T0073104

CORRECT ANTIDIFFERENTIATION
The Change of Variable Well Done

CORRECT ANTIDIFFERENTIATION
The Change of Variable Well Done

Antonio Martínez-Abejón
University of Oviedo, Spain

World Scientific

NEW JERSEY · LONDON · SINGAPORE · BEIJING · SHANGHAI · HONG KONG · TAIPEI · CHENNAI · TOKYO

Published by

World Scientific Publishing Co. Pte. Ltd.

5 Toh Tuck Link, Singapore 596224

USA office: 27 Warren Street, Suite 401-402, Hackensack, NJ 07601

UK office: 57 Shelton Street, Covent Garden, London WC2H 9HE

Library of Congress Cataloging-in-Publication Data
Names: Martínez-Abejón, Antonio, author.
Title: Correct antidifferentiation : the change of variable well done /
    Antonio Martínez-Abejón, University of Oviedo, Spain.
Description: New Jersey : World Scientific, [2021] | Includes bibliographical references and index.
Identifiers: LCCN 2020040370 | ISBN 9789811227455 (hardcover) |
    ISBN 9789811227462 (ebook for institutions) | ISBN 9789811227479 (ebook for individuals)
Subjects: LCSH: Functions of real variables.
Classification: LCC QA331.5 .M286 2021 | DDC 515/.8--dc23
LC record available at https://lccn.loc.gov/2020040370

British Library Cataloguing-in-Publication Data
A catalogue record for this book is available from the British Library.

For any available supplementary material, please visit
https://www.worldscientific.com/worldscibooks/10.1142/12021#t=suppl

To my son Robert

Preface

This book is primarily concerned with the correct calculation of antiderivatives of elementary real-valued functions of a real variable and the correct understanding of these calculations, with particular emphasis on the technique of change of variable.

The motivation for this book is that, in my opinion, most of the circulating materials do not treat the subject of antidifferentiation with sufficient strength or generality. Indeed, when an antiderivative of a given function $f(x)$ is needed and a change of variable $x = g(t)$ is available, the substitution formulas $x = g(t)$ and $dx = g'(t)\,dt$ transform the expression $\int f(x)\,dx$ into $\int (f \circ g)(t) \cdot g'(t)\,dt$, and if an antiderivative $H(t)$ of $(f \circ g)(t) \cdot g'(t)$ is known, the function $F(x) := H \circ g^{-1}(x)$ is usually claimed to be an antiderivative of $f(x)$. But this is only partially true. In fact, there are standard cases where F' coincides with f but locally, that is, $F'(x) = f(x)$ for $x \in A$ where A is a certain subset properly contained in the domain of f, whereas $F(x)$ is not even continuous at the points $x \notin A$, or f is differentiable at $x \notin A$ but $F'(x) \neq f(x)$. That means some modifications still need to be made on $F(x)$ to manage an antiderivative $\widetilde{F}(x)$ of $f(x)$ on the entire domain of f.

Notice that most software packages for Symbolic Calculus stop at the local antiderivative $F(x)$ as a definitive solution for $\int f(x)\,dx$, so they reach a correct solution only when, by chance, $F(x) = \widetilde{F}(x)$ for all x. Undergraduate texts, which have to treat a lot of topics in a short space, deal only with those functions $f(x)$ for which the identity $F(x) = \widetilde{F}(x)$ holds; but they do not generally discuss how to know whether or not $F(x)$ is a true antiderivative of $f(x)$ and, eventually, this is misleading. Higher level books usually refer all issues about elementary antidifferentiation back to undergraduate texts. This book fills the gap by explaining how to reshape $F(x)$ to obtain a true antiderivative $\widetilde{F}(x)$ of $f(x)$. Let us insist on the fact that the process of passing from $F(x)$ to $\widetilde{F}(x)$ cannot be done by means of an algorithm, so it must be learnt by example. However, all actions required for this process are simple and fit perfectly into an undergraduate programme.

It is worth pointing out that the interest of Real Mathematical Analysis in Antidifferentiation has evolved towards very sophisticated theories on the existence of antiderivatives for certain classes of functions ([Bruckner (1978)], [Ciesielski et

Seoane (2018)], [Petruska (1974)]), while every theoretical or practical aspect concerning the calculation of antiderivatives has moved into Computational Algebra as witnessed by the software derived from the Risch algorithm [Risch (1969)], [Risch (1970)] (see also [Bronstein (1997)]). In a certain sense, this book is saying that Real Mathematical Analysis still has a say in the calculation of antiderivatives.

This book is organized into three parts. Part 1 goes from Chapter 1 to 4 and covers the theory of Mathematical Analysis, which sustains the results concerning antidifferentiation of real-valued functions of a real variable (henceforth, functions for short) that will be applied in Part 2. Part 2 goes from Chapter 5 to 12 and focuses on the practice of antidifferentiation. Part 3 consists of Chapter 13 and the appendices. These appendices cover those aspects that do not fit well in Parts 1 or 2 but that are still necessary to make this book reasonably self-contained.

Since this book is oriented to the practice of antidifferentiation, Part 1 does not include a list of examples or exercises. Indeed, most calculus handbooks do an excellent job in that, so we cannot offer any improvement. However, each chapter of the Part 2 contains plenty of examples along with a list of exercises and solutions.

Let us now give a brief summary of each chapter in Part 1.

Chapter 1 consists of a quick survey on real numbers, limit points of sets, limits of real sequences and compactness in the real line.

Chapter 2 covers the topics of function limits, continuity, differentiability and integrability. Most of the results given in Chapter 2 are developed by means of sequential characterizations, which reduce the multiplicity of cases and, in my opinion, are more intuitive than the typical 'ε-δ' arguments. Matters like convexity, extrema of functions and such have been omitted since they are not relevant in antidifferentiation.

A theory of integration is needed to manage results of existence of antiderivatives for continuous functions on intervals. Probably the best integral for this purpose is that of Henstock–Kurzweil, but our choice is the integral of Riemann since it is simpler and is perfectly adapted to the scope of this book.

Chapter 3 introduces the fundamental functions from the core of elementary antidifferentiation: polynomials, n-th roots, the exponential function, the natural logarithm, the circular functions and the hyperbolic functions. The functions $\log x$, e^x, $\cos x$ and $\sin x$ have been constructed by means of integral functions, which is the most appropriate way when dealing with antidifferentiation.

A few comments about the construction of these functions have been postponed to Sections A.1, A.2, A.3 and A.4 in the appendices.

Chapter 4 collects the general theorems of antidifferentiation: linearity, antidifferentiation by parts and change of variable, with stress on this last one. Practical tips in implementing a change of variable are explained in detail in Section 4.2. Section 4.3 describes how to deal with the constants of integration in order to build an antiderivative of a given function from its local antiderivatives. Indeed, the constants of integration stop playing their usual walk-on role assigned in most Calculus handbooks and become actively involved in the process of antidifferentiation. Section 4.5 includes a list of elementary functions which have non-elementary antiderivative and Section 4.6 provides a table of basic primitives.

Each chapter of Part 2 puts the techniques of elementary antidifferentiation into practice for some standard types of functions. We do not pretend to offer an exhaustive list of these types of functions, but to explain how to solve them correctly. For more types of functions, the reader may consult other books, such as [Brychkov et al. (1998)]. All the chapters in Part 2 share a similar structure. A preamble introduces the type of function being dealt with. A first section describes the domain of these functions. Next, and only in the pertinent chapters, a section is devoted to seeking the functions which determine the changes of variable to be used. A further section (or sections) explains in a general manner the way to find the antiderivatives of the considered functions. A last section provides examples. If the reader misses some detail in reading an example, they should go back to the general explanations of the corresponding section. The examples have been presented in increasing order of difficulty. Since a picture is worth a thousand words, some of these examples include graphic representations. Each chapter includes a list of exercises to be solved by the reader.

The contents of the chapters of Part 2 are as follows.

Chapter 5 is devoted to the technique of antidifferentiation by parts.

Chapter 6 is fundamental for the following chapters. It is focused on the antidifferentiation of fractions of polynomials.

Chapter 7 deals with fractions of polynomials over square roots of second degree polynomials.

Chapter 8 is devoted to functions of the form $1/(\alpha x + \beta)^p\sqrt{ax^2 + bx + c}$. From this chapter to Chapter 12, the technique of change of variable is extensively applied.

Chapter 9 deals with rational functions whose arguments are x and rational powers of fractions $(ax + c)/(cx + d)$.

Chapter 10 focuses on binomial differentials. Their antiderivatives are linked to the famous beta functions and were studied by Chebyshev.

Chapter 11 is involved with rational functions whose arguments are the circular functions.

Chapter 12 continues the work initiated in Chapters 7 and 8 about rational functions whose arguments are x and square roots of second degree polynomials.

The chapters in Part 3 can be summarised as follows.

Chapter 13 condenses the procedures of Chapters 5–12 into three tables. A list of useful trigonometric identities is also included.

For this book to be as self-contained as possible, the appendix collects all complementary results, which are usually studied in branches other than Mathematical Analysis, whose proof is too long or sophisticated.

The first section of the appendix shows a discussion which recommends the limitation of the domain of $\sqrt[p]{\cdot}$ to the interval $[0, \infty)$ for all rational, non-integer number p.

Mundane matters such as the division of polynomials and the technique of completing squares, or deeper issues like a proof for the Fundamental Theorem of Algebra or the partial decomposition of a fraction of polynomials, are also included (Section A.10).

There are not many prerequisites for reading this book. In order to keep the language as plain as possible, set theory subtleties such as choice axiom or induction have not been specifically used or cited, and I am sure that the reader will not be hurt by this decision. It is imperative that the reader be skillful in calculating limits of sequences, and derivatives and limits of functions and their sided versions. It is also desirable but not necessary that the reader knows how to plot a function.

This book does not use Complex Analysis, but the reader needs to know the sum and product of complex numbers.

In a few spots, the notions of linear space and basis of a finite dimensional linear space are needed. In general, a first course in Real Analysis should be sufficient, so this book should be accessible for undergraduate and graduate students of any scientific or technical branch. Any teacher who includes antidifferentiation among their lectures would also benefit from this book.

Acknowledgments:

The advice given by many people has been very important in order to produce this book successfully. In particular, I am very indebted to Jesús Araújo, Richard Aron, Javier Pello, Gilles Godefroy, Manuel González and Pfeffer Washek. All of them supplied me with corrections, advice and encouragement. Ricardo Ramos provided important technical support. Sam Hatburn and Ian Hutchinson have done excellent copy-editing work. The facilities and nice atmosphere of the Department of Mathematics of the University of Oviedo made this work possible. I do not forget my students of the Science Faculty of Oviedo: their comments and criticisms pushed me to shape this book. I am also indebted to World Scientific for publishing this book; in particular, I am thankful to Tan Rok Ting, Yolande Koh and Rajesh Babu for their technical assistance.

Oviedo, May 2020

Antonio Martínez-Abejón

Contents

Chapter 1

Background

This chapter introduces the language and basic tools to be used throughout the book. Section 1.1 is a brief description of the numerical systems $\mathbb{N}$, $\mathbb{Z}$, $\mathbb{Q}$, $\mathbb{R}$ and $\mathbb{C}$, and Section 1.2 is devoted to the sequences of real numbers and their limits. Compactness in $\mathbb{R}$ is introduced in Section 1.3 in terms of real-valued sequences.

1.1 Notation and basic concepts

The set of all positive integers, the set of all integers, the set of the rational numbers, the set of the real numbers and the set of all complex numbers are respectively denoted by $\mathbb{N}$, $\mathbb{Z}$, $\mathbb{Q}$, $\mathbb{R}$ and $\mathbb{C}$, so

$$\mathbb{N} = \{1,\, 2,\, 3,\, \ldots\ldots\}$$
$$\mathbb{Z} = \{\ldots,\, \ldots -3,\, -2,\, -1,\, 0, 1,\, 2,\, 3,\, \ldots\ldots\}$$
$$\mathbb{Q} = \left\{ \frac{z}{n} : z \in \mathbb{Z},\ n \in \mathbb{N} \right\}$$

The set $\mathbb{R}$ is endowed with the usual sum and product of real numbers and with the total order $\leqslant$, called *smaller or equal to*. Given two real numbers x and y, the notation $x < y$ (or $y > x$) means that $x \leqslant y$ but $x \neq y$.

The rational numbers satisfy the *density property*: for every pair of real numbers a and b such that $a < b$, there exists a number $q \in \mathbb{Q}$ such that $a < q < b$.

Usually, each real number x is geometrically identified with a point on a line in such a way that the absolute value $|x|$ is the length of the segment whose endpoints are the points x and 0, and as the position of the number 0 divides that line into two halves, if $0 < x$ then x is situated to the right of 0, in which case, the sign of x is defined as $\mathrm{sgn}(x) = 1$; and if $x < 0$ then the position of x is situated to the left and the sign of x is defined as $\mathrm{sgn}(x) = -1$.

All the inclusions $\mathbb{N} \subset \mathbb{Z} \subset \mathbb{Q} \subset \mathbb{R}$ are proper.

The set of complex numbers is formally defined as

$$\mathbb{C} = \{a + bi : (a, b) \in \mathbb{R} \times \mathbb{R}\}$$

where each number $a + bi$ is geometrically identified with the point (a, b) on the plane $\mathbb{R} \times \mathbb{R}$, which is why $\mathbb{C}$ is sometimes referred to as the *complex plane*. The set

$\mathbb{R}$ is properly contained in $\mathbb{C}$ by identifying each real number x with the complex number $x + 0 \cdot i$. This identification is coherent with the geometrical representation of $\mathbb{R}$ as a line (the axis of abscissas) on the complex plane.

The sum and product of $\mathbb{R}$ are extended to $\mathbb{C}$ as follows:

$$(a + bi) + (c + di) := (a + c) + (b + d)i$$
$$(a + bi) \cdot (c + di) := (ac - bd) + (ad + bc)i. \tag{1.1}$$

In particular,

$$i^2 = -1. \tag{1.2}$$

Note that $x^2 \geqslant 0$ for all $x \in \mathbb{R}$.

The sets $\mathbb{Q}$, $\mathbb{R}$ and $\mathbb{C}$ endowed with their respective sums and products are algebraic fields. Clearly, the identity elements of $\mathbb{C}$ with respect to the sum and product are $0 + 0 \cdot i$ and $1 + 0 \cdot i$.

Given a complex number $z = a + bi$, the real numbers a and b are called the *real part* and the *imaginary part* of z respectively and are denoted $a = \Re z$ and $b = \Im z$; the number $\bar{z} := a - bi$ is called the *conjugate of* z; if $z \neq 0$, it is immediate from (1.1) that its inverse with respect to the product is

$$z^{-1} = \frac{\bar{z}}{a^2 + b^2}.$$

Given two complex numbers x and y, it is immediate that

$$\overline{x \cdot y} = \bar{x} \cdot \bar{y}.$$

Since this book deals only with real-valued functions, complex numbers will be used as a mere algebraic tool for factoring polynomials.

A subset A of $\mathbb{R}$ is said to be *bounded above* (resp. *bounded below*) if there exists $r \in \mathbb{R}$ such that $x \leqslant r$ (resp. $r \leqslant x$) for all $x \in A$, in which case r is said to be an *upper bound* (resp. *lower bound*) of A; a subset of $\mathbb{R}$ is said to be *bounded* if it is both bounded above and bounded below. Every non-empty, bounded above (resp. bounded below) subset A of $\mathbb{R}$ has a least upper bound (largest lower bound) which is called the *supremum* (resp. *infimum*) of A, and which is denoted $\sup A$ (resp. $\inf A$). Note that $\sup A$ may belong to A or not; in the first case, $\sup A$ is the largest element of A, also called the *maximum* of A. Analogously, if $\inf A \in A$ then $\inf A$ is the *minimum* of A, that is, the least element of A.

The set $\mathbb{R}$ is extended by adjoining two elements denoted ∞ and $-\infty$ and called *infinity* and *minus infinity* respectively. The new set so obtained is denoted

$$\overline{\mathbb{R}} := \mathbb{R} \cup \{-\infty, \infty\}.$$

The order $\leqslant$ of $\mathbb{R}$ is extended to $\overline{\mathbb{R}}$ as follows:

$$x \leqslant \infty, \quad \text{for all } x \in \overline{\mathbb{R}}$$
$$-\infty \leqslant x, \quad \text{for all } x \in \overline{\mathbb{R}}.$$

As usual, the expression $x < y$ means that $x \leqslant y$ but $x \neq y$.

Once the order $\leqslant$ has been extended to the whole of $\overline{\mathbb{R}}$, the notions of *supremum* and *infimum* can also be extended to the unbounded subsets of $\mathbb{R}$: let A be a non-empty, non-bounded above subset A of $\mathbb{R}$; considering A as a subset of $\overline{\mathbb{R}}$, its only upper bound in $\overline{\mathbb{R}}$ is ∞, and consequently, ∞ is its smallest upper bound in $\overline{\mathbb{R}}$. For this reason, we assign

$$\sup A = \infty;$$

and if A is non-empty, non-bounded below, a similar argument leads to the assignation

$$\inf A = -\infty.$$

Only the subset $\emptyset$ of $\mathbb{R}$ remains without any assignation for its supremum and infimum. This is usually overcome by adopting the identities

$$\sup \emptyset = -\infty \quad \text{and} \quad \inf \emptyset = \infty,$$

which are sustained by the fact that every real number is both an upper bound and a lower bound of $\emptyset$.

The sum and the product of real numbers is partially extended on $\overline{\mathbb{R}}$ as follows:

$$\begin{aligned}
x + \infty = \infty + x &:= \infty, \quad \text{for all } -\infty < x \leqslant \infty \\
x + (-\infty) = (-\infty) + x &:= -\infty, \quad \text{for all } -\infty \leqslant x < \infty \\
x \cdot \infty = \infty \cdot x &:= \infty, \quad \text{for all } 0 < x \leqslant \infty \\
x \cdot \infty = \infty \cdot x &:= -\infty, \quad \text{for all } -\infty \leqslant x < 0 \\
x \cdot (-\infty) = (-\infty) \cdot x &:= -\infty, \quad \text{for all } 0 < x \leqslant \infty \\
x \cdot (-\infty) = (-\infty) \cdot x &:= \infty, \quad \text{for all } -\infty \leqslant x < 0 \\
\frac{x}{\infty} = \frac{x}{(-\infty)} &:= 0, \quad \text{for all } -\infty < x < \infty;
\end{aligned} \qquad (1.3)$$

whereas the forms

$$\infty + (-\infty), \ (\pm\infty) \cdot 0, \ \frac{\pm\infty}{\pm\infty}, \ \frac{x}{0}, \quad \text{for all } x \in \overline{\mathbb{R}} \qquad (1.4)$$

remain undefined because there is no congruent way to define them. For instance, if $\infty + (-\infty)$ were defined to be 0, then $1 = 1 + (\infty + (-\infty)) = (1 + \infty) + (-\infty) = \infty + (-\infty) = 0$, which is a contradiction. However, the forms defined in (1.3) respect the associative, commutative and distributive laws of the sum and the product.

A subset I of $\overline{\mathbb{R}}$ is said to be an *interval* if for every pair of elements x and y of I with $x \leqslant y$, and for every element $x \leqslant z \leqslant y$ then $z \in I$. In particular, the sets $\emptyset$, $\mathbb{R}$, $\overline{\mathbb{R}}$ and any singleton $\{x\}$ with $x \in \overline{\mathbb{R}}$ are intervals of $\overline{\mathbb{R}}$. Every interval I belongs to one of the following types:

$$\begin{aligned}
[a, b] &:= \{x \in \overline{\mathbb{R}} \colon a \leqslant x \leqslant b\} \\
[a, b) &:= \{x \in \overline{\mathbb{R}} \colon a \leqslant x < b\} \\
(a, b] &:= \{x \in \overline{\mathbb{R}} \colon a < x \leqslant b\} \\
(a, b) &:= \{x \in \overline{\mathbb{R}} \colon a < x < b\}
\end{aligned}$$

where a and b are a pair of elements in $\overline{\mathbb{R}}$ denominated the *endpoints* of I. In particular, $\varnothing = (0,0)$, $\mathbb{R} = (-\infty, \infty)$, $\overline{\mathbb{R}} = [-\infty, \infty]$ and $\{x\} = [x, x]$.

An interval I is said to be a *real interval* if $I \subset \mathbb{R}$. In the case where I is empty or a singleton, I is said to be a *degenerate interval*, otherwise I is said to be *non-degenerate*.

A real interval is *open* if it is of the form (a, b). Given an interval I, I° denotes the set

$$I^\circ := I \setminus \{\inf I, \sup I\};$$

I° is an open interval. If $A = \cup_{j \in K} I_j$, where each I_j is an interval, A° denotes the set $\cup_{j \in K} I_j^\circ$.

Given a non-empty subset D of $\mathbb{R}$, a number $p \in \mathbb{R}$ is said to be a *limit point* of D if $(p - \varepsilon, p + \varepsilon) \cap D \setminus \{p\} \neq \varnothing$ for all $\varepsilon > 0$. Roughly speaking, a limit point of D is an element $p \in \mathbb{R}$ approachable by elements of D other than p. If $p \in D$ is not a limit point of D, then it is said to be an *isolated point* of D. The set of all limit points of D is a subset of $\mathbb{R}$ denoted by D'. Note that $I \subset I'$ if I is a non-degenerate interval I.

Given a pair of non-empty subsets D and E of $\mathbb{R}$, a *function f from D into E*, denoted $f \colon D \longrightarrow E$, is an application that associates each element $x \in D$ with a unique element $y \in E$ called the *value* or *image of f at x* and denoted $f(x)$; it is also said that f maps or sends x to $f(x)$. The subset D is named the *domain of f* and the subset $f(D) := \{f(x) \colon x \in D\}$ is named the *image* or *range* of f. Given any subset A of E, the *preimage* of A by f is the subset $f^{-1}(A) := \{x \in D \colon f(x) \in A\}$; given a singleton $\{y\}$ of E, its preimage by f is also denoted as $f^{-1}(y)$ for short. If the preimage of every singleton in E is a singleton, f is said to be *injective*, in which case, given $y \in F(D)$, $f^{-1}(y)$ also denotes the unique element $x \in D$ such that $f(x) = y$; f is said to be *surjective* if $f(D) = E$; if $f \colon D \longrightarrow E$ is injective and surjective, f is said to be *bijective*, in which case the function $f^{-1} \colon E \longrightarrow D$ that maps each $y \in E$ to $f^{-1}(y)$ is called the *inverse function* of f. A function $f \colon D \longrightarrow E$ for which $f(D)$ is a singleton is called *constant*. The *graph* of f is the subset $\{(x, f(x)) \colon x \in D\}$ of $D \times E$.

Given a function $f \colon D \longrightarrow E$ and a subset G of D, the *restriction* of f to G is the function from G into E that maps each $x \in G$ to $f(x)$. This restriction is denoted as $f|_G$.

Given two functions $g \colon D \longrightarrow A$ and $f \colon B \longrightarrow C$ such that $g(D) \subset B$, the *composition* of g with f is the function $f \circ g \colon D \longrightarrow C$ that maps each $x \in D$ to $f(g(x))$.

A function $f \colon D \longrightarrow \mathbb{K}$ is said to be *real-valued* (alt. *complex-valued*) if $\mathbb{K} = \mathbb{R}$ ($\mathbb{K} = \mathbb{C}$), and if D is a subset of $\mathbb{R}$ (alt. of $\mathbb{C}$), then f is said to be a function of a *real variable* (alt., of *a complex variable*).

A subset D of $\mathbb{R}$ is said to be *symmetric* if $-x \in D$ whenever $x \in D$. A real-valued function f defined on a symmetric subset D is said to be *even* if $f(x) = f(-x)$ for all $x \in D$, and is said to be *odd* if $f(-x) = -f(x)$ for all $x \in D$. If $\mathbb{K}$ denotes either $\mathbb{R}$ or $\mathbb{C}$, a function $f \colon D \longrightarrow \mathbb{K}$ is said to be *identically null* if its range is $\{0\}$, in which case it will be denoted $f = 0$ or $f(x) \equiv 0$ in order to distinguish it from the notation $f(a) = 0$, which is reserved to mean that f vanishes at the number $a \in D$. Since we are mainly concerned with real valued functions of one real variable, henceforth we will simply refer to them as *functions*.

Any time a function f is given, its domain must be explicitly given too; otherwise its domain is assumed to be the largest subset of $\mathbb{R}$ where f is well defined, in which case its domain is denoted D_f and called the *natural domain* of f. For instance, if the function $f(x) = \sqrt{x-1}$ is given without mentioning its domain, we are given to understand that its domain is $[1, \infty)$, the largest subset of $\mathbb{R}$ where the square root of $x-1$ makes sense. This is somewhat informal but it avoids the use of superfluous notation.

A function $f \colon D \longrightarrow \mathbb{R}$ is said to be *bounded below* (resp. *bounded above*) if the subset $f(D)$ of $\mathbb{R}$ is bounded below (above); if $f(D)$ is bounded both below and above then f is said to be *bounded*.

A function $f \colon D \longrightarrow \mathbb{R}$ with $D \subset \mathbb{R}$ is said to be *increasing* if $f(x) < f(y)$ whenever $x < y$, and f is said to be *decreasing* if $f(x) > f(y)$ whenever $x < y$; if $f(x) \leqslant f(y)$ when $x < y$ then f is said to be *non-decreasing*, and if $f(x) \geqslant f(y)$ when $x < y$, then f is said to be *non-increasing*. If f is either non-decreasing or non-increasing, f is said to be *monotone*; if f is either increasing or decreasing then f is called *strictly monotone*.

Given a subset D of $\mathbb{R}$, the set of all functions from D into $\mathbb{R}$ is equipped with the usual sum and product of functions. Given two non-empty subsets A and B of real-valued functions defined on D, we denote

$$A + B := \{f + g \colon f \in A, \ g \in B\}$$
$$A \cdot B := \{f \cdot g \colon f \in A, \ g \in B\}.$$

In particular, given $\lambda \in \mathbb{R}$, λA denotes the set $\{\lambda \cdot f \colon f \in A\}$.

It must be emphasized that $A - A$ needs not be $\{0\}$ or $\varnothing$. In fact, we may come across the following situation: let $f \colon D \longrightarrow \mathbb{R}$ a function. Consider the subset $A := \{f(x) + K \colon K \in \mathbb{R}\}$ of functions defined on D. In that case $A - A = \{K \colon K \in \mathbb{R}\}$, the set of constant functions on D.

Let X be a vector space (also called linear space) over $\mathbb{K}$, where $\mathbb{K}$ denotes either $\mathbb{R}$ or $\mathbb{C}$. Given two vector subspaces V and W of X, X is said to be the *direct sum of V and W* if for every $x \in X$ there exists a unique pair $(v, w) \in V \times W$ such that $x = v + w$, in which case, it is denoted $X = V \oplus W$; equivalently, $X = V \oplus W$ if and only if $X = V + W$ and $V \cap W = \{0\}$, and if moreover X is finite-dimensional, then $X = V \oplus W$ if and only if $\dim X = \dim V + \dim W$ and $V \cap W = \{0\}$.

Given a vector space X over $\mathbb{K}$ of dimension n, a finite sequence $\{x_i\}_{i=1}^n \subset X$ is said to be an *algebraic basis of X*, or *basis of X* for short, if for every $x \in X$ there exists a unique sequence $(\lambda_i)_{i=1}^n$ in $\mathbb{K}$ for which $x = \sum_{i=1}^n \lambda_i x_i$.

1.2   Sequences of real numbers

The notion of sequence is a cornerstone of Mathematical Analysis. Since the reader is assumed to be already acquainted with real sequences, only the essential facts will be covered. For more detailed information, see [Ross (1986)].

Definition 1.2.1. A *sequence of real numbers*(i.e., a *real-valued sequence*) is a function $f\colon \mathbb{N} \longrightarrow \mathbb{R}$.

Henceforth, real-valued sequences will be referred to as *sequences* for short.

Roughly speaking, a sequence $f\colon \mathbb{N} \longrightarrow \mathbb{R}$ is an ordered list of infinitely many real numbers

$$(a_n)_{n=1}^\infty = (a_1, a_2, a_3, \ldots^\infty \ldots)$$

where $a_n = f(n)$ is the *n-th coefficient* of the sequence. It is customary to denote a sequence as $(a_n)_{n=1}^\infty$ rather than using the formal notation $f\colon \mathbb{N} \longrightarrow \mathbb{R}$. The sum and product of a pair of sequences are defined as $(a_n)_{n=1}^\infty + (b_n)_{n=1}^\infty := (a_n + b_n)_{n=1}^\infty$ and $(a_n)_{n=1}^\infty \cdot (b_n)_{n=1}^\infty := (a_n \cdot b_n)_{n=1}^\infty$.

Given a sequence $(a_n)_{n=1}^\infty$, $\{a_n\}_{n=1}^\infty$ denotes the set of all its coefficients.

A sequence $(x_n)_{n=1}^\infty$ is said to be *bounded above* (resp. *bounded below*) if the set $\{x_n\colon n \in \mathbb{N}\}$ is bounded above (resp. bounded below). If $(x_n)_{n=1}^\infty$ is bounded both above and below then it is said to be *bounded*.

The concept of limit of a sequence must be now discussed.

Definition 1.2.2. Let $(a_n)_{n=1}^\infty$ be a sequence of real numbers and consider the following statements:

(i) there is $\lambda \in \mathbb{R}$ such that for every $\varepsilon > 0$ there exists $n_\varepsilon \in \mathbb{N}$ such that for all $n \geqslant n_\varepsilon$, we have $a_n \in (\lambda - \varepsilon, \lambda + \varepsilon)$; equivalently, the set $\{n\colon a_n \notin (\lambda - \varepsilon, \lambda + \varepsilon)\}$ is finite;

(ii) for every $K \in \mathbb{R}$ there exists $n_K \in \mathbb{N}$ such that for all $n \geqslant n_K$, we have $a_n > K$; equivalently, the set $\{n\colon a_n \leqslant K\}$ is finite;

(iii) for every $K \in \mathbb{R}$ there exists $n_K \in \mathbb{N}$ such that for all $n \geqslant n_K$, we have $a_n < K$; equivalently, the set $\{n\colon a_n \geqslant K\}$ is finite.

If statement (i) (respectively (ii), (iii)) holds true then λ (resp. ∞, $-\infty$) is said to be the *limit* of $(a_n)_{n=1}^\infty$.

If l is the limit of $(a_n)_{n=1}^\infty$, we write $l = \lim_n a_n$ or $a_n \xrightarrow{n} l$. If $l \in \mathbb{R}$ then $(a_n)_{n=1}^\infty$ is said to be *convergent*, and if $l = 0$ then $(a_n)_{n=1}^\infty$ is said to be *infinitesimal*; if l equals ∞ or $-\infty$ then $(a_n)_{n=1}^\infty$ is said to be *divergent*.

Roughly speaking, $(a_n)_{n=1}^\infty$ converges to a if the coefficients a_n are as close to a as we please for all $n \geqslant n_0$ where n_0 is sufficiently large. For instance, the sequence of successive digits of π,

$$(x_n)_{n=1}^\infty = (3, 3.1, 3.14, 3.141, \ldots)$$

converges to π, and $0 < \pi - x_n < 10^{1-n}$ for all n. In a similar way, a divergent sequence to ∞ can be roughly understood as a sequence whose terms are closer to ∞ the larger n is, where *closer to* ∞ must be understood in the sense that $x \in \mathbb{R}$ is closer to ∞ than any $y \in (-\infty, x)$.

It is evident from the definition of convergence that a sequence $(a_n)_{n=1}^{\infty}$ is convergent if and only if it can be decomposed into a sum of a constant sequence and an infinitesimal one as represented by the equation:

$$(a_n)_{n=1}^{\infty} = (a)_{n=1}^{\infty} + (\varepsilon_n)_{n=1}^{\infty}$$

where the constant coefficient is $a = \lim_n a_n$ and $\varepsilon_n = a_n - a \xrightarrow{n} 0$.

There are sequences, like $\left((-1)^n\right)_{n=1}^{\infty}$, with no limit. Important examples of sequences with a limit are the constant sequence $(\lambda)_{n=1}^{\infty} = (\lambda, \lambda, \lambda, \ldots)$ which converges to λ, and the sequences

$$\frac{1}{n} \xrightarrow{n} 0, \; n \xrightarrow{n} \infty, \; -n \xrightarrow{n} -\infty.$$

There is a connection between the limit points of sets and the limits of sequences:

Proposition 1.2.3. *Let* $x_n \xrightarrow{n} \lambda \in \overline{\mathbb{R}}$:

(i) *if* $\lambda \in \mathbb{R}$ *then* $(x_n)_{n=1}^{\infty}$ *is bounded;*
(ii) *if* $\lambda = \infty$ *then* $(x_n)_{n=1}^{\infty}$ *is bounded below but is not bounded above;*
(iii) *if* $\lambda = -\infty$ *then* $(x_n)_{n=1}^{\infty}$ *is bounded above but is not bounded below.*

Proof. (i) Let $x_n \xrightarrow{n} \lambda \in \mathbb{R}$. From Definition 1.2.2, for $\varepsilon = 1$ there exists $n_0 \in \mathbb{N}$ such that $|x_n - \lambda| < 1$ for all $n \geqslant n_0$, and as the triangular inequality gives $|x_n| \leqslant 1 + |\lambda|$, it follows that the maximum of the finite set

$$\{|x_1|, |x_2|, \ldots, |x_{n_0-1}|, 1 + |x_{n_0}|\}$$

is an upper bound for all coefficients $|x_n|$, hence $(x_n)_{n=1}^{\infty}$ is bounded.

(ii) The fact that $(x_n)_{n=1}^{\infty}$ is not bounded above follows directly from the definition of divergence to ∞.

In order to verify that $(x_n)_{n=1}^{\infty}$ is bounded below, taking $K = 0$ in Definition 1.2.2(ii), we have $x_n > 0$ for all $n \geqslant n_0$ for certain $n_0 \in \mathbb{N}$. Clearly, the minimum of the finite set

$$\{x_1, x_2, \ldots, x_{n_0-1}, 0\}$$

is a lower bound of $(x_n)_{n=1}^{\infty}$.

(iii) The proof parallels that of (ii). $\qquad\qquad\square$

Proposition 1.2.4. *Let* $(x_n)_{n=1}^{\infty}$ *be a sequence convergent to* λ *and let* $X := \{x_n\}_{n=1}^{\infty}$. *Then one and only one of the following statements holds true:*

(i) X *is infinite, in which case* $X' = \{\lambda\}$;
(ii) X *is finite, in which case* $X' = \varnothing$.

Proof. Since $x_n \xrightarrow[n]{} \lambda$, for every $\varepsilon > 0$ there exists $n_\varepsilon \in \mathbb{N}$ such that

$$x_n \in (\lambda - \varepsilon, \lambda + \varepsilon) \quad \text{for all } n \geqslant n_\varepsilon. \tag{1.5}$$

Let μ be a real number such that $\mu \neq \lambda$. Choose a real number $\varepsilon > 0$ such that $\varepsilon < |\lambda - \mu|/2$. Hence,

$$(\mu - \varepsilon, \mu + \varepsilon) \cap (\lambda - \varepsilon, \lambda + \varepsilon) = \varnothing$$

and in combination with (1.5), it follows that $(\mu - \varepsilon, \mu + \varepsilon)$ contains at most a finite number of terms of X, which will be denoted as $x_{k_1}, \ldots x_{k_m}$. Thus the subset $S := \{x_{k_i} : 1 \leqslant i \leqslant m, \ x_{k_i} \neq \mu\}$ is also finite. If $S = \varnothing$, then it is evident that $X \cap (\mu - \varepsilon, \mu + \varepsilon) \backslash \{\mu\} = \varnothing$, so $\mu \notin X'$. And in the case where $S \neq \varnothing$, choosing $\varepsilon_1 := \min\{|\mu - x_{k_i}| : 1 \leqslant i \leqslant m\} > 0$ it is immediate that $X \cap (\mu - \varepsilon_1, \mu + \varepsilon_1) \backslash \{\mu\} = \varnothing$, hence $\mu \notin X'$. This proves that the only possible limit point of X is λ.

If X is infinite, then there is $k > n_\varepsilon$ such that $x_k \neq \lambda$. Hence, (1.5) shows that

$$x_k \in (\lambda - \varepsilon, \lambda + \varepsilon) \backslash \{\lambda\} \neq \varnothing$$

which proves that λ is the only element of X'.

In the case where X is finite, the set $T := X \backslash \{\lambda\}$ is finite, and arguing with T and λ as we did before with S and μ, it can be proved that λ is not a limit point of X. Hence, $X' = \varnothing$ and the proof is done. □

The following result states the uniqueness of the limit of a convergent (or divergent) sequence.

Theorem 1.2.5. *A sequence has at most one limit.*

Proof. Assume $(x_n)_{n=1}^\infty$ has two limits, λ and μ. From Proposition 1.2.3, both λ and μ must be finite, and so, from Proposition 1.2.4, $\lambda = \mu$. □

Definition 1.2.6. A sequence $(x_n)_{n=1}^\infty$ is said to be

(i) *increasing* if $x_n < x_{n+1}$ for all n;
(ii) *non-decreasing* if $x_n \leqslant x_{n+1}$ for all n;
(iii) *decreasing* if $x_n > x_{n+1}$ for all n;
(iv) *non-increasing* if $x_n \geqslant x_{n+1}$ for all n.

Any sequence satisfying any of the four statements above is said to be *monotone*, and if it satisfies (i) or (iii) then it is said to be *strictly monotone*.

The relation between order and limits of sequences is reflected in the following result, also known as the *sandwich rule*.

Theorem 1.2.7. *Given sequences* $(a_n)_{n=1}^\infty$, $(b_n)_{n=1}^\infty$ *and* $(c_n)_{n=1}^\infty$ *such that* $a_n \leqslant b_n \leqslant c_n$ *for all* n, *the following three statements hold true:*

(i) *if* $a_n \xrightarrow[n]{} \lambda$, $c_n \xrightarrow[n]{} \lambda$ *and* $\lambda \in \mathbb{R}$ *then* $b_n \xrightarrow[n]{} \lambda$,

(ii) if $a_n \xrightarrow[n]{} \infty$ then $b_n \xrightarrow[n]{} \infty$,

(iii) if $c_n \xrightarrow[n]{} -\infty$ then $b_n \xrightarrow[n]{} -\infty$.

Proof. The proof follows from Definition 1.2.2. □

Theorem 1.2.8. *If $a_n \xrightarrow[n]{} a$ and $a_n \geqslant 0$ for all n then $a \geqslant 0$.*

Proof. Assume $a < 0$. Then either a is finite or $a = -\infty$. If a is finite, fix $\varepsilon := -a$. Thus there exists n_0 such that $a_n \in (a - \varepsilon, a + \varepsilon)$ for all $n \geqslant n_0$. In particular, $a_{n_0} < 0$, which is a contradiction.

A similar contradiction is reached if it is assumed that $a = -\infty$. □

The relation between the sum, product, and limits of sequences is further developed in the following results.

Proposition 1.2.9. *The following statements hold true:*

(i) *if $(a_n)_{n=1}^{\infty}$ and $(b_n)_{n=1}^{\infty}$ are infinitesimal, so is $(a_n + b_n)_{n=1}^{\infty}$;*

(ii) *if $(a_n)_{n=1}^{\infty}$ is infinitesimal and $(b_n)_{n=1}^{\infty}$ is bounded, then $(a_n b_n)_{n=1}^{\infty}$ is infinitesimal.*

Proof. (i) Given $\varepsilon > 0$, by hypothesis there exists a pair of natural numbers n_1 and n_2 such that $|a_n| < \varepsilon/2$ for all $n \geqslant n_1$ and $|b_n| < \varepsilon/2$ for all $n \geqslant n_2$. Thus, given $n_0 := \max\{n_1, n_2\}$, we have that for all $n \geqslant n_0$,

$$|a_n + b_n| \leqslant |a_n| + |b_n| < \frac{\varepsilon}{2} + \frac{\varepsilon}{2} = \varepsilon$$

hence $a_n + b_n \xrightarrow[n]{} 0$.

(ii) Assume $(b_n)_{n=1}^{\infty}$ is bounded and take $M \in (0, \infty)$ such that $|b_n| \leqslant M$ for all $n \in \mathbb{N}$. Fix $\varepsilon > 0$. As $a_n \xrightarrow[n]{} 0$, there exists $n_0 \in \mathbb{N}$ such that $|a_n| < \varepsilon/M$ for all $n \geqslant n_0$. Thus, for every $n \geqslant n_0$,

$$|a_n b_n| \leqslant |a_n||b_n| \leqslant \frac{\varepsilon}{M} \cdot M = \varepsilon$$

so $a_n b_n \xrightarrow[n]{} 0$. □

Theorem 1.2.10. *Let $(a_n)_{n=1}^{\infty}$ and $(b_n)_{n=1}^{\infty}$ be a pair of sequences such that $a_n \xrightarrow[n]{} a$ and $b_n \xrightarrow[n]{} b$. Let $\odot$ be one of the operations $+$, $-$, $\cdot$ or $\div$ (in the case where $\odot$ is division, all the terms b_n are assumed to be non-null). Thus, if $a \odot b$ is defined in $\mathbb{R}$, then $a_n \odot b_n \xrightarrow[n]{} a \odot b$.*

Proof. The proof is presented in six different parts:

(a) *Both a and b are finite and $\odot$ is one of the operations $+$ or $\cdot$.* Consider the decompositions $a_n = a + \varepsilon_n$ and $b_n = b + \delta_n$, so $\varepsilon_n \xrightarrow[n]{} 0$ and $\delta_n \xrightarrow[n]{} 0$.

Since

$$a_n + b_n = a + b + \varepsilon_n + \delta_n,$$
$$a_n b_n = ab + a\delta_n + \varepsilon_n b + \varepsilon_n \delta_n, \quad \text{for all } n \in \mathbb{N},$$

and as Propositions 1.2.3 and 1.2.9 give that $(\varepsilon_n + \delta_n)_{n=1}^{\infty}$, $(a\delta_n)_{n=1}^{\infty}$, $(\varepsilon_n b)_{n=1}^{\infty}$ and $(\varepsilon_n \delta_n)_{n=1}^{\infty}$ are infinitesimal, it follows that $a_n + b_n \xrightarrow[n]{} a + b$ and $a_n b_n \xrightarrow[n]{} ab$.

(b) *If $b_n \xrightarrow[n]{} b \in (0, \infty)$ then $1/b_n \xrightarrow[n]{} 1/b$.* First, note that $(1/b_n)_{n=1}^{\infty}$ is bounded: in fact, for $\varepsilon := b/2 > 0$, there exists n_0 such that $b/2 = b - \varepsilon < b_n < b + \varepsilon$ for all $n \geqslant n_0$. Thus $0 < 1/b_n < 2/b$, hence $(1/b_n)_{n=1}^{\infty}$ is bounded.

Next, as

$$\frac{1}{b_n} = \frac{1}{b} + \frac{b - b_n}{b_n b} \text{ for all } n \in \mathbb{N}$$

and as $(1/b_n)_{n=1}^{\infty}$ is bounded and $(b - b_n)_{n=1}^{\infty}$ is infinitesimal, Proposition 1.2.9 gives $1/b_n \xrightarrow[n]{} 1/b$.

(c) *If $a = \infty$ and $b \in (-\infty, \infty]$ then $a_n + b_n \xrightarrow[n]{} \infty$.* By Proposition 1.2.3, we may choose a lower bound $M \in \mathbb{R}$ for $(b_n)_{n=1}^{\infty}$. Thus, given any $K \in \mathbb{R}$, as $a_n \xrightarrow[n]{} \infty$, it follows that there exists n_0 such that for every $n \geqslant n_0$, $a_n > K - M$. Hence $a_n + b_n > K$, and therefore, $a_n + b_n \xrightarrow[n]{} \infty$.

(d) *If $a = \infty$ and $b \in (0, \infty]$ then $a_n b_n \xrightarrow[n]{} \infty$.* Fix a real number $K > 0$. On the one hand, as $b > 0$, there exists a real number $H > 0$ and a natural number n_1 such that $b_n > H$ for all $n \geqslant n_1$ (if b is finite, take $H := b/2$, and if $b = \infty$, take $H := 1$). On the other hand, as $a_n \xrightarrow[n]{} \infty$, there exists n_2 such that $a_n > K/H$ for all $n \geqslant n_2$.

Thus for $n_0 := \max\{n_1, n_2\}$, we have that for every $n \geqslant n_0$, $a_n b_n > K$. This proves that $a_n b_n \xrightarrow[n]{} \infty$.

(e) *If $a_n \xrightarrow[n]{} \infty$ then $-a_n \xrightarrow[n]{} -\infty$, and if $a_n \xrightarrow[n]{} -\infty$ then $-a_n \xrightarrow[n]{} \infty$.* This follows from Definition 1.2.2.

(f) *If $b_n \xrightarrow[n]{} \infty$ then $1/b_n \xrightarrow[n]{} 0$.* Given $\varepsilon > 0$, take $K := 1/\varepsilon$. Thus, as $b_n \xrightarrow[n]{} \infty$, there exists n_0 such that $b_n > K$ for all $n \geqslant n_0$, and therefore, $0 < 1/b_n < 1/K = \varepsilon$, hence $1/b_n \xrightarrow[n]{} 0$.

All the remaining cases follow from combinations of the proven cases. □

If $l \odot \lambda$ is one of the undefined forms given in (1.4) then $(a_n \odot b_n)_{n=1}^{\infty}$ may have a limit or may not, and that depends on the particular sequences $(a_n)_{n=1}^{\infty}$ and $(b_n)_{n=1}^{\infty}$ involved. This assertion is illustrated by the following result.

Theorem 1.2.11. *Let $(a_n)_{n=1}^{\infty}$ be an infinitesimal sequence such that $a_n \neq 0$ for all n:*

(i) *if $a_n > 0$ for all n, then $1/a_n \xrightarrow[n]{} \infty$;*
(ii) *if $a_n < 0$ for all n, then $1/a_n \xrightarrow[n]{} -\infty$;*

Proof. Case (i) is simply the inverse argument of case (f) in the demonstration of Theorem 1.2.10. □

We end this section with a result that identifies sequences with a limit by examination of some of its subsequences.

Proposition 1.2.12. *Let A and B be two disjoint, infinite subsets of $\mathbb{N}$ such that $\mathbb{N} = A \cup B$ and let us label the elements of A and B as $A := \{k_1 < k_2 < \dots\}$ and $B := \{m_1 < m_2 < \dots\}$.*

Given a sequence $(x_n)_{n=1}^{\infty}$ for which $x_{k_n} \xrightarrow[n]{} \lambda$ and $x_{m_n} \xrightarrow[n]{} \lambda$, then $x_n \xrightarrow[n]{} \lambda$.

Proof. Assume $\lambda \in \mathbb{R}$ (the cases where $\lambda = \pm\infty$ admit a similar proof). Take $\varepsilon > 0$ and denote $I := (\lambda - \varepsilon, \lambda + \varepsilon)$. Since $x_{k_n} \xrightarrow[n]{} \lambda$ and $x_{m_n} \xrightarrow[n]{} \lambda$ then both subsets $A_\varepsilon := \{n \in A : x_{k_n} \notin I\}$ and $B_\varepsilon := \{n : x_{m_n} \notin I_\varepsilon\}$ are finite. Clearly, the subset $\{n \in \mathbb{N} : x_n \notin I_\varepsilon\}$ equals $A_\varepsilon \cup B_\varepsilon$, hence it is finite too, and that means that $x_n \xrightarrow[n]{} \lambda$. □

1.3 Compactness in $\mathbb{R}$

Thanks to the peculiar structure of $\mathbb{R}$, the language of sequences is sufficient to deal with all the necessary facts concerning compactness in $\mathbb{R}$.

Definition 1.3.1. A subset K of $\mathbb{R}$ is said to be *compact* if it is bounded and $K' \subset K$.

In particular, any finite subset F of $\mathbb{R}$, including the empty subset, is compact because it is bounded and $F' = \varnothing$.

The only compact real intervals are those of the form $[a, b]$ with $a \leqslant b$. In fact, $(a, b)' = (a, b]' = [a, b)' = [a, b]' = [a, b]$.

Bearing in mind that the supremum of a non-empty, non-bounded above (resp. non-bounded below) subset of $\mathbb{R}$ is ∞ (resp. $-\infty$), we can state the first result:

Proposition 1.3.2. *Let $(x_n)_{n=1}^{\infty}$ be a monotone sequence:*

(i) if $(x_n)_{n=1}^{\infty}$ is non-decreasing, then there exists $\lim_n x_n = \sup\{x_n\}_{n\in\mathbb{N}}$;
(ii) if $(x_n)_{n=1}^{\infty}$ is non-increasing, then there exists $\lim_n x_n = \inf\{x_n\}_{n\in\mathbb{N}}$.

Proof. (i) Take $X := \{x_n\}_{n\in\mathbb{N}}$. If $(x_n)_{n=1}^{\infty}$ is bounded above, then $s := \sup X \in \mathbb{R}$. In order to prove that $x_n \xrightarrow[n]{} s$, take $\varepsilon > 0$. As $s - \varepsilon$ is strictly smaller than the least upper bound of X, it cannot be an upper bound of X, so there exists $x_{n_0} \in X$ such that $s - \varepsilon < x_{n_0} \leqslant s$. But $(x_n)_{n=1}^{\infty}$ is non-decreasing, so

$$s - \varepsilon < x_{n_0} \leqslant x_n \leqslant s < s + \varepsilon \text{ for all } n \geqslant n_0$$

which proves that $x_n \xrightarrow[n]{} s$.

The case where $\{x_n\}_{n=1}^{\infty}$ is non-bounded above is similar: given $K \in \mathbb{R}$ there exists a term x_{n_0} greater than K, and therefore $K < x_{n_0} \leqslant x_n$ for all $n \geqslant n_0$, hence $x_n \xrightarrow[n]{} \infty$.

(ii) is similar to (i). □

Proposition 1.3.3. *Every sequence of real numbers contains a monotone subsequence.*

Proof. Given a sequence $(x_n)_{n=1}^{\infty}$, let $P := \{k \in \mathbb{N} \colon x_k \geq x_n \text{ for all } n \geq k\}$.

We face two possible cases: either P is finite (sometimes possibly empty) or P is infinite.

In the first case, there is $k_1 \in \mathbb{N}$ such that $n \notin P$ for all $n \geq k_1$. Thus we select an increasing subsequence $(x_{k_n})_{n=1}^{\infty}$ as follows: the first element is x_{k_1}. As $k_1 \notin P$, there exists $k_2 > k_1$ such that $x_{k_1} < x_{k_2}$. But $k_2 \notin P$, so there exists $k_3 > k_2$ such that $x_{k_2} < x_{k_3}$, and continuing with this procedure indefinitely, we obtain a sequence of positive integers

$$k_1 < k_2 < k_3 < \ldots\ldots\overset{\infty}{.}$$

such that the subsequence $(x_{k_n})_{n=1}^{\infty}$ is increasing.

In the second case, when P is infinite, labelling all the elements of P in ascendent order as

$$p_1 < p_2 < p_3 < \ldots\ldots\overset{\infty}{.}$$

since each p_n belongs to P, it follows that $(x_{p_n})_{n=1}^{\infty}$ is a non-increasing subsequence of $(x_n)_{n=1}^{\infty}$. The proof is done. $\square$

Among the many characterizations of compact subsets of $\mathbb{R}$, we need only the following one.

Theorem 1.3.4. *Given a subset K of $\mathbb{R}$, the following statements are equivalent:*

(a) K is compact;
(b) for every infinite subset A of K, we have $A' \cap K \neq \emptyset$.

Proof. (a)$\Rightarrow$(b) Assume K is compact and let A be an infinite subset of K. Pick a sequence $(x_n)_{n=1}^{\infty}$ of pairwise different elements of A (the reader might want to consult Section A.13 at this point). By Proposition 1.3.3, we may assume $(x_n)_{n=1}^{\infty}$ is monotone. Moreover, as K is bounded, then so is $(x_n)_{n=1}^{\infty}$ and from Proposition 1.3.2, $x_n \xrightarrow{n} p$ for some $p \in \mathbb{R}$. Now Proposition 1.2.4 shows $p \in A'$.

Next, $A \subset K$ implies $A' \subset K'$, and as K is compact, $K' \subset K$, hence $p \in K$. This proves that $p \in A' \cap K \neq \emptyset$.

(b)$\Rightarrow$(a) Assume K is not compact. Then K is not bounded or $K' \not\subset K$. In the case where K is not bounded, we assume K is not bounded above (if K is not bounded below, the proof is analogous). Thus, for every $n \in \mathbb{N}$ we may pick $x_n \in K$ so that $n < x_n$. Thus Theorem 1.2.7 gives $n < x_n \xrightarrow{n} \infty$, and therefore, the set $A := \{x_n\}_{n \in \mathbb{N}}$ is infinite and does not have any limit point, hence (b) does not hold.

In the case where $K' \not\subset K$, pick $p \in K \backslash K'$. For every $n \in \mathbb{N}$ there exists $x_n \in K \backslash \{p\}$ such that $p - 1/n < x_n < p + 1/n$. Then, the sequence $(x_n)_{n=1}^{\infty}$ converges to p, and by virtue of Proposition 1.2.4, p is the only limit point of $A := \{x_n\}_{n \in \mathbb{N}}$. But $p \notin K$, so $A' \cap K = \emptyset$ and therefore (b) does not hold either. $\square$

Chapter 2

Real-valued functions of a real variable

Chapter 2 covers the basics about limits, continuity, differentiability and integrability of real-valued functions of a real variable in antidifferentiation calculus from a theoretical and a practical point of view. In particular, we insist on the notions of one-sided limits, one-sided continuity and one-sided differentiability.

Although we adopt the classical 'ε-δ' definitions for limit and continuity of a function at a point in Sections 2.1 and 2.2, we proceed as soon as possible to their respective characterizations in terms of sequences, which is a more intuitive way of addressing the problem of continuity and saves a lot of time.

Differentiability is studied in Section 2.3 but issues like convexity are not included because they are not relevant to antidifferentiation calculus.

Section 2.4 collects some of the results that form the core of the theory of the Riemann integral for real-valued functions of one real variable which are essential for proving the existence of antiderivatives of continuous functions on intervals. Results that are important in integral calculus but superfluous in antidifferentiation, like the linearity of the Riemann integral, are not included. The only exception is the change of variable for the Riemann integral.

2.1   Limits of functions

Roughly speaking, a number $\lambda \in \overline{\mathbb{R}}$ is the left-hand limit of a function f at a point $x_0 \in \mathbb{R}$ if the values $f(x)$ approach λ as closely as we please when x approaches x_0 on the left-hand side of x_0.[1] In technical terms:

Definition 2.1.1. Let D be a non-empty subset of $\mathbb{R}$, let $x_0 \in \big((-\infty, x_0) \cap D\big)'$ and let $f\colon D \longrightarrow \mathbb{R}$. A number $\lambda \in \overline{\mathbb{R}}$ is said to be the *left-hand limit* of f at x_0 if one of the following three statements holds:

(i) λ is finite and for every $\varepsilon > 0$ there exists $\delta > 0$ such that for every $x \in D \cap (x_0 - \delta, x_0)$ we have $|f(x) - \lambda| < \varepsilon$;

---

[1]If $\lambda = \infty$, proximity to ∞ must be understood in the sense that the larger a real number r is, the closer r is to ∞. Thus, if $r < s$, then r is closer to $-\infty$ than s is.

13

(ii) $\lambda = \infty$ and for every $K \in \mathbb{R}$ there exists a real number $\delta > 0$ such that for every $x \in D \cap (x_0 - \delta, x_0)$ we have $f(x) > K$;

(iii) $\lambda = -\infty$ and for every $K \in \mathbb{R}$ there exists a real number $\delta > 0$ such that for every $x \in D \cap (x_0 - \delta, x_0)$ we have $f(x) < K$.

The left-hand limit of f at x_0 is denoted as $\lim_{x \to x_0^-} f(x)$ or as $f(x_0^-)$.

A similar definition is given when the variable x approaches x_0 on its right-hand side.

Definition 2.1.2. Let D be a non-empty subset of $\mathbb{R}$, let $x_0 \in \big((x_0, \infty) \cap D\big)'$ and let $f : D \longrightarrow \mathbb{R}$. A number $\lambda \in \overline{\mathbb{R}}$ is said to be the *right-hand limit* of f at x_0 if one of the following three statements holds:

(i) λ is finite and for every $\varepsilon > 0$ there exists a real number $\delta > 0$ such that for every $x \in D \cap (x_0, x_0 + \delta)$, we have $|f(x) - \lambda| < \varepsilon$;

(ii) $\lambda = \infty$ and for every $K \in \mathbb{R}$ there exists a real number $\delta > 0$ such that for every $x \in D \cap (x_0, x_0 + \delta)$, we have $f(x) > K$;

(iii) $\lambda = -\infty$ and for every $K \in \mathbb{R}$ there exists a real number $\delta > 0$ such that for every $x \in D \cap (x_0, x_0 + \delta)$, we have $f(x) < K$.

The right-hand limit of f at x_0 is denoted as $\lim_{x \to x_0^+} f(x)$ or as $f(x_0^+)$.

The left-hand limit and right-hand limit of a function at a point are indistinctly called the *one-sided limits*.

Merging the definitions of both one-sided limits at a point, we obtain the notion of two-sided limit of $f(x)$ at x_0, or limit of $f(x)$ at x_0 for short:

Definition 2.1.3. Let D be a non-empty subset of $\mathbb{R}$, let $x_0 \in D'$ and let $f : D \longrightarrow \mathbb{R}$. A number $\lambda \in \overline{\mathbb{R}}$ is said to be the *limit* of f at x_0 if one of the following three statements holds:

(i) λ is finite and for every $\varepsilon > 0$ there exists a real number $\delta > 0$ such that for every $x \in D \cap (x_0 - \delta, x_0 + \delta) \backslash \{x_0\}$, we have $|f(x) - \lambda| < \varepsilon$;

(ii) $\lambda = \infty$ and for every $K \in \mathbb{R}$ there exists a real number $\delta > 0$ such that for every $x \in D \cap (x_0 - \delta, x_0 + \delta) \backslash \{x_0\}$, we have $f(x) > K$;

(iii) $\lambda = -\infty$ and for every $K \in \mathbb{R}$ there exists a real number $\delta > 0$ such that for every $x \in D \cap (x_0 - \delta, x_0 + \delta) \backslash \{x_0\}$, we have $f(x) < K$.

The limit of f at x_0 is denoted as $\lim_{x \to x_0} f(x)$.

It follows from the definitions that if $x_0 \in \big((-\infty, x_0) \cap D\big)' \cap \big((x_0, \infty) \cap D\big)'$ then the limit of f at x_0 exists if and only if both one-sided limits of f at x_0 exist and coincide.

In the case where $x_0 \in \big((-\infty, x_0] \cap D\big)'$ but $x_0 \notin \big([x_0, \infty) \cap D\big)'$, then the existence of the limit of f at x_0 is equivalent to the existence of the left-hand limit of f at x_0. A similar assertion can be claimed for right-hand limits.

The notations $f(x) \xrightarrow{x \to x_0^-} \lambda$ (resp. $f(x) \xrightarrow{x \to x_0^+} \lambda$; $f(x) \xrightarrow{x \to x_0} \lambda$) are also commonly used to denote that λ is the left-hand limit (right-hand limit; limit) of f at x_0.

The behavior of a function $f(x)$ at increasingly large values of x is studied by applying the notions of limits at ∞.

Definition 2.1.4. Let D be a non-bounded above subset of $\mathbb{R}$ and let $f: D \longrightarrow \mathbb{R}$. A number $\lambda \in \overline{\mathbb{R}}$ is said to be the *limit* of f at ∞ if one of the following three statements holds:

(i) λ is finite and for every $\varepsilon > 0$ there exists $H \in \mathbb{R}$ such that for every $x \in D \cap (H, \infty)$, we have $|f(x) - \lambda| < \varepsilon$;

(ii) $\lambda = \infty$ and for every $K \in \mathbb{R}$ there exists $H \in \mathbb{R}$ such that for every $x \in D \cap (H, \infty)$, we have $f(x) > K$;

(iii) $\lambda = -\infty$ and for every $K \in \mathbb{R}$ there exists $H \in \mathbb{R}$ such that for every $x \in D \cap (H, \infty)$, we have $f(x) < K$.

The limit of f at ∞ is denoted as $\lim_{x \to \infty} f(x)$ or as $f(\infty)$.

The notation $f(x) \xrightarrow{x \to \infty} \lambda$ is also usual when λ is the limit of f at ∞.

Definition 2.1.5. Let D be a non-bounded below subset of $\mathbb{R}$ and let $f: D \longrightarrow \mathbb{R}$. A number $\lambda \in \overline{\mathbb{R}}$ is said to be the *limit* of f at $-\infty$ if one of the following three statements holds:

(i) λ is finite and for every $\varepsilon > 0$ there exists $H \in \mathbb{R}$ such that for every $x \in D \cap (-\infty, H)$, we have $|f(x) - \lambda| < \varepsilon$;

(ii) $\lambda = \infty$ and for every $K \in \mathbb{R}$ there exists $H \in \mathbb{R}$ such that for every $x \in D \cap (-\infty, H)$, we have $f(x) > K$;

(iii) $\lambda = -\infty$ and for every $K \in \mathbb{R}$ there exists $H \in \mathbb{R}$ such that for every $x \in D \cap (-\infty, H)$, we have $f(x) < K$.

The limit of f at $-\infty$ is denoted as $\lim_{x \to -\infty} f(x)$ or as $f(-\infty)$.

The notation $f(x) \xrightarrow{x \to -\infty} \lambda$ is also used to say that λ is the limit of f at $-\infty$.

Let us proceed to the sequential characterizations of limits of functions.

Proposition 2.1.6. *Let D be a non-empty subset of $\mathbb{R}$, let $x_0 \in \left((-\infty, x_0) \cap D \right)'$, let $f: D \longrightarrow \mathbb{R}$ and let $\lambda \in \overline{\mathbb{R}}$. Then the following statements are equivalent:*

(a) λ is the left-hand limit of f at x_0;

(b) if $(x_n)_{n=1}^{\infty} \subset (-\infty, x_0) \cap D$ and $x_n \xrightarrow{n} x_0$ then $f(x_n) \xrightarrow{n} \lambda$;

(c) if $(x_n)_{n=1}^{\infty} \subset (-\infty, x_0) \cap D$ is an increasing sequence such that $x_n \xrightarrow{n} x_0$, then $f(x_n) \xrightarrow{n} \lambda$.

Proof. We will prove the equivalence between statements (a), (b) and (c) when $\lambda \in \mathbb{R}$. The cases where $\lambda = \pm\infty$ are left for the reader.

(a)$\Rightarrow$(b) Assume (a) holds and take any sequence $(x_n)_{n=1}^{\infty}$ in $D \cap (-\infty, x_0)$ such that $x_n \xrightarrow[n]{} x_0$. We just need to prove that $f(x_n) \xrightarrow[n]{} \lambda$. To do this, fix $\varepsilon > 0$. By hypothesis, there exists $\delta > 0$ such that

$$|f(x) - \lambda| < \varepsilon \text{ for all } x \in (x_0 - \delta, x_0) \cap D. \tag{2.1}$$

As $x_n \xrightarrow[n]{} x_0$, there exists $n_0 \in \mathbb{N}$ such that for each $n \geq n_0$, x_n belongs to $(x_0 - \delta, x_0)$. Therefore, by virtue of (2.1), $|f(x_n) - \lambda| < \varepsilon$ for all $n \geq n_0$, which proves that $f(x_n) \xrightarrow[n]{} \lambda$, as was needed.

(b)$\Rightarrow$(c) Trivial.

(c)$\Rightarrow$(a) Assume (a) does not hold. Then there exists $\varepsilon > 0$ such that for every $\delta > 0$ there exists $x \in (x_0 - \delta, x_0) \cap D$ such that $|f(x) - \lambda| \geq \varepsilon$. The proof will be done as soon as we find an increasing sequence $(x_n)_{n=1}^{\infty}$ in $D \cap (-\infty, x_0)$ such that $x_n \xrightarrow[n]{} x_0$ but $f(x_n)$ does not converge to λ as $n \to \infty$. This sequence will be obtained recursively:

For $n = 1$, take $\delta_1 := 1$ and choose $x_1 \in (x_0 - \delta, x_0) \cap D$ so that $|f(x_1) - \lambda| \geq \varepsilon$.

Once x_1 has been chosen, take $\delta_2 := \min\{1/2, x_0 - x_1\}$ and choose $x_2 \in (x_0 - \delta, x_0) \cap D$ so that $|f(x_2) - \lambda| \geq \varepsilon$. Clearly, $x_1 < x_2 < x_0$ and $x_0 - x_2 < 1/2$.

Once x_2 has been chosen, take $\delta_3 := \min\{1/3, x_0 - x_2\}$ and choose $x_3 \in (x_0 - \delta, x_0) \cap D$ so that $|f(x_3) - \lambda| \geq \varepsilon$. Thus, $x_1 < x_2 < x_3 < x_0$ and $x_0 - x_3 < 1/3$.

Repeating the above procedure indefinitely, the sequence $(x_n)_{n=1}^{\infty}$ in $D \cap (-\infty, x_0)$ thus obtained is clearly increasing and satisfies $0 < x_0 - x_n < 1/n$ and $|f(x_n) - \lambda| \geq \varepsilon$ for all n. Thus $(x_n)_{n=1}^{\infty}$ converges to x_0 but $\big(f(x_n)\big)_{n=1}^{\infty}$ does not converge to λ. The proof is done. $\qquad\square$

Proposition 2.1.7. *Let D be a non-bounded above subset of $\mathbb{R}$, let $f : D \longrightarrow \mathbb{R}$ and let $\lambda \in \overline{\mathbb{R}}$. Then the following statements are equivalent:*

(a) λ is the limit of f at ∞;
(b) if $(x_n)_{n=1}^{\infty} \subset D$ and $x_n \xrightarrow[n]{} \infty$ then $f(x_n) \xrightarrow[n]{} \lambda$;
(c) if $(x_n)_{n=1}^{\infty} \subset D$ is an increasing sequence such that $x_n \xrightarrow[n]{} \infty$ then $f(x_n) \xrightarrow[n]{} \lambda$.

The proof is similar to that of Proposition 2.1.6.

Proposition 2.1.8. *Let D be a non-empty subset of $\mathbb{R}$, let $x_0 \in \big((x_0, \infty) \cap D\big)'$, let $f : D \longrightarrow \mathbb{R}$ and let $\lambda \in \overline{\mathbb{R}}$. Then the following statements are equivalent:*

(a) λ is the right-hand limit of f at x_0;
(b) if $(x_n)_{n=1}^{\infty} \subset (x_0, \infty) \cap D$ and $x_n \xrightarrow[n]{} x_0$ then $f(x_n) \xrightarrow[n]{} \lambda$;
(c) if $(x_n)_{n=1}^{\infty} \subset (x_0, \infty) \cap D$ is a decreasing sequence such that $x_n \xrightarrow[n]{} x_0$ then $f(x_n) \xrightarrow[n]{} \lambda$.

Proof. The proof parallels that of Proposition 2.1.6. $\qquad\square$

Proposition 2.1.9. *Let D be a non-bounded below subset of $\mathbb{R}$, let $f : D \longrightarrow \mathbb{R}$ and let $\lambda \in \overline{\mathbb{R}}$. Then the following statements are equivalent:*

(a) λ *is the limit of* f *at* $-\infty$;
(b) if $(x_n)_{n=1}^{\infty} \subset D$ *and* $x_n \xrightarrow[n]{} -\infty$ *then* $f(x_n) \xrightarrow[n]{} \lambda$;
(c) if $(x_n)_{n=1}^{\infty} \subset D$ *is a decreasing sequence such that* $x_n \xrightarrow[n]{} -\infty$ *then* $f(x_n) \xrightarrow[n]{} \lambda$.

The proof is very similar to that of Proposition 2.1.7.

Proposition 2.1.10. *Let D be a non-empty subset of $\mathbb{R}$, let $x_0 \in D'$, let $f\colon D \longrightarrow \mathbb{R}$ and let $\lambda \in \mathbb{R}$. Then the following statements are equivalent:*

(a) λ *is the limit of* f *at* x_0;
(b) if $(x_n)_{n=1}^{\infty} \subset D\backslash\{x_0\}$ *and* $x_n \xrightarrow[n]{} x_0$ *then* $f(x_n) \xrightarrow[n]{} \lambda$.

Proof. The proof parallels that of Proposition 2.1.6. □

Note that the uniqueness of the limit of a function at a point is guaranteed by the sequential characterizations (Propositions 2.1.7 to 2.1.10) and Theorem 1.2.5.

By means of Propositions 2.1.6, 2.1.7, 2.1.8, 2.1.9 and 2.1.10, the different notions of a limit of a function at a point can be presented in a unified way as follows:

Theorem 2.1.11. *Let $f\colon D \longrightarrow \mathbb{R}$ be a function and let $x_0 \in \overline{\mathbb{R}}$ satisfying one of the following conditions:*

(i) $x_0 \in D'$;
(ii) $x_0 = \infty$ *and D is non-bounded above;*
(iii) $x_0 = -\infty$ *and D is non-bounded below.*

Then the two following statements are equivalent for a given number $\lambda \in \overline{\mathbb{R}}$:

(a) there exists $\lim_{x \to x_0} f(x) = \lambda$;
(b) $f(x_n) \xrightarrow[n]{} \lambda$ *whenever* $x_n \xrightarrow[n]{} x_0$ *and* $(x_n)_{n=1}^{\infty} \subset D\backslash\{x_0\}$.

The proof is left for the reader.

Note that Theorem 2.1.11 also includes the cases of the one-sided limits at a finite point x_0 by resorting to suitable restrictions of f: the trick is that there exists $\lim_{x \to x_0^-} f(x) = \lambda$ if and only if there exists $\lim_{x \to x_0} f|_{D \cap (-\infty, x_0)}(x) = \lambda$. Indeed, by applying Theorem 2.1.11(b) to $f|_{D \cap (-\infty, x_0)}$, only sequences $(x_n)_{n=1}^{\infty}$ in $D \cap (-\infty, x_0)$ need to be considered. In a similar way, there exists $\lim_{x \to x_0^+} = \lambda$ if and only if there exists $\lim_{x \to x_0} f|_{D \cap (x_0, \infty)}(x) = \lambda$.

Proposition 2.1.12. *Let f and g be a pair of functions from a non-empty subset D of $\mathbb{R}$ into $\mathbb{R}$ and let $\odot$ denote one of the operations $+, -, \cdot$ or $\div$ (in the case that $\odot$ is division, it must be assumed that $g(x) \neq 0$ for all x). Let l and λ be a pair of elements in $\overline{\mathbb{R}}$ such that $l \odot \lambda$ is defined in $\overline{\mathbb{R}}$. Thus we have:*

(i) if $x_0 \in \left(D \cap (-\infty, x_0)\right)'$, $l = \lim_{x \to x_0^-} f(x)$ *and* $\lambda = \lim_{x \to x_0^-} g(x)$ *then there exists* $\lim_{x \to x_0^-} f \odot g(x) = l \odot \lambda$;

(ii) if $x_0 \in \left(D \cap (x_0, \infty)\right)'$, $l = \lim_{x \to x_0^+} f(x)$ and $\lambda = \lim_{x \to x_0^+} g(x)$ then there exists the limit $\lim_{x \to x_0^+} f \odot g(x) = l \odot \lambda$;

(iii) if $x_0 \in D'$, $l = \lim_{x \to x_0} f(x)$ and $\lambda = \lim_{x \to x_0} g(x)$ then there exists the limit $\lim_{x \to x_0} f \odot g(x) = l \odot \lambda$.

Proof. This is a direct consequence of Theorems 1.2.10 and 2.1.11.   □

In the following result, a and b are any pair of numbers in $\overline{\mathbb{R}}$.

Proposition 2.1.13. *For every monotone function $f \colon (a, b) \longrightarrow \mathbb{R}$, there exist $\lim_{x \to a} f(x)$ and $\lim_{x \to b} f(x)$.*

Proof. Denote $J := (a, b)$. In the case where f is non-decreasing, it readily follows from the definitions of limit, supremum and infimum that

$$\lim_{x \to b} f(x) = \sup f(J) \tag{2.2}$$

$$\lim_{x \to a} f(x) = \inf f(J). \tag{2.3}$$

Note that if f is not bounded above, then $\sup f(J) = \infty$, and if f is not bounded below, then $\inf f(J) = -\infty$.

In the case where f is non-increasing, a similar reasoning leads to

$$\lim_{x \to b} f(x) = \inf f(J) \tag{2.4}$$

$$\lim_{x \to a} f(x) = \sup f(J). \tag{2.5}$$

□

Corollary 2.1.14. *Given a monotone function $f \colon (a, b) \longrightarrow \mathbb{R}$ and $x_0 \in (a, b)$, there exist $\lim_{x \to x_0^-} f(x)$ and $\lim_{x \to x_0^+} f(x)$ and both limits are finite.*

Proof. The existence of both limits from both sides is a direct consequence of Proposition 2.1.13 combined with the trick described immediately after Theorem 2.1.11.

In the case where f is non-decreasing, identities (2.2) and (2.3) yield

$$-\infty < \lim_{x \to x_0^-} f(x) \leqslant f(x_0) \leqslant \lim_{x \to x_0^+} f(x) < \infty.$$

Analogously, if f is non-increasing, (2.2) and (2.3) show

$$\infty > \lim_{x \to x_0^-} f(x) \geqslant f(x_0) \geqslant \lim_{x \to x_0^+} f(x) > -\infty.$$

and the proof is done.   □

2.2   Continuous functions

Following the same procedure as in Section 2.1, we will state the definitions of continuity in terms of the classic 'ε-δ' definitions and we will immediately proceed to their sequential characterizations.

Definition 2.2.1. Let D be a non-empty subset of $\mathbb{R}$, let $x_0 \in D$ and let $f \colon D \longrightarrow \mathbb{R}$:

(i) f is said to be *left-hand continuous* at x_0 if for every $\varepsilon > 0$ there exists $\delta > 0$ such that for all $x \in (x_0 - \delta, x_0] \cap D$, we have $|f(x) - f(x_0)| < \varepsilon$;

(ii) f is said to be *right-hand continuous* at x_0 if and only if for every $\varepsilon > 0$ there exists $\delta > 0$ such that for all $x \in [x_0, x_0 + \delta) \cap D$, we have $|f(x) - f(x_0)| < \varepsilon$;

(iii) f is said to be *continuous* at x_0 if for every $\varepsilon > 0$ there exists $\delta > 0$ such that for every $x \in (x_0 - \delta, x_0 + \delta) \cap D$, we have $|f(x) - f(x_0)| < \varepsilon$.

It is evident that a function f is continuous at some point x_0 if and only if it is both left-hand and right-hand continuous at x_0. The term *one-sided continuity* is applied indistinctly for both notions, the left-hand continuity and the right-hand continuity.

Of course, with the same notation as in Definition 2.2.1, if x_0 is an isolated point of D then f is trivially continuous at x_0. But if $x_0 \in D \cap D'$, then f is continuous at x_0 if the values $f(x)$ approach $f(x_0)$ when x approaches x_0; to be more precise, if there exists $\lim_{x \to x_0} f(x) = f(x_0)$. Analogously, if $x_0 \in D \cap (D \cap (-\infty, x_0])'$, f is left-hand continuous at x_0 if there exists $\lim_{x \to x_0^-} f(x) = f(x_0)$. A similar argument works for the right-hand continuity.

Proposition 2.2.2. *Given a function $f \colon D \longrightarrow \mathbb{R}$ and $x_0 \in D$, the following statements are equivalent:*

(a) *f is left-hand continuous at x_0;*

(b) *for every sequence $(x_n)_{n=1}^{\infty} \subset (-\infty, x_0] \cap D$ such that $x_n \xrightarrow[n]{} x_0$, one has $f(x_n) \xrightarrow[n]{} f(x_0)$;*

(c) *for every non-decreasing sequence $(x_n)_{n=1}^{\infty} \subset (-\infty, x_0] \cap D$ such that $x_n \xrightarrow[n]{} x_0$, one has $f(x_n) \xrightarrow[n]{} f(x_0)$.*

The proof parallels closely that of Proposition 2.1.6 but with a few slight differences: here, $f(x_0)$ plays the role of λ in Proposition 2.1.6. Another difference is that the sequences $(x_n)_{n=1}^{\infty}$ considered in the statements (b) and (c) of Proposition 2.2.2 do not need to be strictly monotone. In fact, if $\left(D \cap (-\infty, x_0)\right)' = \varnothing$, then the only sequence $(x_n)_{n=1}^{\infty}$ that satisfies the conditions of (b) or (c) is the constant sequence $(x_0)_{n=1}^{\infty}$. For instance, that is the case where $D = [a, b]$ and $x_0 = a$.

The analogous result for the right-handed continuity is the following.

Proposition 2.2.3. *Given a function $f \colon D \longrightarrow \mathbb{R}$ and $x_0 \in D$, the following statements are equivalent:*

(a) *f is right-hand continuous at x_0;*

(b) *for every sequence $(x_n)_{n=1}^{\infty} \subset [x_0, \infty) \cap D$ such that $x_n \xrightarrow[n]{} x_0$, one has $f(x_n) \xrightarrow[n]{} f(x_0)$;*

(c) *for every non-increasing sequence $(x_n)_{n=1}^{\infty} \subset [x_0, \infty) \cap D$ such that $x_n \xrightarrow[n]{} x_0$, one has $f(x_n) \xrightarrow[n]{} f(x_0)$.*

The proof is left for the reader. Similar observations to those made for the proof of Proposition 2.2.2 are also valid here.

Proposition 2.2.4. *Given a function $f\colon D \longrightarrow \mathbb{R}$ and $x_0 \in D$, the following statements are equivalent:*

(a) *f is continuous at x_0;*
(b) *for every sequence $(x_n)_{n=1}^{\infty} \subset D$ such that $x_n \xrightarrow[n]{} x_0$, one has $f(x_n) \xrightarrow[n]{} f(x_0)$.*

The proof of Proposition 2.2.4 is also left for the reader.

A function $f\colon D \longrightarrow \mathbb{R}$ is said to be *continuous on D* (or just *continuous* for short) if f is continuous at each point $x \in D$.

If $x_0 \in D \cap D'$ and f is not continuous at x_0, then x_0 is said to be a point of discontinuity of f, in which case, exactly one of the following three cases holds true:

(i) there exists $\lim_{x \to x_0} f(x) = \lambda \in \mathbb{R}$ but $\lambda \neq f(x_0)$;
(ii) there exist $\lim_{x \to x_0^-} f(x) = l$, $\lim_{x \to x_0^+} f(x) = \lambda$ but $l \neq \lambda$ or one of the two one-sided limits is not finite;
(iii) case (ii) does not hold and f does not have a limit at x_0.

If (i) holds true then it is said that f has a *removable discontinuity* at x_0; if (ii) holds, then it is said that f has a *jump discontinuity* at x_0; and if (iii) holds, then the discontinuity at x_0 is called *essential*.

The following are the main theorems for continuous functions.

Theorem 2.2.5. *Let $f\colon D \longrightarrow \mathbb{R}$ and $g\colon D \longrightarrow \mathbb{R}$ be a pair of continuous functions at a point x_0. Then $f + g$, $f - g$ and fg are continuous at x_0, and if $g(x) \neq 0$ for all $x \in D$ then f/g is continuous at x_0.*

Proof. Let $(x_n)_{n=1}^{\infty} \subset D$ be a sequence such that $x_n \xrightarrow[n]{} x_0$ and let $\odot$ denote one of the operations $+$, $-$, $\cdot$ or $\div$.

Combining Proposition 2.2.4 with Theorem 1.2.10, we have $f(x_n) \xrightarrow[n]{} f(x_0)$ and $g(x_n) \xrightarrow[n]{} g(x_0)$, and therefore, $f(x_n) \odot g(x_n) \xrightarrow[n]{} f(x_0) \odot g(x_0)$, which proves that $f \odot g$ is continuous at x_0. $\qquad\square$

Note that if f and g are left-hand continuous at x_0 then Theorem 2.2.5 proves that $f + g$, $f - g$, $f \cdot g$ and f/g are also left-hand continuous at x_0: to understand why this is so, it is sufficient to apply Theorem 2.2.5 to the restrictions of f and g to $D \cap (-\infty, x_0]$. Similar arguments work for the right-hand continuity at x_0.

Theorem 2.2.6. *Let $g\colon D \longrightarrow \mathbb{R}$ and $f\colon E \longrightarrow \mathbb{R}$ be a pair of functions such that $g(D) \subset E$. If g is continuous at x_0 and f is continuous at $g(x_0)$ then $f \circ g$ is continuous at x_0.*

Proof. Let $(x_n)_{n=1}^{\infty} \subset D$ be a sequence such that $x_n \xrightarrow[n]{} x_0$. Since g is continuous at x_0 and f at $g(x_0)$, two applications of Proposition 2.2.4 yield $g(x_n) \xrightarrow[n]{} g(x_0)$ and $f(g(x_n)) \xrightarrow[n]{} f(g(x_0))$, and the proof is done. $\qquad\square$

Theorem 2.2.7 (Bolzano). *Let f be a continuous function on a compact interval $[a, b]$. If $f(a) < 0$ and $f(b) > 0$ then there exists $\xi \in (a, b)$ such that $f(\xi) = 0$.*

Proof. Let $A := \{x \in [a, b] : f(x) < 0\}$. Since $a \in A$ and A is bounded above by b, it follows that $\xi := \sup A \in [a, b]$.

For every $n \in \mathbb{N}$, $\xi - 1/n$ is not an upper bound of A, hence there exists $a_n \in A$ such that $\xi - 1/n < a_n \leqslant \xi$. By the sandwich rule, $a_n \xrightarrow[n]{} \xi$, and as $f(a_n) < 0$ for all n, the sequential characterization of the continuity and Theorem 1.2.8 give

$$f(a_n) \xrightarrow[n]{} f(\xi) \leqslant 0. \tag{2.6}$$

This entails $a \leqslant \xi < b$ and therefore, there is room enough between ξ and b for a decreasing sequence $(b_n)_{n=1}^{\infty}$ to be put into (ξ, b) with the additional condition that $b_n \xrightarrow[n]{} \xi$. As none of the elements b_n belongs to A, then $f(b_n) \geqslant 0$, so

$$f(b_n) \xrightarrow[n]{} f(\xi) \geqslant 0 \tag{2.7}$$

and it follows from (2.6) and (2.7) that $f(\xi) = 0$. $\qquad\square$

Theorem 2.2.8 (Bolzano–Darboux). *Let f be a continuous function on D. If I is an interval contained in D then $f(I)$ is an interval too.*

Proof. Pick any pair of elements y_1 and y_2 in $f(I)$ such that $y_1 \leqslant y_2$ and let $y_1 \leqslant \zeta \leqslant y_2$. The proof will be done as soon as we see that ζ belongs to $f(I)$. If $y_1 = \zeta$ or $y_2 = \zeta$ then there is nothing to prove. Assume then that $y_1 < \zeta < y_2$ and take x_1 and x_2 in I so that $y_1 = f(x_1)$ and $y_2 = f(x_2)$. Let J denote the compact subinterval of I whose endpoints are x_1 and x_2.

Let $g := f|_J - \zeta$. As g is continuous on J, and as $g(x_1) < 0$ and $g(x_2) > 0$, Theorem 2.2.7 yields $\xi \in J$ such that $0 = g(\xi) = f(\xi) - \zeta$, which means that $\zeta \in f(J) \subset f(I)$, and the proof is done. $\qquad\square$

Theorem 2.2.9 (Weierstrass). *Let f be a continuous function on a set K. If K is compact, so is $f(K)$.*

Proof. By virtue of Theorem 1.3.4, we merely have to prove that $B' \cap f(K) \neq \varnothing$ for all infinite subsets B of $f(K)$. There are two possible cases: either $f(K)$ is infinite or it is not.

If $f(K)$ is finite then it does not contain any infinite set and therefore $f(K)$ is trivially compact.

If $f(K)$ is infinite, take any infinite subset B of $f(K)$. Thus we may take a sequence $(y_n)_{n=1}^{\infty}$ of pairwise different elements of B.[2] For every y_n, take one element $x_n \in f^{-1}(y_n)$. Clearly, $(x_n)_{n=1}^{\infty}$ is a sequence of pairwise different elements in K and therefore, $A := \{x_n : n \in \mathbb{N}\}$ is an infinite subset of K. But K is compact so there exists $x_0 \in A' \cap K$. As $x_0 \in A'$, there exists a subsequence $(x_{k_n})_{n=1}^{\infty}$ of $(x_n)_{n=1}^{\infty}$ such that $x_{k_n} \xrightarrow[n]{} x_0$. But x_0 belongs to K and f is continuous at x_0, so

---

[2] A few notes about the cardinality of sets are given in Section A.13.

$y_{k_n} = f(x_{k_n}) \xrightarrow[n]{} f(x_0)$, and Proposition 1.2.4 gives $f(x_0) \in B' \cap f(K)$, which is what we wanted to prove. $\qquad\square$

Corollary 2.2.10. *If f is a continuous function on $[a, b]$ then $f([a, b]) = [m, M]$.*

Proof. Theorems 2.2.8 and 2.2.9 show that $f([a, b])$ is a compact interval. The result follows from the fact that a non-empty interval is compact if and only if it is of the form $[c, d]$. In this case, $f([a, b]) = [m, M]$ where m and M are respectively the minimum and the maximum of $f([a, b])$. $\qquad\square$

Theorem 2.2.11. *If I is an interval and $f \colon I \longrightarrow \mathbb{R}$ is an injective, continuous function on I, then f is strictly monotone and its inverse function $f^{-1} \colon f(I) \longrightarrow \mathbb{R}$ is continuous on $f(I)$.*

Proof. If f is not monotone then there exist three elements $x_1 < x_2 < x_3$ in I such that either

$$f(x_1) < f(x_2) > f(x_3) \tag{2.8}$$

or

$$f(x_1) > f(x_2) < f(x_3). \tag{2.9}$$

Assume (2.8) holds. First, consider the intervals $J_1 := [x_1, x_2]$, $J_2 := [x_2, x_3]$, let $M := \max\{f(x_1), f(x_3)\}$ and take $M < \xi < f(x_2)$. By virtue of Theorem 2.2.8, there is a pair of elements $c \in (x_1, x_2)$ and $d \in (x_2, x_3)$ such that $f(c) = f(d) = \xi$, but $c \neq d$ and f is injective, which is a contradiction.

If (2.9) holds, a similar argument also leads to a contradiction.

Once we have proved that f is monotone, it only remains to be seen that f^{-1} is continuous on $f(I)$. To do so, we take any $y_0 \in f(I)$ and we prove that f^{-1} is left-hand and right-hand continuous at y_0.

Let us assume f is increasing (the decreasing case is analogous). It is immediate that f^{-1} is also increasing. Let (y_n) be any non-decreasing sequence in $f(I)$ such that $y_n \xrightarrow[n]{} y_0$, and let $x_n := f^{-1}(y_n)$ for all $n = 0, 1, 2, \dots$. Clearly $(x_n)_{n=1}^{\infty}$ is non-decreasing and bounded above by x_0, so

$$x_n \xrightarrow[n]{} s := \sup\{x_n : n \in \mathbb{N}\} \leqslant x_0$$

and as f is continuous on I, then $y_n = f(x_n) \xrightarrow[n]{} f(s)$. Thus

$$f(s) = y_0 = f(x_0)$$

and the injectivity of f yields $s = x_0$. This proves that $f^{-1}(y_n) \xrightarrow[n]{} f^{-1}(y_0)$, and therefore, by virtue of Proposition 2.2.2(c), f^{-1} is left-hand continuous at y_0. A similar argument proves that f^{-1} is right-hand continuous at y_0, and the proof is done. $\qquad\square$

In general, given a continuous function on a subset D of $\mathbb{R}$ and given a fixed $\varepsilon > 0$, the choice of the parameter δ in Definition 2.2.1 depends on every point x_0 of D, but if the same number δ works for all $x_0 \in D$, then f is said to be uniformly continuous. To be more precise:

Definition 2.2.12. A function $f \colon D \longrightarrow \mathbb{R}$ is said to be *uniformly continuous* on D if for every $\varepsilon > 0$ there exists $\delta > 0$ such that for every pair of elements x and z in D such that $|x - z| < \delta$, one has $|f(x) - f(z)| < \varepsilon$.

Theorem 2.2.13 (Heine–Cantor). *Given a compact subset K of $\mathbb{R}$, every continuous function on K is uniformly continuous on K.*

Proof. Assume f is a continuous, non-uniformly continuous function on K.

Since f is not uniformly continuous, there exists $\varepsilon > 0$ such that for every $n \in \mathbb{N}$ there exists a pair of elements x_n and z_n in K such that

$$|x_n - z_n| < 1/n$$
$$|f(x_n) - f(z_n)| \geqslant \varepsilon. \tag{2.10}$$

Combining Propositions 1.3.2 and 1.3.3, we may extract a convergent subsequence $(x_{k_n})_{n=1}^{\infty}$ from $(x_n)_{n=1}^{\infty}$ so that $\lim_n x_{k_n} =: x_0 \in [a, b]$.

Moreover, f is continuous at x_0, so there exists $\delta > 0$ such that for every $z \in K$ such that $|z - x_0| < \delta$, we have

$$|f(z) - f(x_0)| < \varepsilon/2. \tag{2.11}$$

Choose $m \in \mathbb{N}$ such that $1/m < \delta/2$ and take a term x_{k_n} such that $k_n > m$ and such that $|x_{k_n} - x_0| < 1/m$. Then

$$|z_{k_n} - x_0| \leqslant |z_{k_n} - x_{k_n}| + |x_{k_n} - x_0| < \frac{1}{k_n} + \frac{1}{m} < \frac{2}{m} < \delta.$$

Hence $|f(z_{k_n}) - f(x_0)| < \varepsilon/2$ by virtue of (2.11); moreover, as $|x_{k_n} - x_0| < 1/m < \delta$, we also have $|f(x_{k_n}) - f(x_0)| < \varepsilon/2$ and therefore,

$$|f(z_{k_n}) - f(x_{k_n})| \leqslant |f(z_{k_n}) - f(x_0)| + |f(x_0) - f(x_{k_n})| < \frac{\varepsilon}{2} + \frac{\varepsilon}{2}$$

which contradicts (2.10). This proves that f is uniformly continuous. $\qquad\square$

2.3 Differentiable functions

Differentiability is a notion oriented to the study of the variation of a function at a given point. Given a continuous function $y = f(x)$ on a non-degenerate interval I, fix a point x_0 in I. Given a second point $z \in I$ such that $x_0 < z$, the increment of f when the variable x moves from x_0 to z is $\triangle f(x) := f(z) - f(x_0)$, also denoted as $\triangle y$ for short, and the increment of the variable x is $\triangle x := z - x_0$. Therefore the relative increment or ratio of the increment of f with respect to the increment $\triangle x$ of the variable x at x_0 is

$$\frac{\triangle y}{\triangle x} = \frac{f(z) - f(x_0)}{z - x_0}.$$

Note that $\frac{\Delta y}{\Delta x}$ equals exactly the slope of the line $r(x) = \frac{\Delta y}{\Delta x}(x - x_0) + f(x_0)$ that passes through the points $\big(x_0, f(x_0)\big)$ and $\big(z, f(z)\big)$ of the graph of $f(x)$ (see Diagram (2.12)).

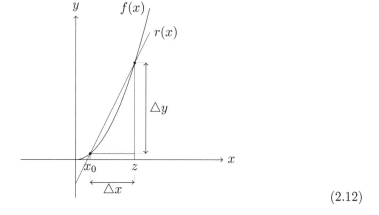

$$(2.12)$$

If we want to know the relative variation of $f(x)$ at x_0, then we need to reduce the distance between x_0 and z as much as possible, but keeping $z \neq x_0$. One way to do this is by taking a decreasing sequence $(x_n)_{n=1}^{\infty}$ convergent to x_0 where the successive terms x_n will play the role of z in the argument above. The line r_n is the one determined by the pair of points $\big(x_0, f(x_0)\big)$ and $\big(x_n, f(x_n)\big)$ (see Diagram 2.14). It can be seen that the larger n is, the closer $\big(x_n, f(x_n)\big)$ is to $\big(x_0, f(x_0)\big)$ and the closer the line r_n is to the tangent t_r to the graph of $f|_{I \cap [x_0, \infty)}(x)$ at $\big(x_0, f(x_0)\big)$, so the value

$$\lim_{n} \frac{f(x_n) - f(x_0)}{x_n - x_0} \qquad (2.13)$$

is the limit of the slopes of the successive lines r_n, that is, the slope of the tangent t_r, and at the same time, it is the relative increment of $f|_{I \cap [x_0, \infty)}(x)$ at x_0, in other words, the relative variation of f at x_0 when x approaches x_0 from its right-hand side.

In Diagram (2.14), the tangent t_l to the graph of $f|_{I \cap (-\infty, x_0]}(x)$ is drawn. In order to obtain its slope, we need to take an increasing sequence $(x_n)_{n=1}^{\infty}$ such that $x_n \xrightarrow{n} x_0$. Obviously, t_l and t_r do not coincide if the graph of $f(x)$ has a corner at

$(x_0, f(x_0))$, but if the graph of f is smooth enough at that point, then $t_l = t_r$.

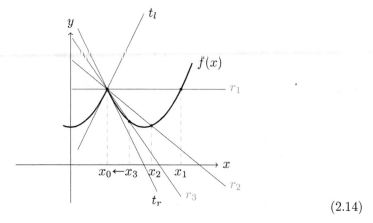

$$(2.14)$$

Of course, it would be desirable for all increasing sequences $(x_n)_{n=1}^{\infty}$ convergent to x_0 to share the same value for the limit given in (2.13), in which case thanks to the sequential characterization for the limit of a function at a point, we would rather use the function limit

$$\lim_{x \to x_0} \frac{f(x) - f(x_0)}{x - x_0}.$$

All these ideas are formalized below.

Definition 2.3.1. Let $f \colon D \longrightarrow \mathbb{R}$ be a function, let $x_0 \in D$ and let λ be a real number:

(i) if $x_0 \in \big(D \cap (-\infty, x_0] \big)'$ and there exists

$$\lim_{x \to x_0^-} \frac{f(x) - f(x_0)}{x - x_0} = \lambda$$

then λ is said to be the *left-hand derivative of f at x_0*;

(ii) if $x_0 \in \big(D \cap [x_0, \infty) \big)'$ and there exists

$$\lim_{x \to x_0^+} \frac{f(x) - f(x_0)}{x - x_0} = \lambda$$

then λ is said to be the *right-hand derivative of f at x_0*;

(iii) if $x_0 \in D'$ and there exists

$$\lim_{x \to x_0} \frac{f(x) - f(x_0)}{x - x_0} = \lambda$$

then λ is said to be the *derivative of f at x_0*.

The left-hand derivative (resp. right-hand derivative; derivative) of f at x_0 is denoted as $f'(x_0^-)$ (resp. $f'(x_0^+)$; $f'(x_0)$).

The term *one-sided differentiability* is applied indistinctly for both notions, the left-hand differentiability and the right-hand differentiability.

It is clear that if x_0 belongs to $\left(D \cap (-\infty, x_0]\right)'$ and to $\left(D \cap [x_0, \infty)\right)'$, then f is differentiable at x_0 if and only if $f'(x_0^-) = f'(x_0^+)$.

In order to round up this informal introduction to Section 2.3, let us say that $f'(x_0)$ measures the variation of f at x_0, and at the same time, it is the slope of the tangent line to the graph of f at $(x_0, f(x_0))$, which is expressed by the equation $g(x) = f'(x_0) \cdot (x - x_0) + f(x_0)$: the greater the variation of f at x_0, the steeper the slope of the line $y = g(x)$.

Similarly, $f'(x_0^-)$ and $f'(x_0^+)$ are the respective variations of $f|_{D \cap (-\infty, x_0]}$ and of $f|_{D \cap [x_0, \infty)}$ at x_0, and the slopes of the respective tangent lines to the graphs of $f|_{D \cap (-\infty, x_0]}$ and of $f|_{D \cap [x_0, \infty)}$ at $(x_0, f(x_0))$.

Differentiability implies continuity:

Proposition 2.3.2. *If* $f \colon D \longrightarrow \mathbb{R}$ *is differentiable at* x_0, *then it is continuous at* x_0.

Proof. Let $(x_n)_{n=1}^{\infty} \subset D \backslash \{x_0\}$ be any sequence convergent to x_0. Since f is differentiable at x_0,

$$\varepsilon_n := \frac{f(x_n) - f(x_0)}{x_n - x_0} - f'(x_0) \xrightarrow[n]{} 0.$$

Thus $f(x_n) - f(x_0) = (f'(x_0) + \varepsilon_n)(x_n - x_0) \xrightarrow[n]{} 0$. Hence f is continuous at x_0. $\quad\square$

Corollary 2.3.3. *If* $f \colon D \longrightarrow \mathbb{R}$ *is left-hand (resp. right-hand) differentiable at* x_0, *then it is left-hand (resp. right-hand) continuous at* x_0.

Proof. Apply Proposition 2.3.2 to $f|_{D \cap (-\infty, x_0]}$ (resp. to $f|_{D \cap [x_0, \infty)}$). $\quad\square$

As a consequence of Corollary 2.3.3, if both one-sided derivatives of f at x_0 exist, then f is continuous at x_0, irrespective of whether f is differentiable or not at x_0. A typical example of this is the function $f(x) = |x|$: in fact, the one-sided derivatives are $f'(0^-) = -1$ and $f'(0^+) = 1$.

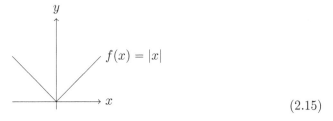

$$(2.15)$$

Definition 2.3.4. If $f \colon D \longrightarrow \mathbb{R}$ is differentiable at each $x \in D$, then f is said to be *differentiable on* D. Then the function from D into $\mathbb{R}$ that maps each x to $f'(x)$ is called the *derivative of* f *on* D and f is said to be an *antiderivative of* f'.

Proposition 2.3.5. *Let f and g be a pair of functions with domain D, both differentiable at x_0. Then the following statements hold:*

(i) $f + g$ *is differentiable at* x_0 *and* $(f + g)'(x_0) = f'(x_0) + g'(x_0)$;
(ii) fg *is differentiable at* x_0 *and* $(fg)'(x_0)$ $f'(x_0)g(x_0) + f(x_0)g'(x_0)$;
(iii) if $g(x) \neq 0$ *for all* x, *then* $1/g$ *is differentiable at* x_0 *and*

$$(1/g)'(x_0) = -\frac{g'(x_0)}{g(x_0)^2}.$$

Proof. Bearing in mind Proposition 2.1.12, the three statements are proved as follows:
(i)

$$\frac{(f + g)(x) - (f + g)(x_0)}{x - x_0} = \frac{f(x) - f(x_0)}{x - x_0} + \frac{g(x) - g(x_0)}{x - x_0} \xrightarrow{x \to x_0} f'(x_9) + g'(x_0).$$

(ii) As $(fg)(x) - (fg)(x_0) = f(x)g(x) - f(x_0)g(x) + f(x_0)g(x) - f(x_0)g(x_0)$ we have

$$\frac{(fg)(x) - (fg)(x_0)}{x - x_0} = \frac{f(x) - f(x_0)}{x - x_0} \cdot g(x) + f(x_0) \cdot \frac{g(x) - g(x_0)}{x - x_0}$$
$$\xrightarrow{x \to x_0} f'(x_0)g(x_0) + f(x_0)g'(x_0).$$

(iii)

$$\frac{1/g(x) - 1/g(x_0)}{x - x_0} = \frac{1}{g(x)g(x_0)} \cdot \frac{g(x_0) - g(x)}{x - x_0} \xrightarrow{x \to x_0} -\frac{g'(x_0)}{g(x_0)^2}.$$

$\square$

Proposition 2.3.6 (chain rule). *Let* $g\colon D \longrightarrow \mathbb{R}$ *and* $f\colon E \longrightarrow \mathbb{R}$ *be a pair of functions such that* $g(D) \subset E$ *and let* $x_0 \in D$. *If* g *is differentiable at* x_0 *and* f *is differentiable at* $g(x_0)$, *then* $f \circ g$ *is differentiable at* x_0 *and* $(f \circ g)'(x_0) = f'(g(x_0)) \cdot g'(x_0)$.

Proof. Let $(x_n)_{n=1}^\infty \subset D \backslash \{x_0\}$ be a sequence convergent to x_0 and denote $y_0 := g(x_0)$ and $y_n := g(x_n)$ for all $n \in \mathbb{N}$. By virtue of Proposition 2.3.2, g is continuous at x_0, so $y_n \xrightarrow{n} y_0$. Let $A := \{n \in \mathbb{N}: y_n = y_0\}$, $B := \mathbb{N} \backslash A$ and let their elements be relabeled as $A = \{k_1 < k_2 < \ldots\}$ and $B = \{m_1 < m_2 < \ldots\}$. At least one of these sets is infinite. In the case where A is infinite, then the sequential characterization of function limits yields

$$g'(x_0) = \lim_n \frac{y_{k_n} - y_0}{x_{k_n} - x_0} = \lim_n 0 = 0$$

and moreover,

$$\frac{f \circ g(x_{k_n}) - f \circ g(x_0)}{x_{k_n} - x_0} = 0 \xrightarrow{n} 0 \qquad (2.16)$$

and if B is infinite, then

$$\frac{f \circ g(x_{m_n}) - f \circ g(x_0)}{x_{m_n} - x_0} \tag{2.17}$$

$$= \frac{f(y_{m_n}) - f(y_0)}{y_{m_n} - y_0} \cdot \frac{y_{m_n} - y_0}{x_{m_n} - x_0} \xrightarrow[n]{} f'(y_0) g'(x_0).$$

Thus, the statement is proved by (2.16) in the case where A is infinite and B is not, and by (2.17) in the case where B is infinite and A is not. In the case where both A and B are infinite, the statement is proved by both (2.16) and (2.17) in combination with Proposition 1.2.12. ◻

Let us explore the differentiability of inverse functions.

Proposition 2.3.7. *Let f be an injective, continuous function on an interval I. If f is differentiable at a point $x_0 \in I$, then only one of the following statements holds:*

(i) $f'(x_0) \neq 0$, in which case f^{-1} is differentiable at $f(x_0)$ and $(f^{-1})'(f(x_0)) = 1/f'(x_0)$;

(ii) $f'(x_0) = 0$, in which case f^{-1} is not differentiable at $f(x_0)$.

Proof. Denote $y_0 := f(x_0)$ and let $(y_n)_{n=1}^{\infty} \subset f(I) \backslash \{y_0\}$ be a convergent sequence to y_0. For each $n \in \mathbb{N}$, let $x_n \in I$ be the unique element in I such that $f(x_n) = y_n$. Recall that f^{-1} is strictly monotone and continuous on $f(I)$ by virtue of Theorem 2.2.11, hence $x_n \xrightarrow[n]{} x_0$. Without loss of generality, we assume that f is increasing. Thus $x_n \neq x_0$ for all $n \in \mathbb{N}$ and

$$\frac{f^{-1}(y_n) - f^{-1}(y_0)}{y_n - y_0} = \frac{x_n - x_0}{f(x_n) - f(x_0)} > 0 \quad \text{for all } n \in \mathbb{N}. \tag{2.18}$$

If we assume now that $f'(x_0) \neq 0$, then (2.18) and Theorem 1.2.10 yield

$$\frac{f^{-1}(y_n) - f^{-1}(y_0)}{y_n - y_0} \xrightarrow[n]{} \frac{1}{f'(x_0)}$$

and statement (i) is proved.

But in the case that $f'(x_0) = 0$, then (2.18) and Theorem 1.2.11 give

$$\frac{f^{-1}(y_n) - f^{-1}(y_0)}{y_n - y_0} \xrightarrow[n]{} \infty.$$

Hence f^{-1} is not differentiable at y_0, which fulfils statement (ii). ◻

The following mean value theorems are crucial for studying the variation of a given function and for proving L'Hôpital's rule.

Theorem 2.3.8 (Rolle). *Let $a < b$ and let $f : [a, b] \longrightarrow \mathbb{R}$ be a continuous function on $[a, b]$ and differentiable at each point of (a, b). If $f(a) = f(b)$, then there exists a point $c \in (a, b)$ such that $f'(c) = 0$.*

Proof. Since $[a, b]$ is compact, by Corollary 2.2.10 there exist c_1 and c_2 in $[a, b]$ such that $f(c_1) \leqslant f(x) \leqslant f(c_2)$ for all $x \in [a, b]$.

If $\{c_1, c_2\} = \{a, b\}$, then f is constant and consequently, $f'(x) = 0$ for all x. Otherwise, at least one of the points c_i lays on (a, b). If $c_2 \in (a, b)$ (if $c_1 \in (a, b)$ the proof is analogous), then, for every $x \in [a, c_2)$, we have $(f(x) - f(c_2))/(x - c_2) \geqslant 0$. Hence

$$f'(c_2) = f'(c_2^-) = \lim_{x \to c_2^-} \frac{f(x) - f(c_2)}{x - c_2} \geqslant 0,$$

and for every $x \in (c_2, b]$, $(f(x) - f(c_2))/(x - c_2) \leqslant 0$. Hence

$$f'(c_2) = f'(c_2^+) = \lim_{x \to c_2^+} \frac{f(x) - f(c_2)}{x - c_2} \leqslant 0,$$

which proves that $f'(c_2) = 0$, as desired. $\qquad\square$

The following theorem is known as Cauchy's mean value theorem.

Theorem 2.3.9. *Let $a < b$ and let f and g be a pair of continuous functions on $[a, b]$ and differentiable on (a, b). Then there exists a point $c \in (a, b)$ such that*

$$f'(c) \cdot [g(b) - g(a)] = g'(c) \cdot [f(b) - f(a)].$$

Proof. The proof is merely an application of Theorem 2.3.8 to the function $h(x) := f(x)[g(b) - g(a)] - g(x)[f(b) - f(a)]$. $\qquad\square$

Cauchy's theorem is known as the Lagrange mean value theorem when g is the identity.

Proposition 2.3.10. *Let I be a non-degenerate interval and let $f : I \longrightarrow \mathbb{R}$ be a differentiable function on I. Then $f' = 0$ if and only if f is constant.*

Proof. Assume f is constant, that is, there is a real number k such that $f(x) = k$ for all $x \in I$. Hence, for each $x_0 \in I$,

$$f'(x_0) = \lim_{x \to x_0} \frac{f(x) - f(x_0)}{x - x_0} = \lim_{x \to x_0} \frac{0}{x - x_0} = 0$$

which proves that $f'(x) = 0$ for all $x \in I$.

Conversely, let us assume that $f'(x) = 0$ for all x and fix $x_0 \in I$. Given any $x \in I \backslash \{x_0\}$, Theorem 2.3.9 provides a point c_x between x and x_0 such that

$$f'(c_x) = \frac{f(x) - f(x_0)}{x - x_0}.$$

But by hypothesis, $f'(c_x) = 0$ so $f(x) = f(x_0)$ for all $x \in I$, which proves that f is constant. $\qquad\square$

The following result clearly states that $f'(x)$ is a numerical indicator of the variation of f at the point x.

Proposition 2.3.11. *Let I be a non-degenerate, real interval and let f be a differentiable function on I. Then the following two statements hold:*

(i) f is non-decreasing if and only if $f'(x) \geqslant 0$ for all $x \in I$;
(ii) if $f'(x) > 0$ for all $x \in I$, then f is increasing.

Proof. (i) Assume f is non-decreasing. Thus, for fixed $x_0 \in I$, it is evident that $(f(x) - f(x_0))/(x - x_0) \geqslant 0$ for all $x \in I\backslash\{x_0\}$. Hence

$$f'(x_0) = \lim_{x \to x_0} \frac{f(x) - f(x_0)}{x - x_0} \geqslant 0.$$

Conversely, assume $f'(x) \geqslant 0$ for all x. Then, given a pair of elements $x < z$ in I, Theorem 2.3.9 provides an element $c \in (x, z)$ such that

$$f(z) - f(x) = f'(c) \cdot (z - x) \geqslant 0$$

which proves that f is non-decreasing.

(ii) Using the same line of reasoning given for the converse of (i), if $f'(u) > 0$ for all u, we obtain

$$f(z) - f(x) = f'(c) \cdot (z - x) > 0,$$

that is, f is increasing.   □

It is immediate that under the same hypotheses of Proposition 2.3.11, f is non-increasing if and only if $f'(x) \leqslant 0$ for all $x \in I$, and if $f'(x) < 0$ for all x then f is decreasing.

Statement (i) fails if the domain of f is not an interval. In fact, the function $f(x) := 1/x$ defined on $(-\infty, 0) \cup (0, \infty)$ is not monotone but $f'(x) > 0$ at each x.

The converse of (ii) also fails. For instance, $f(x) = x^3$ is increasing but $f'(0) = 0$. Of course, analogous statements can be given for non-increasing functions.

L'Hôpital's rule is is useful for calculating certain types of antiderivatives. Only the following restricted version is needed here.

Theorem 2.3.12 (L'Hôpital's rule). *Let $f(x)$ and $g(x)$ be a pair of differentiable functions on (a, b) such that $-\infty < a < b < \infty$ and let $\lambda \in \mathbb{R}$ such that the following conditions are satisfied:*

(i) $\lim_{x \to b} f(x) = \lim_{x \to b} g(x) = 0$,
(ii) $g(x) \neq 0$ and $g'(x) \neq 0$ for all x
(iii) there exists $\lim_{x \to b} f'(x)/g'(x) = \lambda$;

then there exists $\lim_{x \to b} f(x)/g(x) = \lambda$.

Proof. Extend f and g to b as $f(b) = g(b) = 0$. Let $(x_n)_{n=1}^{\infty} \subset (a, b)$ be a sequence such that $x_n \xrightarrow[n]{} b$. For every x_n, Theorem 2.3.9 applied to the interval $[x_n, b]$ provides an element $c_n \in (x_n, b)$ such that

$$\frac{f(x_n)}{g(x_n)} = \frac{f'(c_n)}{g'(c_n)}. \qquad (2.19)$$

Hence, as $x_n \xrightarrow[n]{} b$, Theorem 1.2.7 yields $c_n \xrightarrow[n]{} b$. But by hypothesis, $f'(c_n)/g'(c_n) \xrightarrow[n]{} \lambda$ so identity (2.19) proves $f(x)/g(x) \xrightarrow[x \to b]{} \lambda$.   □

Of course, L'Hôpital's rule also works if each instance of $x \longrightarrow b$ is replaced by $x \longrightarrow a$.

The general version of L'Hôpital's rule includes the cases $-\infty \leqslant a < b \leqslant \infty$ and permits the replacement of condition (i) by the alternative condition that $\lim_{x \to b} g(x) = \infty$.

There are many ways to approximate a function by polynomials. We are interested in the approximation by Taylor polynomials. In the following, $f^{(k)}(x)$ denotes the k-th derivative of f at x if this derivative exists, and $f^{(0)}(x)$ denotes $f(x)$.

Definition 2.3.13. Given a function $f \colon D \longrightarrow \mathbb{R}$ n times differentiable at a, the n-th Taylor polynomial of f at a is the polynomial given by

$$P_{a,n}(x) := f(a) + \frac{f^{(1)}(a)}{1!}(x - a) + \frac{f^{(2)}(a)}{2!}(x - a)^2 + \ldots + \frac{f^{(n)}(a)}{n!}(x - a)^n.$$

The reason $P_{a,n}(x)$ approximates quite well to $f(x)$ near a is that $f^{(k)}(a) = P_{a,n}^{(k)}(a)$ for all $k = 0, 1, \ldots, n$. This approximation can be measured by means of the following result, known as Taylor's theorem.

Theorem 2.3.14. *Let f be a function $n + 1$ times differentiable on an open interval I. Let a and x be a pair of different elements of I and let $I_{a,x}$ be the closed interval whose endpoints are a and x. Then there exists a number $\xi \in I_{a,x}^{\circ}$ for which*

$$f(x) = P_{a,n}(x) + \frac{f^{(n+1)}(\xi)}{(n+1)!}(x - a)^{n+1}.$$

Proof. Consider the functions

$$G(t) := P_{t,n}(x)$$
$$= f(t) + \frac{f'(t)}{1!}(x - t) + \frac{f^{(2)}(t)}{2!}(x - t)^2 + \ldots + \frac{f^{(n)}(t)}{n!}(x - t)^n, \ t \in I$$
$$g(t) := (x - t)^{n+1}, \ t \in I.$$

Since $f(t)$ is $n + 1$ times differentiable, it follows that $G(t)$ is differentiable on I. Thus, an application of Theorem 2.3.9 to the functions $G(t)$ and $g(t)$ on the interval $I_{a,x}$ shows the existence of a number $\xi \in I_{a,x}^{\circ}$ such that

$$\frac{G'(\xi)}{g'(\xi)} = \frac{G(x) - G(a)}{g(x) - g(a)}. \tag{2.20}$$

Clearly, the right side of (2.20) is

$$\frac{G(x) - G(a)}{g(x) - g(a)} = -\frac{f(x) - P_{a,n}(x)}{(x - a)^{n+1}}. \tag{2.21}$$

In order to evaluate the left side of (2.20), realize that for each $k = 1, 2, \ldots, n$, the derivative of $f^{(k)}(t)(x - t)^k/k!$ with respect to t is

$$-\frac{1}{(k-1)!}f^{(k)}(t) \cdot (x - t)^{k-1} + \frac{1}{k!}f^{(k+1)}(t) \cdot (x - t)^k, \ t \in I.$$

Thus, it readily follows that

$$G'(t) = \frac{f^{(n+1)}(t)}{n!}(x-t)^n, \quad t \in I.$$

As $g'(t) = -(n+1)(x-t)^n$ for all t, the evaluation of $G'(t)/g'(t)$ at ξ yields $-f^{(n+1)}(\xi)/(n+1)!$. Hence, from (2.21) and (2.20),

$$\frac{f(x) - P_{a,n}(x)}{(x-a)^{n+1}} = \frac{f^{(n+1)}(\xi)}{(n+1)!}$$

and the proof is done by solving $f(x)$ in this last identity.   □

2.4   Integrable functions

The purpose of the Riemann integral is to measure the area determined by the graph of a function $f \geqslant 0$ defined on a compact interval $[a,b]$ and the axis of abscissas. An estimate of that area can be obtained by choosing a finite amount of consecutive terms $a = t_0 < t_1 < \ldots < t_{n-1} < t_n = b$ and summing the areas of all the rectangles, each of which is based on the subinterval $I_i := [t_{i-1}, t_i]$ and whose height, m_i, is the greatest real number such that $m_i \leqslant f(x)$ for all $x \in I_i$ so that the rectangle does not surpass the graph of f (see Diagram 2.22). Of course, the value thus obtained is an underestimate of the area A determined for f. In order to obtain an overestimate for A, we consider the sum of the areas of a collection of intervals each of which is based on I_i but whose height, M_i, is the least real number such that $f(x) \leqslant M_i$ for all $x \in I_i$. Hence these new rectangles will contain all of the area determined by f. A glance at Diagram 2.22 shows that the error made by any of these two estimations of A will be smaller than the sum of the areas of all the yellow rectangles. Thus, the smaller the distance between any pair of consecutive terms t_{i-1} and t_i is, the smaller the error will be.

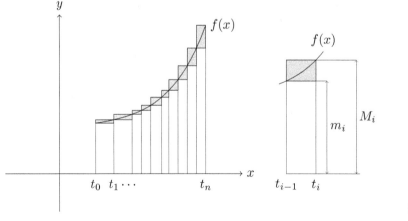

$$(2.22)$$

Note that if $f(x) < 0$ for all x, then the height of every rectangle inscribed in the graph of $f(x)$ is negative, which produces negative areas. If the reader feels uncomfortable with the idea of a negative area, then the integral of f can be understood as being an average of f on $[a, b]$ instead of as the area as determined by f.

All these ideas are formalized below.

A *partition* $\{t_i\}_{i=0}^n$ of a compact interval $[a, b]$ (with $a < b$) is a finite collection of consecutively ordered points $\{a = t_0 < t_1 < \ldots < t_n = b\}$. Partitions are denoted by capital letters P, Q, etc., and the set of all partitions of $[a, b]$ is denoted as $\mathcal{P}([a, b])$. Given a pair of partitions P and Q of $[a, b]$, Q is said to be a *refinement* of P if each element of P belongs to Q.

Given a bounded function $f \colon [a, b] \longrightarrow \mathbb{R}$ and a partition $P = \{t_i\}_{i=0}^n$ of $[a, b]$, the *upper* and the *lower sums of Darboux* of f associated with P are respectively defined by

$$U(f, P) := \sum_{i=1}^n \sup f([t_{i-1}, t_i]) \cdot (t_i - t_{i-1})$$

$$L(f, P) := \sum_{i=1}^n \inf f([t_{i-1}, t_i]) \cdot (t_i - t_{i-1})$$

Proposition 2.4.1. *For every bounded function* $f \colon [a, b] \longrightarrow \mathbb{R}$ *and for every pair of partitions* P *and* Q *of* $[a, b]$ *where* Q *is a refinement of* P, *we have*

$$L(f, P) \leqslant L(f, Q) \leqslant U(f, Q) \leqslant U(f, P).$$

Proof. Let $P = \{x_i\}_{i=0}^m$ and $Q = \{t_i\}_{i=0}^n$, and let $\{t_{k_0} < t_{k_1} < \ldots < t_{k_m}\}$ be the points of P that belong to Q so that $x_i = t_{k_i}$ for $i = 0, 1, 2, \ldots, m$.

Denote $M_i := \sup f([t_{i-1}, t_i])$ for $1 \leqslant i \leqslant n$, and $M_j' := \sup f([x_{j-1}, x_j])$ for $1 \leqslant j \leqslant m$.

Note that for each $1 \leqslant j \leqslant m$ and each $k_{j-1} + 1 \leqslant i \leqslant k_j$, $f([t_{i-1}, t_i])$ is a subset of $f([x_{j-1}, x_j])$. Hence M_j' is an upper bound of $f([t_{i-1}, t_i])$ and consequently,

$$M_i \leqslant M_j', \quad \text{for all } 1 \leqslant j \leqslant m \text{ and all } k_{j-1} + 1 \leqslant i \leqslant k_j. \tag{2.23}$$

It follows from (2.23) that

$$U(f, Q) = \sum_{i=1}^n M_i(t_i - t_{i-1}) = \sum_{j=1}^m \sum_{i=k_{j-1}+1}^{k_j} M_i(t_i - t_{i-1})$$

$$\leqslant \sum_{j=1}^m \sum_{i=k_{j-1}+1}^{k_j} M_j'(t_i - t_{i-1}) = \sum_{j=1}^m M_j' \sum_{i=k_{j-1}+1}^{k_j} (t_i - t_{i-1})$$

$$= \sum_{j=1}^m M_j'(t_{k_j} - t_{k_{j-1}}) = U(f, P).$$

A symmetric argument shows that $L(f, Q) \geqslant L(f, P)$.

In order to prove $L(f, Q) \leqslant U(f, Q)$, by denoting $m_i := \inf f([t_{i-1}, t_i])$ for $i = 1, 2 \ldots, n$, it is immediate that $m_i \leqslant M_i$, so

$$L(f, Q) = \sum_{i=1}^{n} m_i(t_i - t_{i-1}) \leqslant \sum_{i=1}^{n} M_i(t_i - t_{i-1}) \leqslant U(f, Q)$$

and the proof is done.   □

Proposition 2.4.1 tells us that the greatest number smaller than any upper sum of Darboux is also an overestimate that measures the area limited by f with more precision than any upper sum. In a similar manner, the smallest number greater than any lower sum is a better underestimate than any lower sum of Darboux. These estimates are named *upper* and *lower Darboux integrals* of f, respectively, and are formally defined as

$$\sup \int_a^b f := \inf\{U(f, P) \colon P \in \mathcal{P}([a, b])\}$$

$$\inf \int_a^b f := \sup\{L(f, P) \colon P \in \mathcal{P}([a, b])\}.$$

Clearly,

$$\inf \int_a^b f \leqslant \sup \int_a^b f.$$

Definition 2.4.2. A bounded function $f \colon [a, b] \longrightarrow \mathbb{R}$ is said to be *Riemann-integrable* on $[a, b]$ if $\inf \int_a^b f = \sup \int_a^b f$, in which case, the *Riemann integral of f on $[a, b]$* is the real number denoted as

$$\int_a^b f := \inf \int_a^b f.$$

Henceforth, we will just say that f is *integrable* on $[a, b]$ whenever f is Riemann-integrable on $[a, b]$ and the number $\int_a^b f$ will just be referred to as the *integral of f on $[a, b]$* for short.

Proposition 2.4.3. *If f and g are integrable on $[a, b]$ and $f(x) \leqslant g(x)$ for all x, then $\int_a^b f \leqslant \int_a^b g$.*

Proof. Given any partition P of $[a, b]$, from the hypothesis that $f(x) \leqslant g(x)$ it immediately follows that $L(f, P) \leqslant L(g, P)$, and the result follows from Definition 2.4.2.   □

Proposition 2.4.4 (Cauchy's criterium). *A bounded function $f \colon [a, b] \longrightarrow \mathbb{R}$ is integrable on $[a, b]$ if and only if for every $\varepsilon > 0$ there exists a partition P of $[a, b]$ such that $U(f, P) - L(f, P) < \varepsilon$.*

Proof. Assume f is integrable on $[a, b]$ and take $\varepsilon > 0$. Then there exists a pair of partitions P_1 and P_2 of $[a, b]$ such that $U(f, P_1) - \int_a^b f < \varepsilon/2$ and $\int_a^b f - L(f, P_2) < \varepsilon/2$. Thus, given any refinement $P \in \mathcal{P}([a, b])$ of P_1 and of P_2, according to Proposition 2.4.1, it follows that

$$U(f, P) - L(f, P) \leqslant U(f, P_1) - \int_a^b f + \int_a^b f - L(f, P_2) < \frac{\varepsilon}{2} + \frac{\varepsilon}{2} = \varepsilon.$$

Conversely, assume that for every $\varepsilon > 0$ there exists a partition P_ε such that $U(f, P_\varepsilon) - L(f, P_\varepsilon) < \varepsilon$. Then

$$0 \leqslant \sup \int_a^b f - \inf \int_a^b f \leqslant U(f, P_\varepsilon) - L(f, P_\varepsilon) < \varepsilon, \qquad (2.24)$$

and since (2.24) works for all $\varepsilon > 0$, then $0 = \sup \int_a^b f - \inf \int_a^b f$, which proves that f is integrable. $\qquad \square$

The following result identifies the two most important classes of integrable functions:

Theorem 2.4.5. *Let $f \colon [a, b] \longrightarrow \mathbb{R}$ be a function for which at least one of the following two conditions holds:*

(i) f is monotone on $[a, b]$;
(ii) f is continuous on $[a, b]$.

Then f is integrable on $[a, b]$.

Proof. Fix $\varepsilon > 0$. By virtue of Proposition 2.4.4, the proof will be done for both (i) and (ii) as soon as we find a partition P of $[a, b]$ such that $U(f, P) - L(f, P) < \varepsilon$.

(i) Suppose f is non-decreasing. If $f(a) = f(b)$, then f is constant and therefore, $U(f, Q) - L(f, Q) = 0$ for all partitions Q. Hence f is trivially integrable. Assume then that $f(a) < f(b)$. Then it is sufficient to choose a partition $P = \{t_i\}_{i=0}^n$ such that $t_i - t_{i-1} < \varepsilon/(f(b) - f(a))$ for all i, so that $\sup f([t_{i-1}, t_i]) = f(t_i)$ and $\inf f([t_{i-1}, t_i]) = f(t_{i-1})$. Therefore

$$U(f, P) - L(f, P) = \sum_{i=1}^n (f(t_i) - f(t_{i-1}))(t_i - t_{i-1})$$

$$< \frac{\varepsilon}{f(b) - f(a)} \sum_{i=1}^n (f(t_i) - f(t_{i-1})) = \varepsilon$$

as desired.

(ii) Let f be a continuous function on $[a, b]$. As $[a, b]$ is compact, f is uniformly continuous by virtue of Theorem 2.2.13, and therefore there exists $\delta > 0$ such that for all x and z in $[a, b]$ that satisfy $|x - z| < \delta$, we have $|f(x) - f(z)| < \varepsilon/(b - a)$. Hence, we just need to take a partition $P = \{t_i\}_{i=0}^n$ such that $|t_i - t_{i-1}| < \delta$ for all i. Indeed, as each subinterval $[t_{i-1}, t_i]$ is compact, by virtue of Corollary 2.2.10 there is a pair of elements c_i and d_i in $[t_{i-1}, t_i]$ where $f|_{[t_{i-1}, t_i]}$ reaches its minimum and its maximum, that is,

$$f(c_i) \leqslant f(x) \leqslant f(d_i) \quad \text{for all } x \in [t_{i-1}, t_i].$$

Thus, as $|c_i - d_i| < \delta$, we get $|f(c_i) - f(d_i)| < \varepsilon/(b-a)$ and consequently,

$$U(f, P) - L(f, P) = \sum_{i=1}^{n} (f(d_i) - f(c_i))(t_i - t_{i-1})$$

$$< \frac{\varepsilon}{b-a} \sum_{i=1}^{n} (t_i - t_{i-1}) = \varepsilon$$

and by Cauchy's criterium, f is integrable. $\square$

Henceforth, when $-\infty < a < b < \infty$ and f is integrable on $[a, b]$, it is convenient to agree

$$\int_a^a f := 0$$

$$\int_b^a f := -\int_a^b f,$$

in order to construct the integral functions of a given function as follows: Let I be any real interval and assume $f: I \longrightarrow \mathbb{R}$ is an integrable function on every compact interval contained in I. Fix any element $a \in I$; then the function $F_a: I \longrightarrow \mathbb{R}$ defined as

$$F_a(x) := \int_a^x f, \quad x \in I$$

is said to be an *integral function of f* or an *indefinite integral of f*. Of course, each choice of a means a different integral function of f.

The following results are related to the effective computation of integrals and to the properties of integral functions.

Theorem 2.4.6 (mean value theorem). *The following statements hold for every bounded function $f: [a, b] \longrightarrow \mathbb{R}$:*

(i) *if f is integrable on $[a, b]$, there exists $\mu \in [m, M]$ such that $\int_a^b f = \mu \cdot (b - a)$, where m and M are respectively a lower bound and an upper bound of $f([a, b])$;*
(ii) *if, moreover, f is continuous on $[a, b]$, there exists a point $c \in [a, b]$ such that $\int_a^b f = f(c) \cdot (b - a)$.*

Proof. (i) By virtue of Proposition 2.4.3, $m(b - a) \leqslant \int_a^b f \leqslant M(b - a)$, and consequently it follows that $\mu := \int_a^b f/(b - a) \in [m, M]$.
  (ii) By virtue of Corollary 2.2.10, there are two points ζ and ξ in $[a, b]$ such that

$$[f(\zeta), f(\xi)] = f([a, b]). \tag{2.25}$$

Thus, from (i), there exists $\mu \in [f(\zeta), f(\xi)]$ such that $\int_a^b f = \mu \cdot (b - a)$, and from (2.25), $\mu = f(c)$ for some $c \in [a, b]$, and the proof is done. $\square$

Theorem 2.4.7. *Let I be a real interval and let $f : I \longrightarrow \mathbb{R}$ be an integrable function on every compact interval contained in I. Fix $a \in I$ and consider the integral function $F(x) := \int_a^x f$ defined on I. Then the following two statements hold:*

(i) $F(x)$ is continuous on I;
(ii) if f is continuous on I, then $F(x)$ is differentiable on I and $F'(x) = f(x)$ for all $x \in I$.

Proof. (i) Fix $x_0 \in I$ and let $(x_n)_{n=1}^{\infty} \subset I$ be any sequence such that $x_n \xrightarrow[n]{} x_0$ where $x_n \neq x_0$ for all n. Let $[a, b]$ be a compact interval contained in I large enough so that x_0 and all x_n belong to $[a, b]$. Note that f is bounded on $[a, b]$. Hence there is a pair of real numbers m and M such that $m \leqslant f(x) \leqslant M$ for all $x \in [a, b]$. Thus, for every $n \in \mathbb{N}$, statement (i) of Theorem 2.4.6 provides a number $\mu_n \in [m, M]$ such that

$$\int_{x_0}^{x_n} f = \mu_n(x_n - x_0). \tag{2.26}$$

As $(\mu_n)_{n=1}^{\infty}$ is bounded, Proposition 1.2.9 and (2.26) yield

$$F(x_n) - F(x_0) = \int_{x_0}^{x_n} f \xrightarrow[n]{} 0$$

which proves that there exists $\lim_{x \to x_0} F(x) = F(x_0)$, so $F(x)$ is continuous at x_0.

(ii) Fix $x_0 \in I$ and a sequence $(x_n)_{n=1}^{\infty}$ as in statement (i). For every x_n, statement (ii) of Theorem 2.4.6 provides a point c_n between x_n and x_0 for which

$$\frac{F(x_n) - F(x_0)}{x_n - x_0} = \frac{\int_{x_0}^{x_n} f}{x_n - x_0} = f(c_n).$$

Thus, as $x_n \xrightarrow[n]{} x_0$, then $c_n \xrightarrow[n]{} x_0$ as well, and as f is continuous at x_0, then $f(c_n) \xrightarrow[n]{} f(x_0)$, so

$$\frac{F(x_n) - F(x_0)}{x_n - x_0} \xrightarrow[n]{} f(x_0).$$

This proves that F is differentiable at x_0 and $F'(x_0) = f(x_0)$. $\qquad \square$

Theorem 2.4.7 entails the existence of antiderivatives for a continuous function f on an interval I, and moreover, it provides an effective way to calculate its integral function if an antiderivative G of f is known:

$$\int_a^x f = G(x) - G(a), \quad x \in I. \tag{2.27}$$

Indeed, given $a \in I$, Theorem 2.4.7 proves that F_a is an antiderivative of f. Thus $(F_a - G)'$ is null on the interval I, and by virtue of Proposition 2.3.10, there exists a constant K such that $G(x) = F_a(x) + K$ for all $x \in I$. Clearly, $G(a) = K$, so $F_a(x)$ can be calculated as $G(x) - G(a)$ which proves (2.27).

If f is integrable and has an antiderivative F, the identity $\int_a^b f = F(b) - F(a)$ still holds even if f is not continuous:

Theorem 2.4.8 (Barrow's rule). *Let f be a differentiable function on a compact interval $[a,b]$ such that f' is integrable on $[a,b]$. Then $\int_a^b f' = f(b) - f(a)$.*

Proof. Let $P = \{t_i\}_{i=0}^n$ be any partition of $[a,b]$ and let us denote $m_i :=$ $\inf f'([t_{i-1}, t_i])$ and $M_i := \sup f'([t_{i-1}, t_i])$.

By virtue of Theorem 2.3.9, for every $i = 1, 2, \ldots, n$ there exists $c_i \in [t_{i-1}, t_i]$ such that $f'(c_i) = \big(f(t_i) - f(t_{i-1})\big)/(t_i - t_{i-1})$, and as $m_i \leqslant f'(c_i) \leqslant M_i$, it follows that

$$\sum_{i=1}^n m_i(t_i - t_{i-1}) \leqslant \sum_{i=1}^n f(t_i) - f(t_{i-1}) \leqslant \sum_{i=1}^n M_i(t_i - t_{i-1})$$

that is,

$$L(f', P) \leqslant f(b) - f(a) \leqslant U(f', P) \tag{2.28}$$

and as (2.28) holds for every partition P, it follows that $\int_a^b f' = f(b) - f(a)$. $\qquad\square$

Corollary 2.4.9. *Let g be a differentiable function on $[a,b]$ such that g' is continuous on $[a,b]$, and let f be a continuous function on $I := g([a,b])$. Then $f \circ g$ is integrable on $[a,b]$ and $\int_{g(a)}^{g(b)} f = \int_a^b (f \circ g) \cdot g'$.*

Proof. By virtue of Theorem 2.4.7, the integral function $F(y) := \int_{g(a)}^y f$ defined at each $y \in I$ is an antiderivative of f and

$$\int_{g(a)}^{g(b)} f = F\big(g(b)\big) - F\big(g(a)\big).$$

And by following the same line of reasoning,

$$\int_a^b (f \circ g) \cdot g' = \int_a^b (F \circ g)' = F \circ g(b) - F \circ g(a),$$

and the proof is done. $\qquad\square$

Note that the function g of Corollary 2.4.9 is not required to be injective.

Chapter 3

Elementary real-valued functions

We call the following functions *fundamental functions*: x^n, $\sqrt[n]{x}$ (with $n \in \mathbb{N}$), e^x, $\log x$, $\sin x$, $\cos x$, $\arcsin x$ and $\arccos x$. Any other function defined in finite terms by composition, sum, product or quotient of the fundamental functions will henceforth be called *elementary*.

Continuity, differentiability and all the relevant properties of the fundamental functions concerning antidifferentiation are listed in this chapter. A formal construction of $\log x$, e^x, $\sin x$ and $\cos x$ by means of integration techniques is given in Sections 3.3 and 3.5.

Of course, there are other ways to introduce the fundamental functions and some of them are discussed in Sections A.2 to A.4.

Some elementary functions like polynomials, $\tan x$ or the hyperbolic functions, and also the fundamental functions $\sqrt[n]{x}$, $\log x$, $\arcsin x$ and $\arccos x$ have been studied in detail. Convexity and other features which are not important in regards to antidifferentiation have not been included. Instead, the graphs of many functions have been plotted.

3.1 Power functions and root functions

For every positive integer n, the n-th power function is defined as

$$p_n(x) := x^n, \ x \in \mathbb{R}.$$

It is immediate that all functions p_n are increasing on $[0, \infty)$, and that

$$p_{2n-1}(-x) = -p_{2n-1}(x) \tag{3.1}$$
$$p_{2n}(-x) = p_{2n}(x) \tag{3.2}$$

for all $x \in \mathbb{R}$ and all $n \in \mathbb{N}$. Hence only the odd power functions are invertible.

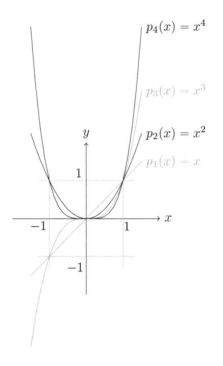

$$(3.3)$$

All power functions are differentiable at each $a \in \mathbb{R}$. In fact, as

$$x^n - a^n = (x - a)(x^{n-1} + ax^{n-2} + a^2 x^{n-3} + \ldots + a^{n-2}x + a^{n-1}) \qquad (3.4)$$

it follows that

$$\frac{x^n - a^n}{x - a} = \sum_{k=0}^{n-1} a^k x^{n-1-k} \xrightarrow[x \to a]{} na^{n-1}$$

which proves that

$$p_n'(x) = nx^{n-1}, \quad \text{for all } n \in \mathbb{N} \text{ and all } x \in \mathbb{R}. \qquad (3.5)$$

Since each power function is increasing and continuous on $[0, \infty)$, and as $p_n(0) = 0$ and $\lim_{n \to \infty} p_n(x) = \infty$, Theorem 2.2.8 implies that p_n maps $[0, \infty)$ onto itself. Hence $p_n|_{[0,\infty)}$ admits an inverse function denoted

$$\sqrt[n]{\cdot} := p_n|_{[0,\infty)}^{-1} \qquad (3.6)$$

which also maps $[0, \infty)$ onto itself. In particular, this implies that $\lim_{x \to \infty} \sqrt[n]{x} = \infty$.

For every $x \in [0, \infty)$, the image $y := \sqrt[n]{x}$ is called the *n-th root* of x and it is the only non-negative real number satisfying the identity $y^n = x$.

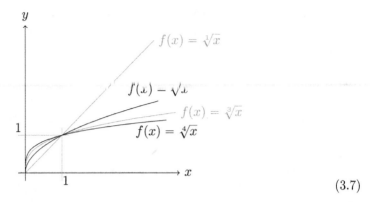

$$(3.7)$$

By virtue of (3.5) and of Proposition 2.3.7 for the differentiation of inverse functions, $\sqrt[n]{\cdot}$ is differentiable on $(0, \infty)$ and its derivative is

$$\sqrt[n]{x}' = \sqrt[n]{x}^{1-n}, \qquad x \in (0, \infty)$$

but it is not differentiable at $x = 0$ if $n \geqslant 2$.

It is customary to write $\sqrt{x}$ instead of $\sqrt[2]{x}$. It is also customary to denote

$$x^{\frac{p}{q}} := \sqrt[q]{x^p} = \sqrt[q]{x}^p$$

for $x \in (0, \infty)$, $p \in \mathbb{N} \cup \{0\}$ and $q \in \mathbb{N}$. Under this notation, it is immediate that

$$x^{\frac{p}{q}} = (x^p)^{\frac{1}{q}} = (x^{\frac{1}{q}})^p$$

$$x^{\frac{p_1}{q_1}} x^{\frac{p_2}{q_2}} = x^{\frac{p_1}{q_1} + \frac{p_2}{q_2}}$$

$$(x^{\frac{1}{p_1}})^{\frac{1}{p_2}} = x^{\frac{1}{p_1 p_2}} = (x^{\frac{1}{p_2}})^{\frac{1}{p_1}}$$

where p, p_1 and p_2 are non-negative integers, q, q_1 and q_2 are natural numbers and $x \in (0, \infty)$. As usual, $0^r = 0$ if r is a positive rational number.

3.2 Polynomials

A *polynomial over the real numbers* is a function of the form

$$P(x) = a_n x^n + a_{n-1} x^{n-1} + \ldots + a_2 x^2 + a_1 x + a_0, \qquad x \in \mathbb{R} \qquad (3.8)$$

where each a_k is a real number, n is a non-negative integer, a_n is non-null and the domain of $P(x)$ is $\mathbb{R}$. The numbers a_k are called the *coefficients* of f and $a_k x^k$ is its *term of degree* k; the coefficients a_n and a_0 are respectively called the *leading coefficient* and the *constant term* of $P(x)$; the positive integer n associated with the highest power of x is called the *degree of* $P(x)$ and is denoted $\deg P(x)$. A constant polynomial $P(x) = a_0$ with $a_0 \neq 0$ has degree zero; the *null polynomial* $P(x) \equiv 0$ is not assigned a degree.

The main issue of polynomials concerning antidifferentiation is the problem of their factorization, and that means the introduction of polynomials over $\mathbb{C}$.

We say that the function in (3.8) is a *polynomial over the complex numbers* if its coefficients a_k are complex and its domain is $\mathbb{C}$. Thus, a polynomial over $\mathbb{R}$ can be viewed as a polynomial over $\mathbb{C}$ whose coefficients are real and whose domain is restricted to $\mathbb{R}$. A number $z \in \mathbb{C}$ is said to be a *root* or a *zero* of a polynomial $P(x)$ over $\mathbb{R}$ or over $\mathbb{C}$ if $P(z) = 0$.

A polynomial $P(x)$ over $\mathbb{C}$ is said to be *divisible* by a polynomial $Q(x)$ if there exists a polynomial $R(x)$ such that $P(x) = Q(x) \cdot R(x)$; note that if the coefficients of $P(x)$ and those of $Q(x)$ are real numbers, so are the coefficients of $R(x)$, as shown by means of the long division algorithm (see Section A.7).

Let $\mathbb{K}$ denote either $\mathbb{R}$ or $\mathbb{C}$: if $P(x)$ is a non-null polynomial over $\mathbb{K}$ divisible by a polynomial $Q(x)$ over $\mathbb{K}$ such that $1 \leqslant \deg Q(x) < \deg P(x)$ then $P(x)$ is said to be *reducible over* $\mathbb{K}$ and $Q(x)$ is called a *factor* of $P(x)$, otherwise $P(x)$ is said to be *irreducible over* $\mathbb{K}$. A pair of polynomials $P(x)$ and $Q(x)$ over $\mathbb{K}$ are said to be *mutually irreducible over* $\mathbb{K}$ if no polynomial $D(x)$ over $\mathbb{K}$ with $\deg D(x) \geqslant 1$ is a factor of both $P(x)$ and $Q(x)$.

The next result about polynomials over $\mathbb{C}$ requires the introduction of modulus of complex numbers. That is now possible, thanks to the fact that square roots of non-negative real numbers have been well founded in Section 3.1. Given a complex number $z = a + bi$ (with a and b the real part and the imaginary part of z respectively), its *modulus* is defined as the real number

$$|z| := \sqrt{a^2 + b^2}.$$

Geometrically, $|z|$ is identified with the length of the segment in the plane $\mathbb{R}^2$ whose enpoints are the points $(0,0)$ and (a,b). It is immediate from (1.1) that for every pair of complex numbers z_1 and z_2,

$$|z_1 \cdot z_2| = |z_1| \cdot |z_2|. \tag{3.9}$$

Let us look further into the linear structure of polynomials.

Proposition 3.2.1. *Let $\mathbb{K}$ denote either $\mathbb{R}$ or $\mathbb{C}$, let $n \in \mathbb{N}$ and let $\mathcal{P}_n$ denote the set of all polynomials over $\mathbb{K}$ whose degree is equal to or smaller than n. Then $\mathcal{P}_n$ is a vector space over $\mathbb{K}$ and $\{1, x, x^2, \ldots, x^n\}$ is a basis of $\mathcal{P}_n$.*

Proof. Clearly, if $p_\alpha(x) = \sum_{k=0}^{n} \alpha_k x^k$ and $p_\beta = \sum_{k=0}^{n} \beta_k x^k$ are two polynomials in $\mathcal{P}_n$ and λ and μ are a pair of elements of $\mathbb{K}$ then $\lambda p_\alpha(x) + \mu p_\beta(x) = \sum_{k=0}^{n} (\lambda \alpha_k + \mu \beta_k) x^k$ is a polynomial over $\mathbb{K}$ whose degree is at most n. Hence it belongs to $\mathcal{P}_n$, and that proves $\mathcal{P}_n$ is a vector space over $\mathbb{K}$.

Since $\{1, x, \ldots, x^n\}$ trivially spans $\mathcal{P}_n$, we only need to check that it is linearly independent in order to prove that it is a basis of $\mathcal{P}_n$. To do that, consider any linear combination

$$Q(x) = \lambda_m x^m + \lambda_{m-1} x^{m-1} + \ldots + \lambda_1 x + \lambda_0$$

with $\lambda_m \neq 0$ and $1 \leqslant m \leqslant n$. For $M := \max_{1 \leqslant k \leqslant m-1} |\lambda_k|$ and for values of $x \in \mathbb{K}$ such that $|x| \geqslant \max\{1, 2Mn/|\lambda_m|\}$, we have

$$|Q(x)| \geqslant |x|^m \left(|\lambda_m| - \frac{|\lambda_{m-1}|}{|x|} - \frac{|\lambda_{m-2}|}{|x|^2} - \cdots - \frac{|\lambda_0|}{|x|^m} \right)$$

$$\geqslant |x|^m \left(|\lambda_m| - \frac{Mn}{|x|} \right) \geqslant |x|^m \frac{|\lambda_m|}{2} > 0.$$

Hence $Q(x) \neq 0$, which proves that $\{x^k\}_{k=0}^n$ is linearly independent and, therefore, a basis of $\mathcal{P}_n$. $\qquad\square$

Corollary 3.2.2. *Let $\mathbb{K}$ denote $\mathbb{R}$ or $\mathbb{C}$ and let $P(x) = \sum_{k=0}^n \alpha_k x^k$ be a polynomial over $\mathbb{K}$. If $P(t) = 0$ for all $t \in \mathbb{R}$, then $\alpha_k = 0$ for all $0 \leqslant k \leqslant n$.*

Proof. If $\mathbb{K} = \mathbb{R}$ then $P(x)$ can be regarded as the identically null polynomial over $\mathbb{R}$, and the result follows readily from the fact that $\{x^k\}_{k=1}^n$ is a basis of $\mathcal{P}_n$.

Assume now that $\mathbb{K} = \mathbb{C}$. Denote $\beta_k := \Re\alpha_k$ and $\gamma_k := \Im\alpha_k$ for all k and consider the polynomials $Q(x) := \sum_{k=0}^n \beta_k x^k$ and $R(x) := \sum_{k=0}^n \gamma_k x^k$ over $\mathbb{R}$. Thus,

$$0 = P(t) = Q(t) + R(t)i, \quad \text{for all } t \in \mathbb{R}.$$

Hence both $Q(x)$ and $R(x)$ vanish on $\mathbb{R}$, and, as proved in the case $\mathbb{K} = \mathbb{R}$, that means $\beta_k = \gamma_k = 0$ for all k, and in turn, $\alpha_k = 0$ too. $\qquad\square$

The following result identifies all bases of the $\mathbb{K}$-vector space $\mathcal{P}_n$:

Proposition 3.2.3. *Let $\mathbb{K}$ denote either $\mathbb{R}$ or $\mathbb{C}$, and let $\{B_k(x)\}_{k=0}^n$ be a finite collection of polynomials over $\mathbb{K}$ such that $\deg B_k(x) = k$ for all k. Then $\{B_k(x)\}_{k=0}^n$ is a basis of $\mathcal{P}_n$.*

Proof. As in Proposition 3.2.1, $\mathcal{P}_n$ denotes the vector space over $\mathbb{K}$ of all polynomials over $\mathbb{K}$ with degree equal to or smaller than n. Since $\dim \mathcal{P}_n = n + 1$ and $\{B_k(x)\}_{k=0}^n$ has exactly $n + 1$ elements the proof will be done as soon as we prove that $\{B_k(x)\}_{k=0}^n$ is linearly independent. To do so, consider a linear combination

$$B(x) = \mu_n B_n(x) + \mu_{n-1} B_{n-1}(x) + \ldots + \mu_1 B_1(x) + \mu_0 B_0(x) \qquad (3.10)$$

such that $B(x) \equiv 0$. Let b_k be the leading coefficient of each $B_k(x)$. Of course, $B_0(x)$ is a non-null constant $b_0 \in \mathbb{K}$. Expanding the polynomials $B_k(x)$ in (3.10), it is clear that $B(x) = \mu_n b_n x^n + Q_{n-1}(x)$ where $Q_{n-1}(x)$ is a polynomial of degree at most $n - 1$, and since $\{x^k\}_{k=0}^n$ is a basis of $\mathcal{P}_n$, it follows that $\mu_n b_n = 0$, hence $\mu_n = 0$. That implies $B(x) = \mu_{n-1} B_{n-1}(x) + \ldots + \mu_1 B_1(x) + \mu_0 B_0(x)$. Repeating the same argument, $B(x) = \mu_{n-1} b_{n-1} x^{n-1} + Q_{n-2}(x)$ with $\deg Q_{n-2}(x) \leqslant n - 2$, which leads to $\mu_{n-1} = 0$. Keeping this procedure $n - 2$ more times, we successively prove that $0 = \mu_{n-2} = \ldots = \mu_1 = \mu_0$. This shows that $\{B_k(x)\}_{k=0}^n$ is linearly independent and, therefore, a basis of $\mathcal{P}_n$. The proof is done. $\qquad\square$

The most important results about the existence of roots and factorization of polynomials are included in the next theorem.

Theorem 3.2.4. *The following statements hold for every polynomial $P(x)$ over $\mathbb{C}$:*

(i) $P(x)$ has a root in $\mathbb{C}$;
(ii) given $a \in \mathbb{C}$, $P(x)$ is divisible by $x - a$ if and only if a is a root of $P(x)$.

Statement (i), known as the Fundamental Theorem of Algebra, does not admit elementary proof so it will be postponed to Theorem A.10.4. However, part (ii) is very simple: if $x - a$ divides $P(x)$ then $P(a) = 0$ trivially. For the converse, if $P(a) = 0$ then

$$P(x) = P(x) - P(a) = \sum_{k=1}^{n} a_k(x^k - a^k) \tag{3.11}$$

and as formula (3.4) proves that every summand $a_k(x^k - a^k)$ in (3.11) is divisible by $x - a$, it follows that $x - a$ is a factor of $P(x)$.

Note that Theorem 3.2.4 implies that a complex polynomial is irreducible if and only if its degree is 1 and, therefore, any polynomial $P(x)$ of degree $n \geqslant 1$ with leading coefficient α can be factored as

$$P(x) = \alpha(x - z_1)^{m_1}(x - z_2)^{m_2} \ldots (x - z_n)^{m_n} \tag{3.12}$$

where the numbers z_k are pairwise different and the numbers m_i are natural. Obviously, the numbers z_k are the complex roots of $P(x)$ and the *multiplicity of each root z_k* is m_i; if $m_i = 1$ then z_i is called a *simple root* of $P(x)$, otherwise it is called a *multiple root* of $P(x)$; sometimes, if $a \in \mathbb{C}$ is not a root of $P(x)$, we refer to a as a root of *multiplicity zero* of $P(x)$. Formula (3.12) is called the *irreducible factorization of $P(z)$ over $\mathbb{C}$*.

After Theorem 3.2.4, the next problem is how to find out the roots of a given polynomial. Only polynomials with degree smaller than five admit a general formula to find their roots, and, with a few exceptions, the roots of polynomials of fifth or larger degree can be approached only by means of numerical methods.

The formulas for the roots of third and fourth degree polynomials are certainly intricate (see Sections A.6 and A.11 in the Appendices). The roots for a second degree polynomial $P(x) = ax^2 + bx + c$ over $\mathbb{R}$ are usually given by means of the informal formula

$$\frac{-b \pm \sqrt{b^2 - 4ac}}{2a};$$

the next result restates that formula in an appropriate manner.[1] The real number $\Delta P(x) := b^2 - 4ac$ is termed the *discriminant of $P(x)$*.

---

[1]The formula $x = (-b \pm \sqrt{b^2 - 4ac})/(2a)$ for the roots of $ax^2 + bx + c$ use an abuse of terminology as explained in Section A.1.

Proposition 3.2.5. *Given a polynomial $P(x) = ax^2 + bx + c$ over $\mathbb{R}$ with $a \neq 0$, one and only one of the following statements hold:*

(i) if $\Delta\, P(x) = 0$ then $P(x)$ has a unique real root, its multiplicity is two and its value is $x_1 = -b/2a$;

(ii) if $\Delta\, P(x) > 0$ then $P(x)$ has two real roots, both of multiplicity one and their values are

$$x_1 = \frac{-b + \sqrt{b^2 - 4ac}}{2a} \quad and \quad x_2 = \frac{-b - \sqrt{b^2 - 4ac}}{2a}$$

(iii) if $\Delta\, P(x) < 0$ then $P(x)$ has two complex roots, both of multiplicity one and their values are

$$x_1 = \frac{-b + i\sqrt{|b^2 - 4ac|}}{2a} \quad and \quad x_2 = \frac{-b - i\sqrt{|b^2 - 4ac|}}{2a}$$

Proof. Since the roots of $P(x)$ and of $-P(x)$ are the same and their discriminants coincide, we may assume that $a > 0$. Let $\Delta := b^2 - 4ac$ be the discriminant of $P(x)$. Completing squares, we get

$$P(x) = ax^2 + bx + c = \left(\sqrt{a}x + \frac{b}{2\sqrt{a}} \right)^2 - \frac{\Delta}{4a}. \tag{3.13}$$

(i) if $\Delta = 0$, then (3.13) gives the irreducible decomposition of $P(x)$ over $\mathbb{C}$ as

$$P(x) = \left(\sqrt{a}x + \frac{b}{2\sqrt{a}} \right)^2 = a\left(x + \frac{b}{2a} \right)^2$$

which immediately gives that $x_1 := -b/(2a)$ is the only root of $P(x)$, is real and has multiplicity two.

(ii) If $\Delta > 0$ then the irreducible decomposition of $P(x)$ over $\mathbb{C}$ is managed from (3.13) as

$$P(x) = \left[\sqrt{a}x + \frac{b}{2\sqrt{a}} - \frac{\sqrt{\Delta}}{2\sqrt{a}} \right]\left[\sqrt{a}x + \frac{b}{2\sqrt{a}} + \frac{\sqrt{\Delta}}{2\sqrt{a}} \right]$$

$$= a\left[x + \frac{b}{2a} - \frac{\sqrt{\Delta}}{2a} \right]\left[x + \frac{b}{2a} + \frac{\sqrt{\Delta}}{2a} \right],$$

which clearly shows that the roots of $P(x)$ are

$$x_1 = -\frac{b}{2a} - \frac{\sqrt{\Delta}}{2a} \quad and \quad x_2 = -\frac{b}{2a} + \frac{\sqrt{\Delta}}{2a}$$

both real and simple.

(iii) If $\Delta < 0$, the irreducible decomposition of $P(x)$ over $\mathbb{C}$ is also obtained from (3.13) as

$$P(x) = \left[\sqrt{a}x + \frac{b}{2\sqrt{a}} - \frac{\sqrt{|\Delta|}}{2\sqrt{a}}i \right]\left[\sqrt{a}x + \frac{b}{2\sqrt{a}} + \frac{\sqrt{|\Delta|}}{2\sqrt{a}}i \right]$$

$$= a\left[x + \frac{b}{2a} - \frac{\sqrt{|\Delta|}}{2a}i \right]\left[x + \frac{b}{2a} + \frac{\sqrt{|\Delta|}}{2a}i \right]$$

which shows that the two roots of $P(x)$ are

$$x_1 = -\frac{b}{2a} - \frac{\sqrt{|\Delta|}}{2a}i \text{ and } x_2 = -\frac{b}{2a} + \frac{\sqrt{|\Delta|}}{2a}i,$$

both in $\mathbb{C}\backslash\mathbb{R}$ and simple. □

The following result is very useful in finding the rational roots of a polynomial of any degree with rational coefficients.

Proposition 3.2.6. *Let* $P(x) = a_n x^n + \ldots + a_1 x + a_0$ *be a polynomial where every coefficient* a_k *is an integer. If* p/q *is a root of* $P(x)$ *with* $p \in \mathbb{Z}$, $q \in \mathbb{N}$ *and* g.c.d.$(p,q) = 1$ *then* p *is a divisor of* a_0 *and* q *is a divisor of* a_n.

Proof. Since p/q is a root of $P(x)$, then

$$0 = a_n \frac{p^n}{q^n} + a_{n-1}\frac{p^{n-1}}{q^{n-1}} + \ldots + a_1 \frac{p}{q} + a_0. \tag{3.14}$$

In order to eliminate denominators in (3.14), we multiply by q^n, so

$$0 = a_n p^n + a_{n-1}p^{n-1}q + \ldots + a_1 p q^{n-1} + a_0 q^n.$$

Thus

$$p(-a_n p^{n-1} - a_{n-1}p^{n-2}q + \ldots - a_1 q^{n-1}) = a_0 q^n$$

which implies that p is a divisor of $a_0 q^n$, and as g.c.d.$(p,q) = 1$, it follows that p is a divisor of a_0. A similar argument proves that q is a divisor of a_n. □

Proposition 3.2.7. *Let* $P(x)$ *be a polynomial whose coefficients are all real numbers. Then, if* z *is a root of* $P(x)$ *so is* $\bar{z}$.

Proof. Assume $P(z) = 0$. Since the coefficients of $P(x)$ are real numbers, and as $\overline{z_1 z_2} = \overline{z_1}\,\overline{z_2}$ for any pair of complex numbers z_1 and z_2, it is immediate that $0 = \overline{P(z)} = P(\bar{z})$ so $\bar{z}$ is a root of $P(x)$. □

Given a complex number $z = a + bi$, since

$$(x - z)(x - \bar{z}) = x^2 - 2ax + a^2 + b^2 \tag{3.15}$$

is a polynomial over $\mathbb{R}$, it follows from Proposition 3.2.7 that the degree of an irreducible polynomial over $\mathbb{R}$ is 1 or 2.

Thus, if $P(x)$ is a polynomial over $\mathbb{R}$, its irreducible factorization over $\mathbb{C}$ is

$$P(x) = a \prod_{i=1}^{m}(x - \alpha_i)^{m_i} \cdot \prod_{i=1}^{n}(x - z_i)^{n_i}(x - \bar{z}_i)^{n_i} \tag{3.16}$$

where a is the leading coefficient of $P(x)$, the numbers α_i are its real roots, the numbers z_i and $\bar{z}_i$ are its complex, non-real roots, m_i is the multiplicity of α_i, n_i is the multiplicity of z_i and $\bar{z}_i$ and of course, all α_i, z_i and $\bar{z}_i$ are pairwise different.

Following (3.15), each product $(x - z_i)(x - \bar{z}_i)$ is a polynomial $x^2 + \mu_i x + \nu_i$ over $\mathbb{R}$, so (3.16) becomes

$$P(x) = a \prod_{i=1}^{m}(x - \alpha_i)^{m_i} \cdot \prod_{i=1}^{n}(x^2 + \mu_i x + \nu_i)^{n_i} \tag{3.17}$$

where all factors in (3.17) are mutually irreducible and irreducible over $\mathbb{R}$. The expression (3.17) is called the *irreducible decomposition of* $P(x)$ *over* $\mathbb{R}$.

3.3 The exponential function and the log function

The exponential function e^x and the $\log x$ function and their main properties can be easily obtained by means of integral functions as shown in the following.

The $\log x$ *function*

The log function can be defined as the integral function

$$\log x := \int_1^x \frac{1}{t}\, dt, \quad x \in (0, \infty). \tag{3.18}$$

Clearly, $\log 1 = 0$. As the function $1/x$ is continuous on $(0, \infty)$, the function given in (3.18) is well defined and differentiable on $(0, \infty)$, and according to Theorem 2.4.7, its derivative is

$$\log' x = \frac{1}{x} > 0, \quad x \in (0, \infty).$$

Hence, log is increasing (Proposition 2.3.11) and by Theorem 2.2.8, its range is an interval. More precisely,

$$\log\big((0, \infty)\big) = (\lim_{x \to 0^+} \log x, \lim_{x \to \infty} \log x). \tag{3.19}$$

In order to calculate $\lim_{x \to \infty} \log x$, consider the partition $P := \{0 < 2^0 < 2^1 < 2^2 <\overset{n}{...}< 2^{n-1} < 2^n\}$ of $[0, 2^n]$. Clearly,

$$\log(2^n) = \int_1^{2^n} \frac{dt}{t} \geqslant L(1/t, P) = 1 + \sum_{k=1}^{n} \frac{1}{2} \geqslant \frac{n}{2} \xrightarrow[n]{} \infty$$

and as log is increasing, $\lim_{x \to \infty} \log x = \lim_n \log 2^n = \infty$.

In order to compute $\lim_{x \to 0^+} \log x$, the change of variable $\tau = 1/t$ and Corollary 2.4.9 give

$$\int_1^x \frac{1}{t}\, dt = -\int_1^{1/x} \frac{1}{\tau}\, d\tau;$$

therefore, as $\lim_{x \to \infty} \log x = \infty$, it follows that

$$\lim_{x \to 0^+} \log x = -\lim_{x \to 0^+} \log(1/x) = -\lim_{t \to \infty} \log t = -\infty,$$

which proves after (3.19) that

$$\log\big((0, \infty)\big) = (-\infty, \infty).$$

It only remains to see that the function defined in (3.18) satisfies the identity

$$\log xy = \log x + \log y, \quad \text{all } x, y \in (0, \infty). \tag{3.20}$$

In order to do that, fix any real number $y > 0$ and define the function $L(x) := \log(xy)$ for all $x \in (0, \infty)$. By the chain rule, $L'(x) = 1/x = \log' x$. Thus, by Proposition 2.3.10, there exists a constant K such that $L(x) = \log x + K$. Since $\log 1 = 0$, it follows that $\log y = L(1) = K$, which proves (3.20).

Finally, as 1 belongs to the range of $\log x$, the number e is defined as the only real number for which

$$\log e = 1. \tag{3.21}$$

It is immediate from (3.18) that $e > 1$.[2]

---

[2]The number e has been defined in (3.21) in symbolic terms, and this is sufficient for the Elementary Antidifferentiation theory developed in this book. An estimation of its value is $e = 2.7182\ldots$.

The function e^x

Since log is a bijection from $(0, \infty)$ onto $(-\infty, \infty)$, it has an inverse function $E \colon (-\infty, \infty) \longrightarrow (0, \infty)$. Clearly, $E(x)$ is increasing, and by Proposition 2.3.7, $E(x)$ is differentiable and

$$E'(x) = E(x), \quad x \in (-\infty, \infty).$$

From (3.20), it is not too difficult to see that

$$E(x + y) = E(x) \cdot E(y). \tag{3.22}$$

It is also immediate that $E(0) = 1$ and $E(1) = e$. Moreover, since $E(x)$ is increasing and maps $(-\infty, \infty)$ onto $(0, \infty)$ continuously, it is straightforward that

$$\lim_{x \to -\infty} E(x) = 0 \text{ and } \lim_{x \to \infty} E(x) = \infty.$$

The following result reveals the connections between the number e and $E(x)$:

Proposition 3.3.1. *For every* $p \in \mathbb{Z}$ *and every* $q \in \mathbb{N}$, $E(p/q) = e^{p/q}$.

Proof. By (3.22),

$$e = E(1) = E\left(\frac{1}{q} + \overset{q}{\ldots} + \frac{1}{q}\right) = E\left(\frac{1}{q}\right)^q$$

which yields $E(1/q) = \sqrt[q]{e}$, and therefore, $E(p/q) = \sqrt[q]{e^p}$. $\square$

In order to attribute a meaning to the expression e^y when y is an irrational real number, pick any sequence $(r_n)_{n=1}^{\infty}$ of rational numbers convergent to y. Then, the combination of Proposition 3.3.1 and the continuity of $E(x)$ at y yields

$$e^{r_n} = E(r_n) \xrightarrow[n]{} E(y)$$

irrespectively of the chosen sequence $(r_n)_{n=1}^{\infty}$. Thus, this last identity and the fact that $E(x) = e^x$ for all $x \in \mathbb{Q}$ sustain the universally accepted notation

$$e^x := E(x), \quad x \in \mathbb{R}.$$

In agreement with this notation, $e^{x+y} = e^x e^y$ for all real numbers x and y.

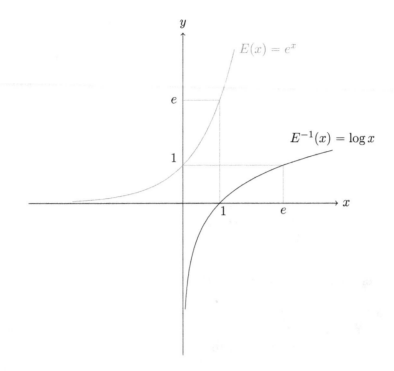

3.4 General exponentiation

Given any pair of numbers a and b, the expression a^b is defined for $a > 0$ as the real number

$$a^b := e^{b \log a} \qquad (3.23)$$

and if $a < 0$, the expression a^b remains undefined except when $b \in \mathbb{Z}$, in which case a^b retains its usual meaning.

The identities

$$\log a^b = b \log a$$
$$a^{b_1} a^{b_2} = a^{b_1 + b_2}$$
$$a_1^b a_2^b = (a_1 a_2)^b$$
$$(a^{b_1})^{b_2} = a^{b_1 b_2}$$

are immediate from (3.23).

For every real number $a > 0$, formula (3.23) allows us to define the exponential function $E_a(x)$ as

$$E_a(x) := a^x, \qquad x \in (-\infty, \infty).$$

Since $E_a(x) = e^{x \log a}$, it is a simple exercise to check that $E_a(x)$ is differentiable on $\mathbb{R}$ and its derivative is

$$E_a'(x) = \log(a) e^x, \qquad x \in \mathbb{R}.$$

Thus, if $0 < a < 1$, then $\log a < 0$ and therefore, $E_a(x)$ is decreasing. And if $1 < a$, a similar argument gives $E_a(x)$ is increasing.

In the case $a > 1$, given any sequence $(x_n)_{n=1}^{\infty}$ divergent to ∞, we have $x_n \log a \xrightarrow[n]{} \infty$, so

$$a^{x_n} = e^{x_n \log a} \xrightarrow[n]{} \infty$$

and if $x_n \xrightarrow[n]{} -\infty$, then a similar argument gives

$$a^{x_n} \xrightarrow[n]{} 0.$$

Hence, $\lim_{x \to \infty} E_a(x) = \infty$ and $\lim_{x \to -\infty} E_a(x) = 0$, so that $E_a(\mathbb{R}) = (0, \infty)$.

And if $0 < a < 1$, as $E_a(x) = 1/E_{1/a}(x)$, it follows that $\lim_{x \to \infty} E_a(x) = 0$ and $\lim_{x \to -\infty} E_a(x) = \infty$ and therefore, $E_a(\mathbb{R}) = (0, \infty)$.

3.5   The circular functions and their inverse functions

The functions $\cos x$ and $\sin x$, named *cosine* and *sine* respectively, are called *circular* because they satisfy the identity $\cos^2 t + \sin^2 t = 1$ for all $t \in \mathbb{R}$, which means that the point $(x, y) = (\cos t, \sin t)$ belongs to the graph of the circumference $x^2 + y^2 = 1$.

The circular functions are usually introduced at an early stage by means of certain graphical constructions and geometrical intuitions. Although these procedures provide a clear picture about the circular functions, a rigorous geometrical approach is a tough job because it requires the introduction of deep concepts like that of length of an arc and orientation, and this is not possible in an elementary setting. Nowadays, when a rigorous construction of the circular functions is needed, the simplicity of an analytic approach is preferred.

The analytic construction of the circular functions given here follows three steps: introduction of the celebrated number π, construction of the auxiliary functions $A(x)$ and $C(x)$, and construction of $\cos x$ and $\sin x$.

The number π

Since $\sqrt{1 - x^2}$ is a continuous function on $[-1, 1]$, it is integrable on the same interval by virtue of Theorem 2.4.5. This allows us to define the number π as[3]

$$\pi := 2 \int_{-1}^{1} \sqrt{1 - x^2} \, dx. \tag{3.24}$$

It is not difficult to realize that $\pi > 0$.

---

[3]Formula (3.24) is a precise definition of the number π, but it is rather inpractical to estimate its numerical value. However, since the antidifferentiation shown in this book is developed in terms of Symbolic Calculus, formula (3.24) is all we need for our purposes. For an effective estimation of π as 3.14159..., we recommend numerical methods.

Auxiliary functions

Let $A: [-1,1] \longrightarrow \mathbb{R}$ be given by

$$A(x) := \frac{1}{2} x \sqrt{1 - x^2} + \int_x^1 \sqrt{1 - t^2}\, dt.$$

The function $A(x)$ is well defined and continuous on $[-1,1]$ because of the continuity of $\sqrt{\cdot}$ and Theorem 2.4.7. The just mentioned theorem also proves that $A(x)$ is differentiable on $(-1,1)$ and

$$A'(x) = -\frac{1}{2\sqrt{1 - x^2}}, \quad x \in (-1,1). \tag{3.25}$$

As $A'(x) < 0$ for all $x \in (-1,1)$, then Proposition 2.3.11 shows $A(x)$ is decreasing on $(-1,1)$. But A is continuous at -1 and at 1, so it is decreasing on the entire interval $[-1,1]$ by virtue of Propositions 2.1.13, 2.2.2 and 2.2.3. Thus $A([-1,1]) = [A(1), A(-1)]$. Clearly, $A(1) = 0$ and by (3.24), $A(-1) = \pi/2$.

Hence, A is a continuous bijection from $[-1,1]$ onto $[0, \pi/2]$, and we can define the function $C: [0, \pi] \longrightarrow [-1,1]$ as

$$C(x) := A^{-1}\left(\frac{x}{2}\right), \quad x \in [0, \pi].$$

Clearly,

$$C(0) = 1, C(\pi/2) = 0 \text{ and } C(\pi) = -1 \tag{3.26}$$

and by Theorem 2.2.11 and Proposition 2.3.7,

$$C(x) \text{ is continuous and decreasing on } [0, \pi] \tag{3.27}$$

$$C'(x) = -\sqrt{1 - C(x)^2}, \quad x \in (0, \pi). \tag{3.28}$$

The functions $\cos x$ and $\sin x$

For each $k \in \mathbb{Z}$, let $I_k := k\pi + [0, \pi)$. Since $\{I_k\}_{k \in \mathbb{Z}}$ is a collection of pairwise disjoint intervals whose union is $\mathbb{R}$, the function $C(x)$ is extended on the entire set $\mathbb{R}$ as follows:

$$\cos x := \begin{cases} C(x - 2k\pi), & x \in I_{2k},\ k \in \mathbb{Z}; \\ C(2k\pi - x), & x \in I_{2k-1},\ k \in \mathbb{Z}. \end{cases} \tag{3.29}$$

It is immediate that

$$\cos x > 0, \ x \in 2k\pi + (-\pi/2, \pi/2),$$
$$\cos x = 0, \ x = \pm\pi/2 + k\pi,$$
$$\cos(x) = (-1)^k, \ x = k\pi,$$
$$\cos x < 0, \ x \in 2k\pi + (\pi/2, 3\pi/2) \tag{3.30}$$

and that $\cos x$ is periodic of period 2π.

Since $C(x)$ is continuous on $[0, \pi]$, the sided limits of $\cos x$ at each $k\pi$ $(k \in \mathbb{Z})$ can be calculated as

$$\lim_{x \to k\pi \pm} \cos x = C(0) = 1, \text{ if } k \text{ is even}$$

$$\lim_{x \to k\pi \pm} \cos x = C(\pi) = -1, \text{ if } k \text{ is odd}$$

which proves that $\cos x$ is continuous at each $k\pi$. The continuity of $\cos x$ at each $x \in \bigcup_{k \in \mathbb{Z}} I_k^\circ$ follows from the continuity of $C(x)$ on $(0, \pi)$. This proves that $\cos x$ is continuous on $\mathbb{R}$.

Next, let us look into the differentiability of $\cos x$. It is straightforward from (3.28) and from the chain rule that $\cos x$ is differentiable at each $x \in \mathbb{R} \backslash \{k\pi : k \in \mathbb{Z}\}$ and its derivative at these points is

$$\cos' x = \begin{cases} -\sqrt{1 - C^2(x - 2k\pi)} = -\sqrt{1 - \cos^2 x}, \, x \in I_{2k}^\circ, \, k \in \mathbb{Z}, \\ \sqrt{1 - C^2(2k\pi - x)} = \sqrt{1 - \cos^2 x}, \quad x \in I_{2k-1}^\circ, \, k \in \mathbb{Z}. \end{cases} \tag{3.31}$$

Formula (3.31), L'Hôpital's rule and the continuity of $\cos x$ on $\mathbb{R}$ allow us to calculate the sided derivatives of $\cos x$ at each $x = k\pi$ $(k \in \mathbb{Z})$ as

$$\cos'(k\pi^-) = \lim_{x \to k\pi^-} \cos'(x) = \lim_{x \to k\pi^-} (-1)^{k+1}\sqrt{1 - \cos^2 x} = 0$$

$$\cos'(k\pi^+) = \lim_{x \to k\pi^+} \cos'(x) = \lim_{x \to k\pi^+} (-1)^k \sqrt{1 - \cos^2 x} = 0$$

so

$$\cos'(k\pi) = 0, \text{ all } k \in \mathbb{Z}. \tag{3.32}$$

The function $\sin x$ can be now defined as

$$\sin x := -\cos'(x), \quad x \in \mathbb{R}. \tag{3.33}$$

By (3.31) and (3.32),

$$\sin x = \begin{cases} \sqrt{1 - \cos^2 x}, \, x \in I_{2k}, \, k \in \mathbb{Z}, \\ -\sqrt{1 - \cos^2 x}, \, x \in I_{2k-1}, \, k \in \mathbb{Z}, \end{cases} \tag{3.34}$$

and it readily follows that

$$\sin x > 0, \, x \in (0, \pi),$$
$$\sin x = 0, \, x \in \{0, \pi\},$$
$$\sin x = (-1)^k, \, x = \pi/2 + k\pi,$$
$$\sin x < 0, \, x \in (-\pi, 0), \tag{3.35}$$

that $\sin x$ is periodic of period 2π and the circular identity

$$\cos^2 x + \sin^2 x = 1. \text{ all } x \in \mathbb{R}. \tag{3.36}$$

Moreover, by virtue of (3.34) and the chain rule, there exists $\sin' x = \cos x$ for each $x \in \mathbb{R} \backslash \{k\pi : k \in \mathbb{Z}\}$. Then, the continuity of $\cos x$ on $\mathbb{R}$ and L'Hôpital's rule show that $\sin x$ is differentiable on $\mathbb{R}$ and

$$\sin'(x) = \cos x, \, x \in \mathbb{R}. \tag{3.37}$$

The main properties of the circular functions managed through their construction are gathered in the following result:

Proposition 3.5.1. *The functions* $\cos x$ *and* $\sin x$ *are differentiable on* $\mathbb{R}$, *periodic of period* 2π, *their range is* $[-1, 1]$ *and satisfy the following properties:*

(i) $\cos^2 x + \sin^2 x = 1$ *for all* $x \in \mathbb{R}$,
(ii) $\sin' x = \cos x$ *and* $\cos' x = -\sin x$ *for all* $x \in \mathbb{R}$,
(iii) $\cos x$ *is even and* $\sin x$ *is odd*,
(iv) $\cos x$ *is decreasing on* $[0, \pi]$, $\cos 0 = 1$, $\cos(\pi/2) = 0$ *and* $\cos \pi = -1$,
(v) $\sin x$ *is increasing on* $[-\pi/2, \pi/2]$, $\sin(-\pi/2) = -1$, $\sin 0 = 0$ *and* $\sin(\pi/2) = 1$.

Proof. We will prove only those parts which have not been explicitly proved before. The fact that $\cos x$ is even follows directly from (3.29); and this and (3.34) imply that $\sin x$ is odd.

Part (v) is straightforward from (iv) and (3.34). $\qquad\square$

The graphs of $\cos x$ and $\sin x$ are plotted next:

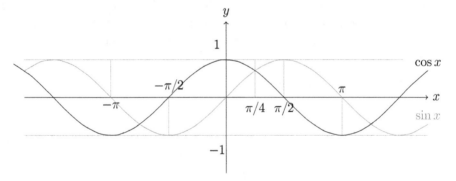

Trigonometric identities

The identities of the following result are often used in antidifferentiation. They will be proved by means of analytic techniques.

Proposition 3.5.2. *The following identities hold for all pairs of real numbers* x *and* y:

(i) $\cos(x + y) = \cos x \cos y - \sin x \sin y$,
(ii) $\sin(x + y) = \sin x \cos y + \cos x \sin y$,
(iii) $\sin x + \sin y = 2 \sin \left(\dfrac{x + y}{2} \right) \cos \left(\dfrac{x - y}{2} \right)$,
(iv) $\cos x + \cos y = 2 \cos \left(\dfrac{x + y}{2} \right) \cos \left(\dfrac{x - y}{2} \right)$,
(v) $\sin x \sin y = \dfrac{1}{2} [\cos(x - y) - \cos(x + y)]$,

(vi) $\cos x \cos y = \dfrac{1}{2}\big[\cos(x-y) + \cos(x+y)\big],$

(vii) $\sin x \cos y = \dfrac{1}{2}\big[\sin(x-y) + \sin(x+y)\big].$

Proof. For (i) and (ii), fix $y \in \mathbb{R}$ and consider the functions

$$f(x) := \sin x \cos(x+y) - \sin(x+y)\cos x$$
$$g(x) := \sin x \sin(x+y) + \cos x \cos(x+y), \quad x \in \mathbb{R}.$$

Since $\cos' x = -\sin x$ and $\sin' x = \cos x$, it follows that $f' \equiv 0$ and $g' \equiv 0$. Hence, f and g are constant and therefore, $f(x) = f(0) = -\sin y$ and $g(x) = g(0) = \cos y$ for all x. Thus

$$\sin x \cos(x+y) - \sin(x+y)\cos x = -\sin y \tag{3.38}$$
$$\sin x \sin(x+y) + \cos x \cos(x+y) = \cos y. \tag{3.39}$$

Formula (i) is managed by multiplying both sides of (3.38) by $\sin x$ and both sides of (3.39) by $\cos x$, and identifying the sum of the left sides of both equations with the sum of their right sides. For formula (ii), we proceed in the same way but multiplying both sides of (3.38) by $-\cos x$ and both sides of (3.39) by $\sin x$.

Once (i) and (ii) are proved, the remaining equations from (iii) to (vii) are a simple exercise. ☐

In particular, the useful identities

$$\cos(x+\pi) = -\cos x,$$
$$\sin(x+\pi) = -\sin x,$$
$$\cos(x+\pi/2) = -\sin x,$$
$$\sin(x+\pi/2) = \cos x$$

follow easily from parts (i) and (ii) of Proposition 3.5.2.

The functions $\arcsin x$ *and* $\arccos x$

Since $\sin x$ is increasing on $[-\pi/2,\, \pi/2]$ and its image on that interval is $[-1,1]$, it is possible to define the function *arcsine* as the inverse function

$$\arcsin x := \sin\big|_{[-\pi/2,\pi/2]}^{-1} : [-1,1] \longrightarrow \left[-\frac{\pi}{2}, \frac{\pi}{2}\right]$$

In a similar manner, as $\cos x$ is decreasing on $[0, \pi]$ and its image on that interval is $[-1,1]$, the *arccosine* is defined as the inverse

$$\arccos x := \cos\big|_{[0,\pi]}^{-1} : [-1,1] \longrightarrow [0,\pi].$$

Since $\sin x$ is differentiable on $\mathbb{R}$, Propositions 3.5.1(ii) and 2.3.7 show that $\arcsin x$ is differentiable on $(-1,1)$ and

$$\arcsin' x = 1/\cos(\arcsin x), \quad x \in (-1,1). \tag{3.40}$$

Plugging $t = \arcsin x$ into the circular identity $\cos^2 t + \sin^2 t = 1$, we get $\cos^2(\arcsin x) + x^2 = 1$, that is, $\cos^2(\arcsin x) = 1 - x^2$. But $-\pi/2 < \arcsin x < \pi/2$ for all $x \in (-1, 1)$, so $\cos(\arcsin x) > 0$. That means $\cos(\arcsin x) = \sqrt{1 - x^2}$, and therefore, formula (3.40) becomes

$$\arcsin' x = \frac{1}{\sqrt{1 - x^2}}, \quad x \in (-1, 1).$$

A similar argument shows that $\arccos x$ is differentiable on $(-1, 1)$ and its derivative is

$$\arccos' x = -\frac{1}{\sqrt{1 - x^2}}, \quad x \in (-1, 1).$$

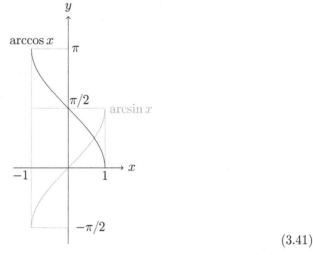

$$(3.41)$$

In a similar manner, given a fixed $k \in \mathbb{Z}$, the functions $\arcsin_k x$ and $\arccos_k x$ are defined on every $x \in [-1, 1]$ as the inverses

$$\arcsin_k x := \sin\big|^{-1}_{\left[k\pi - \frac{\pi}{2}, k\pi + \frac{\pi}{2}\right]} \tag{3.42}$$

$$\arccos_k x := \cos\big|^{-1}_{\left[k\pi, (k+1)\pi\right]}. \tag{3.43}$$

It is straightforward that

$$\arcsin_k x = (-1)^k \arcsin x + k\pi \tag{3.44}$$

$$\arccos_k x = (-1)^k \arccos x + k\pi \tag{3.45}$$

for all $k \in \mathbb{Z}$ and all $x \in [-1, 1]$.

The *tangent* function is defined as

$$\tan x := \frac{\sin x}{\cos x}, \quad x \in \mathbb{R}\backslash\{\pi/2 + n\pi : n \in \mathbb{Z}\}.$$

Following Proposition 2.3.5, $\tan x$ is differentiable on its domain and its derivative is

$$\tan x = 1/\cos^2 x = 1 + \tan^2 x.$$

The function $\tan x$ is periodic with period π and maps each interval $(-\pi/2 + n\pi, \pi/2 + n\pi)$ onto $(-\infty, \infty)$. Moreover, since $\tan x$ is increasing on $(-\pi/2, \pi/2)$, it admits an inverse function called *arctangent*, which maps $(-\infty, \infty)$ onto $(-\pi/2, \pi/2)$:

$$\arctan x := \tan\big|_{(-\frac{\pi}{2}, \frac{\pi}{2})}^{-1} : (-\infty, \infty) \longrightarrow \left(-\frac{\pi}{2}, \frac{\pi}{2}\right).$$

Proposition 2.3.7 shows that $\arctan x$ is differentiable on $\mathbb{R}$ and

$$\arctan' x = \frac{1}{1 + \tan^2(\arctan x)} = \frac{1}{1 + x^2}, \quad x \in \mathbb{R}.$$

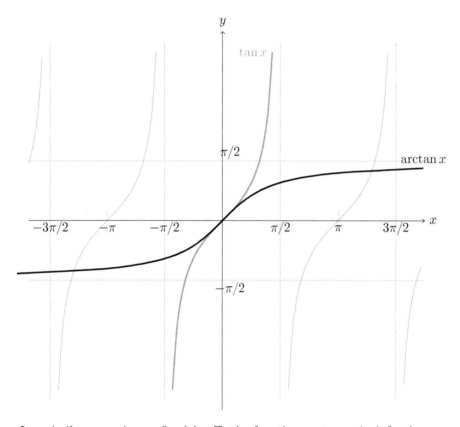

In a similar way, given a fixed $k \in \mathbb{Z}$, the function $\arctan_k x$ is defined as

$$\arctan_k x := \tan\big|_{(k\pi - \frac{\pi}{2}, k\pi + \frac{\pi}{2})}^{-1}, \quad x \in \mathbb{R}. \tag{3.46}$$

It is immediate that $\arctan_k x = \arctan x + k\pi$ for all x and all k.

For other construction of the circular functions, see Section A.4.

In order to prevent any possible ambiguity, henceforth the notation $\cos^{-1}(x)$, $\sin^{-1}(x)$ and $\tan^{-1}(x)$ denote only $1/\cos(x)$, $1/\sin(x)$ and $1/\tan(x)$ respectively. In this way, $\cos^{-2}(x)$ denotes $1/\cos(x)^2$. The inverse functions of $\cos x$, $\sin x$ and $\tan x$ will be exclusively denoted as $\arccos x$, $\arcsin x$ and $\arctan x$ respectively.

3.6 The hyperbolic functions and their inverse functions

The *hyperbolic sine* is defined as

$$\sinh x := \frac{e^x - e^{-x}}{2}, \quad x \in \mathbb{R}$$

and the *hyperbolic cosine* is defined as

$$\cosh x := \frac{e^x + e^{-x}}{2}, \quad x \in \mathbb{R}.$$

These functions are called *hyperbolic* because for every real number t,

$$\cosh^2 t - \sinh^2 t = 1 \tag{3.47}$$

hence the point $(x, y) = (\cosh t, \sinh t)$ belongs to the graph of the hyperbola $x^2 - y^2 = 1$.

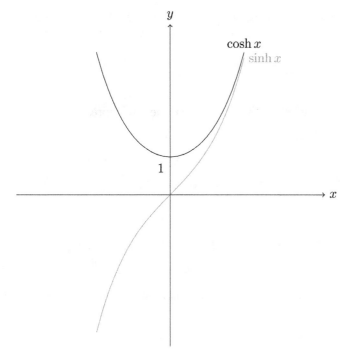

As $\lim_{x \to -\infty} e^x = 0$ and $\lim_{x \to \infty} e^x = \infty$, we have that

$$\lim_{x \to -\infty} \sinh x = -\infty, \quad \lim_{x \to \infty} \sinh x = \infty$$

and

$$\lim_{x \to -\infty} \cosh x = \lim_{x \to \infty} \cosh x = \infty.$$

It is immediate that $\cosh(-x) = \cosh x$ and $\sinh(-x) = -\sinh(x)$ for all x, and that both are increasing on $[0, \infty)$. Thus, since the image of $\sinh x$ is the interval $(-\infty, \infty)$, the *argument of the hyperbolic sine* is defined as the inverse function

$$\arg \sinh x := \sinh^{-1} : (-\infty, \infty) \longrightarrow (-\infty, \infty).$$

And as the image of $[0, \infty)$ by cosh is $[1, \infty)$, the *argument of the hyperbolic cosine* is defined as the inverse function

$$\arg \cosh x := \cosh \Big|_{[0,\infty)}^{-1} : [1, \infty) \longrightarrow [0, \infty).$$

It is immediate that $\sinh' x = \cosh x$ and $\cosh' x = \sinh x$. Thus Proposition 2.3.7 yields that $\arg \arcsin x$ is differentiable on $(-\infty, \infty)$ and

$$\arg \sinh' x = 1/\cosh(\arg \sinh x); \tag{3.48}$$

but $\cosh x > 0$ for all x, so formula (3.47) yields $\cosh(\arg \sinh x) = \sqrt{1 + x^2}$ and by means of formula (3.48) we obtain

$$\arg \sinh' x = \frac{1}{\sqrt{1 + x^2}}, \quad x \in (-\infty, \infty).$$

With a similar argument, since $\arg \cosh' x = 1/\sinh(\arg \cosh x)$ an application of formula (3.47) and the fact that $\sinh(\arg \cosh x) > 0$ for all $x > 1$ yields

$$\arg \cosh' x = \frac{1}{\sqrt{x^2 - 1}}, \quad x \in (1, \infty). \tag{3.49}$$

The $\arg \sinh$ and $\arg \cosh$ functions can be also represented as:

$$\arg \sinh x = \log(x + \sqrt{x^2 + 1}), \quad x \in (-\infty, \infty) \tag{3.50}$$
$$\arg \cosh x = \log(x + \sqrt{x^2 - 1}), \quad x \in [1, \infty). \tag{3.51}$$

Indeed, (3.50) is obtained by solving for y in the identity

$$\frac{e^y - e^{-y}}{2} = \sinh y = x, \quad y \in \mathbb{R}, \; x \in \mathbb{R}. \tag{3.52}$$

To do that, multiply $2e^y$ times both sides of (3.52), so that

$$(e^y)^2 - 2e^y x - 1 = 0.$$

Thus, the formula for the roots of a polynomial of second degree given in Proposition 3.2.5 yields

$$e^y = x \pm \sqrt{x^2 + 1}. \tag{3.53}$$

Since $e^y > 0$ for all $y \in \mathbb{R}$, the solution $e^y = x - \sqrt{x^2 + 1}$ in (3.53) must be discarded. Thus, $e^y = x + \sqrt{x^2 + 1}$, and solving for y in this last equation, we obtain $y = \log(x + \sqrt{x^2 + 1})$, which proves (3.50).

The identity (3.51) is proved in a similar manner: consider the identity

$$\frac{e^y + e^{-y}}{2} = \cosh y = x, \quad y \in \mathbb{R}, \; x \in [1, \infty). \tag{3.54}$$

In order to solve for y in (3.52), multiply both sides by $2e^y$ and then apply Proposition 3.2.5, so that

$$e^y = x \pm \sqrt{x^2 - 1}. \tag{3.55}$$

Since $x \geqslant 1$, (3.55) yields

$$y = \log(x \pm \sqrt{x^2 - 1}), \quad x \in [1, \infty). \tag{3.56}$$

A moment of reflection will reveal that the only continuous functions which satisfy (3.56) are $y_1(x) = \log(x + \sqrt{x^2 - 1})$ and $y_2(x) = \log(x - \sqrt{x^2 - 1})$, and of course, arg cosh x is among them. Since $\lim_{x \to \infty} y_2(x) = 0$ and $\lim_{x \to \infty} \arg\cosh x = \infty$, clearly $y_2(x) \neq \arg\cosh x$. Hence, $y_1(x) \equiv \arg\cosh x$, and this proves (3.51).

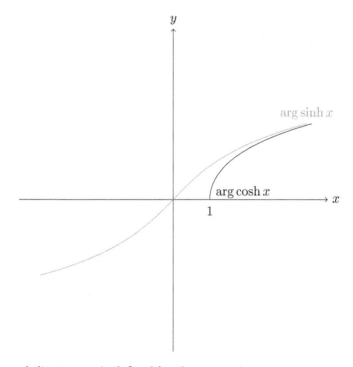

The *hyperbolic tangent* is defined by the expression

$$\tanh x := \frac{\sinh x}{\cosh x}, \quad x \in \mathbb{R}. \tag{3.57}$$

It is straightforward that $\tanh(-x) = -\tanh x$ and that

$$\tanh' x = \frac{1}{\cosh^2 x}, \quad x \in \mathbb{R} \tag{3.58}$$

hence $\tanh x$ is increasing. Thus, as $\tanh 0 = 0$ and $\lim_{x \to \infty} \tanh x = 1$, its image is $(-1, 1)$.

The *argument of the hyperbolic tangent* is defined as the inverse of $\tanh x$:

$$\arg\tanh x := \tanh^{-1} x, \quad x \in (-1, 1); \tag{3.59}$$

its derivative can be easily calculated from Proposition 2.3.7 as

$$\arg\tanh' x = \frac{1}{1 - x^2}, \quad x \in (-1, 1). \tag{3.60}$$

The function arg tanh x can be represented as

$$\arg\tanh x = \frac{1}{2}\log\left(\frac{1+x}{1-x}\right), \quad x \in (-1,1). \tag{3.61}$$

A proof for (3.61) follows easily by solving for y in the identity

$$x = \tanh y = \frac{e^{2y}-1}{e^{2y}+1}. \tag{3.62}$$

To do that, first solve for e^{2y} in (3.62), so that $e^{2y} = (1+x)/(1-x)$, and (3.61) follows immediately.

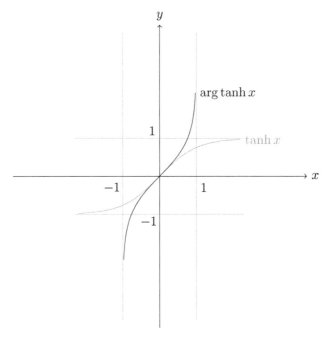

In order to prevent any possible ambiguity, henceforth the notation $\cosh^{-1}(x)$, $\sinh^{-1}(x)$ and $\tanh^{-1}(x)$ denote only $1/cosh(x)$, $1/sinh(x)$ and $1/tanh(x)$ respectively. In this way, $\cosh^{-2}(x)$ denotes $1/\cosh(x)^2$. The inverse functions of $\cosh x$, $\sinh x$ and $\tanh x$ will be exclusively referred to as arg cosh x, arg sinh x and arg tan x respectively.

Chapter 4

Antidifferentiation

This chapter is focused on the techniques of elementary antidifferentiation. Section 4.1 establishes formally the notions of antiderivative, primitive, regular domain, constant of integration and locally constant functions on regular domains.

Section 4.2 describes the basic techniques used in antidifferentiation: linearity, antidifferentiation by parts and change of variable, with special emphasis on the last one. These three techniques are just obtained as a translation into antidifferentiation language of their respective results on differentiation of linear combinations, product and composition of differentiable functions. Hence, Section 4.2 is independent from the theory of integration developed in Section 2.4. However, it must be emphasized that when the aforementioned techniques are put into practice, the Riemann integral plays an implicit but important role. Indeed, this integral ensures the existence of an antiderivative for every continuous function on a regular domain, and this fact makes it easy to find this antiderivative as shown in Section 4.3. To summarize: it is well known by means of Barrow's rule that antidifferentiation is useful in integration; but integration is also useful in antidifferentiation.

Section 4.3 explains how to negotiate the typical difficulties that arise when implementing a change of variable. In particular, the role of the constants of integration is clarified, and Section 4.4 provides a general procedure which explains how to combine the results of Section 4.3 with a change of variable. This procedure will be widely used in Chapters 10 and 11. Section 4.5 includes a list of elementary functions whose antiderivatives are not elementary. Section 4.6 provides a table of primitives which will be useful for solving the exercises in subsequent chapters.

4.1 Antiderivatives and primitives

Let us begin by introducing the central notions formally.

Definition 4.1.1. Let D be a subset of $\mathbb{R}$ with no isolated point. Given a pair of functions f and F defined on D, F is said to be an *antiderivative* of f if F is differentiable on D and $F'(x) = f(x)$ for all $x \in D$.

Definition 4.1.2. The set of all antiderivatives of a given function f is called the *primitive* of f and is denoted as

$$\int f(x)\,dx.$$

Henceforth, if $F(x)$ is an antiderivative of $f(x)$, we will also say that $F(x)$ is a solution of the primitive $\int f(x)\,dx$, and the expression *solving* $\int f(x)\,dx$ means finding out all solutions of $\int f(x)\,dx$.

Given a function $f\colon D \longrightarrow \mathbb{R}$ for which an antiderivative $F(x)$ is known, it is clear that the primitive of f is the set of all functions of the form $F(x)+C(x)$ where $C(x)$ is differentiable on D and $C'(x) = 0$ for all $x \in D$. The simplest case occurs when D is just a non-degenerate interval:

Proposition 4.1.3. *Let I be a non-degenerate interval and let $F(x)$ be an antiderivative of a given function $f\colon I \longrightarrow \mathbb{R}$. Then the primitive of $f(x)$ is*

$$\int f\,dx = \{F(x) + C \colon C \in \mathbb{R}\}.$$

Proof. Let us denote $A := \{F(x) + C \colon C \in \mathbb{R}\}$. Since every constant function is differentiable and its derivative is the null function, it is immediate that $A \subset \int f\,dx$.

For the reverse inclusion, take $G(x) \in \int f\,dx$. Then $G'(x) - F'(x) \equiv 0$, and as the domain of $G - F$ is a non-degenerate interval, Proposition 2.3.10 gives that $G(x) - F(x)$ is a constant function, so $G(x) \in A$. The proof is done. □

According to Theorem 2.4.7, any continuous function on a non-degenerate interval has an antiderivative, hence it fulfils the hypotheses of Proposition 4.1.3. Sometimes, however, the natural domain of a continuous function like $f(x) = 1/x$ is not an interval but a union of pairwise disjoint intervals. These functions are discussed in the following results.

Definition 4.1.4. A non-empty subset D of $\mathbb{R}$ is said to be a *regular domain* if $D = \cup_{j \in A} I_j$ where

(i) each I_j is a non-degenerate interval,
(ii) $D = \cup_{j \in A} I_j$,
(iii) for every $p \in D$ there exists $\delta > 0$ such that $(p - \delta, p + \delta) \cap I_j = \varnothing$ for all but one index $j \in A$,

in which case, $\{I_j\}_{j \in A}$ is said to be (up to relabeling of the intervals I_j) the *regular decomposition* of D.

In order to avoid cumbersome notation, we will say that the regular decomposition of D is $D = \cup_{j \in A} I_j$ instead of using the formal notation $\{I_j\}_{j \in A}$.

Note that statement *(iii)* implies that the intervals I_j of the regular decomposition of D are pairwise disjoint, that is, $I_j \cap I_k = \varnothing$ if and only if $j \neq k$. Also realize that a regular domain has no isolated point.

Example 4.1.5. The set $D = [0,1) \cup \{1\} \cup (1,2) \cup (3,5) \cup (5, \infty)$ is a regular domain and its regular decomposition is $D = [0,2) \cup (3,5) \cup (5, \infty)$.

Example 4.1.6. The set $D := \mathbb{R} \backslash \mathbb{Z}$ is a regular domain and its regular decomposition is $D = \cup_{k \in \mathbb{Z}} (k-1, k)$.

Example 4.1.7. Given the sets

(i) $A := [-1, 0] \cup \left[\bigcup_{n=1}^{\infty} \left(\frac{1}{2n}, \frac{1}{2n-1} \right) \right]$

(ii) $B := [-1, 0) \cup \left[\bigcup_{n=1}^{\infty} \left(\frac{1}{2n}, \frac{1}{2n-1} \right) \right]$

B is a regular domain but A is not.

If D is a regular domain and $D = \bigcup_{i \in I} D_i$ is its regular decomposition, then I is either finite or countably infinite. Indeed, since each D_i is non-degenerate, the density of the rational numbers in $\mathbb{R}$ allows us to choose a rational number $q_i \in D_i$, and since $\mathbb{Q}$ is countable (Corollary A.13.2), then $\{q_i\}_{i \in I}$ is either finite or countable. Moreover, the numbers q_i are pairwise different because the intervals D_i are pairwise disjoint. Hence, there are as many intervals D_i as elements q_i have been chosen. Thus, I is either finite or countable.

Henceforth, we will restrict ourselves to functions whose domain is regular unless otherwise stated.

Proposition 4.1.8. *Let D be a regular domain with regular decomposition $D = \cup_{j \in I} I_j$ and let f be a function defined on D such that each restriction $f|_{I_j}$ has an antiderivative F_j on I_j. Then the function $F: D \longrightarrow \mathbb{R}$ given by $F(x) := F_j(x)$ for every $x \in I_j$ and every $j \in I$ is an antiderivative of f on D.*

Proof. Fix $x_0 \in D$ and let $k \in I$ be the only index such that $x_0 \in I_k$. By definition of regular decomposition, there exists $\delta > 0$ such that

$$D_0 := D \cap (x_0 - \delta, x_0 + \delta) = I_k \cap (x_0 - \delta, x_0 + \delta).$$

Let $(x_n)_{n=1}^{\infty} \subset D \backslash \{x_0\}$ be any convergent sequence to x_0 and take $n_0 \in \mathbb{N}$ such that $x_n \in D_0$ for all $n \geqslant n_0$. Thus, as $D_0 \subset I_k$, it follows that for every $n \geqslant n_0$

$$\frac{F(x_n) - F(x_0)}{x_n - x_0} = \frac{F_k(x_n) - F_k(x_0)}{x_n - x_0} \xrightarrow{n} f(x_0)$$

and from the sequential characterization of Proposition 2.1.10, it follows that there exists $F'(x_0) = f(x_0)$. $\square$

Theorem 4.1.9. *If D is a regular domain, then every continuous function on D has an antiderivative.*

Proof. Let $D = \cup_{j \in I} I_j$ be the regular decomposition of D. Since each I_j is a non-degenerate interval, Theorem 2.4.7 yields an antiderivative $F_j: I_j \longrightarrow \mathbb{R}$ for $f|_{I_j}$, and the result follows from Proposition 4.1.8. $\square$

There are non-continuous functions defined on an interval which possess antiderivatives as shown in Example A.5.2. Moreover, given a continuous function $f: D \longrightarrow \mathbb{R}$, the regularity of its domain is a sufficient but not a necessary condition for f to possess an antiderivative.

The existence of an antiderivative for a continuous function $f: D \longrightarrow \mathbb{R}$ with the sole condition that D has no isolated point is a delicate business far beyond the scope of this book. Eventually, the domain of any function considered in elementary antidifferentiation is regular. Hence, Theorem 4.1.9 will meet all needs throughout this book.

Definition 4.1.10. A function $C: D \longrightarrow \mathbb{R}$ is said to be *locally constant on D* if it is constant on each interval I contained in D.

Let D be a regular domain whose regular decomposition is $D = \cup_{j \in I} D_j$. It readily follows from Definition 4.1.10 that a function $f: D \longrightarrow \mathbb{R}$ is locally constant if f is constant on each interval D_j.

Example 4.1.11. The function C defined on $D := (-\infty, 0) \cup (0, \infty)$ that maps each $x < 0$ to -1 and each $x > 0$ to 1 is locally constant on D.

Proposition 4.1.12. *Let $F: D \longrightarrow \mathbb{R}$ be a function defined on a regular domain. Then, F is differentiable on D and $F' \equiv 0$ if and only if $F(x)$ is locally constant on D.*

Proof. Let $\cup_{j \in I} I_j$ be the regular decomposition of D and assume $F'(x) = 0$ for all $x \in D$. Thus $F'|_{I_j}(x) = 0$ for all $x \in I_j$ and all $j \in I$, and by virtue of Proposition 2.3.10, each $F|_{I_j}$ is constant.

The converse follows from Proposition 4.1.8 and from the fact that every constant function is an antiderivative of the null function on each I_j. □

Proposition 4.1.13. *Let D be a regular domain with regular decomposition $D = \cup_{j \in A} I_j$ and let $f: D \longrightarrow \mathbb{R}$ be a function with an antiderivative $F(x)$. Then the primitive of $f(x)$ is the set which consists of all the functions*

$$F(x) + K(x)$$

where $K(x)$ is locally constant on D.

Proof. It follows from Propositions 4.1.3 and 4.1.12. □

The role played by the regular domains in Proposition 4.1.13 is very important, as the following example shows: for every $n \in \mathbb{N}$, let $I_n := \left[1/(2n), 1/(2n-1)\right)$, so that $0 < \sup I_{n+1} < \inf I_n$ for all n. Let $D := \bigcup_{n=1}^{\infty} I_n$ and $E := D \cup \{0\}$. Clearly D is a regular domain but E is not. After Proposition 4.1.13 it is straightforward that the primitive of the null function on D is the set of all locally constant functions on D.

However, the primitive of the null function on E is the set of all locally constant functions $K(x)$ on E which satisfy

$$\lim_n n[K(1/2n) - K(0)] = 0$$

which is a sufficient and necessary condition for $K(x)$ to be differentiable at 0 with $K'(0) = 0$.

Despite the abuse of notation, when the domain of $f(x)$ is one interval and $F(x)$ is an antiderivative of $F(x)$, it is customary to write

$$\int f(x)\, dx = F(x) + C, \quad C \in \mathbb{R} \tag{4.1}$$

instead of using the formal notation $\int f(x)\, dx = \{F(x) + C : C \in \mathbb{R}\}$, and the unspecific constant C in (4.1) is called the *constant of integration*.

If the domain of $f(x)$ admits a regular decomposition $D = \cup_{j \in A} I_j$ where A has more than one element and $F(x)$ is an antiderivative of $f(x)$, we will maintain this abuse of notation and will write

$$\int f(x)\, dx = F(x) + C(x) \tag{4.2}$$

where $C(x)$ is any locally constant function on D.

Writing the constant of integration C or the function $C(x)$ of formulas (4.1) and (4.2) whenever a primitive occurs might seem fancy and cumbersome but it is necessary in order to obtain a complete solution. In fact, the use of the constant of integration avoids any possible paradox such as that of Example 4.1.14. What is more, its role is absolutely crucial in solving certain primitives such as that of Example 4.3.4.

Example 4.1.14. Solve the primitive $\int x^2 + 2x + 1\, dx$.

Proof. We can antidifferentiate term by term so that

$$\int x^2 + 2x + 1\, dx = \int x^2\, dx + \int 2x\, dx + \int dx = \frac{1}{3}x^3 + x^2 + x + C, \tag{4.3}$$

or we also can proceed globally as follows:

$$\int x^2 + 2x + 1\, dx = \int (x+1)^2\, dx = \frac{1}{3}(x+1)^3 + C$$
$$= \frac{1}{3}x^3 + x^2 + x + \frac{1}{3} + C. \tag{4.4}$$

Although the polynomials $\frac{1}{3}x^3 + x^2 + x$ and $\frac{1}{3}x^3 + x^2 + x + \frac{1}{3}$ in (4.3) and in (4.4) do not coincide, they only differ in a constant. Hence

$$\left\{ \frac{1}{3}x^3 + x^2 + x + C : C \in \mathbb{R} \right\} = \left\{ \frac{1}{3}x^3 + x^2 + x + \frac{1}{3} + C : C \in \mathbb{R} \right\}$$

which shows that the solutions given in (4.3) and in (4.4) are the same. $\qquad\square$

Example 4.1.15. Calculate the primitives of $f\colon \mathbb{R} \longrightarrow \mathbb{R}$ and of $g\colon \mathbb{R}\backslash\{0\} \longrightarrow \mathbb{R}$ where $f(x) = x^2$ and $g = f|_{\mathbb{R}\backslash\{0\}}$.

Proof. Since the domain of $f(x)$ is the interval $(-\infty, \infty)$ and $x^3/3$ is an antiderivative of $f(x)$, from Proposition 4.1.3, the primitive of $f(x)$ is

$$\int f(x)\, dx = \frac{1}{3}x^3 + C,\ C \in \mathbb{R}.$$

Analogously, but applying Proposition 4.1.13 instead of Proposition 4.1.3, we get

$$\int g(x)\, dx = \frac{1}{3}x^3 + K(x),\ \text{where}\ K(x) = \begin{cases} c_1 & \text{if } x > 0 \\ c_2 & \text{if } x < 0 \end{cases}$$

where c_1 and c_2 are any pair of constants. $\square$

4.2   Fundamental theorems of antidifferentiation. Change of variable

This section is concerned with the linearity of antidifferentiation, the antidifferentiation by parts and very particularly, with change of variable.

Let us recall one more time that $\int f(x)\, dx$ does not denote a function but a set of functions.

Theorem 4.2.1 (linearity). *Given a pair of functions f and g defined on a regular domain D, both of which have an antiderivative, it follows that*

$$\int \lambda \cdot f(x) + \mu \cdot g(x)\, dx = \lambda \int f(x)\, dx + \mu \int g(x)\, dx$$

for all pairs of real numbers λ and μ.

Proof. Let F and G be antiderivatives of f and g respectively. Then $F + G$ is an antiderivative of $f + g$ (Proposition 2.3.5). Moreover, as the set of locally constant functions on D is a linear subspace, it follows immediately from the definition of primitive that

$$\int \lambda \cdot f(x) + \mu \cdot g(x)\, dx = \lambda \int f(x)\, dx + \mu \int g(x)\, dx$$

for any pair of real numbers λ and μ. $\square$

Theorem 4.2.2 (integration by parts). *Let f and g be a pair of differentiable functions on a regular domain D. Then, if $f'g$ has an antiderivative, so has fg' and*

$$\int f(x)g'(x)\, dx = f(x)g(x) - \int f'(x)g(x)\, dx.$$

Proof. Let $H(x)$ be an antiderivative of $f'(x)g(x)$. Thus, as $(f\cdot g)'(x) = f'(x)g(x) + f(x)g'(x)$, it follows that $f\cdot g(x) - H(x)$ is an antiderivative of $f\cdot g'(x)$. Theorem 4.2.1 yields $\int f(x)g'(x)\, dx = f(x)g(x) - \int f'(x)g(x)\, dx$ and the proof is done. $\square$

Definition 4.2.3. A function $g\colon I \longrightarrow \mathbb{R}$ defined on a non-degenerate interval I is called a *change of variable (on I)* if it is differentiable on I and $g'(x) \neq 0$ for all $x \in I$.

The main property of a change of variable is its reversibility. This fact is proved in the following proposition and considered in relation to other results in this section.

Proposition 4.2.4. *Every change of variable g on an interval I is invertible and g^{-1} is a change of variable on $g(I)$.*

Proof. Note that g is continuous on I, so according to Theorem 2.2.8, $J := g(I)$ is an interval. In order to see that g is injective, fix $x_0 \in I$. Given any $x \in I\backslash\{x_0\}$, by virtue of Cauchy's mean value theorem (Theorem 2.3.9) there exists $c_x \in I$ situated between x_0 and x such that $g(x) - g(x_0) = g'(c_x) \cdot (x - x_0)$. But by hypothesis, $g'(c_x) \neq 0$, so it follows that $g(x) - g(x_0) \neq 0$, which proves that g is injective, and therefore, strictly monotone by virtue of Theorem 2.2.11. Next, an application of Proposition 2.3.11 shows that either g is increasing and $g'(x) > 0$ for all $x \in I$, or g is decreasing and $g'(x) < 0$ for all $x \in I$. Note that in both cases, J is non-degenerate.

The fact that g^{-1} is differentiable on J and $(g^{-1})'(t) \neq 0$ for all $t \in J$ follows directly from Proposition 2.3.7 and from the hypothesis that $g'(x) \neq 0$ for all $x \in I$. This proves that g^{-1} is a change of variable on J. $\qquad\square$

The following result sometimes eases some calculations for solving a primitive. Note that given a change of variable $g\colon (a, b) \longrightarrow \mathbb{R}$ and any $s \in [a, b] \subset \overline{\mathbb{R}}$, the existence of $\lim_{x \to s} g(x)$ is a direct consequence of Proposition 2.1.13 and Corollary 2.1.14.

Proposition 4.2.5. *Let $g\colon J \longrightarrow \mathbb{R}$ be a change of variable and let $f\colon g(J) \longrightarrow \mathbb{R}$ be a function. Let x_0 and y_0 be a pair of numbers in $\overline{\mathbb{R}}$ such that $y_0 = \lim_{x \to x_0} g(x)$. Then f has limit at y_0 if and only if $f \circ g$ has limit at x_0, in which case, $\lim_{y \to y_0} f(y) = \lim_{x \to x_0} f \circ g(x)$.*

Proof. It is a consequence of Theorem 2.1.11: assume there exists $l = \lim_{y \to y_0} f(y)$ and let $(x_n)_{n=1}^{\infty} \subset J\backslash\{x_0\}$ be a sequence convergent to x_0. As g is injective, it follows that $f \circ g(x_n) \xrightarrow[n]{} l$, which proves that there exists $l = \lim_{x \to x_0} f \circ g(x)$.

Since g^{-1} is a change of variable and $x_0 = \lim_{y \to y_0} g^{-1}(y)$, a similar proof works for the converse implication. $\qquad\square$

Theorem 4.2.6 (change of variable). *Let $g\colon J \longrightarrow \mathbb{R}$ be a change of variable on an interval J and let $f\colon g(J) \longrightarrow \mathbb{R}$ be a function.*

If $H(t)$ is an antiderivative of $(f \circ g)(t) \cdot g'(t)$ on J, then $H \circ g^{-1}(x)$ is an antiderivative of $f(x)$ on $g(J)$.

Proof. By Proposition 4.2.4, g has an inverse g^{-1} that maps $g(J)$ onto J. Denote $I := g(J)$ and $F(x) := H \circ g^{-1}(x)$. From Proposition 2.3.7, $(g^{-1})'(x) = 1/g'\big(g^{-1}(x)\big)$ for all $x \in I$. Thus, by the chain rule, F is differentiable at each $x \in I$ and its derivative is

$$F'(x) = (H \circ g^{-1})'(x) = H'\big(g^{-1}(x)\big) \cdot (g^{-1})'(x)$$

$$= f \circ g\big(g^{-1}(x)\big) \cdot g'\big(g^{-1}(x)\big) \cdot \frac{1}{g'\big(g^{-1}(x)\big)} = f(x).$$

$\square$

Roughly speaking, the theorem of change of variable says that it is sufficient to solve $\int (f \circ g)(t)g'(t)\, dt$ in order to solve $\int f(x)\, dx$.

The following remarks are useful in the application of the theorem of change of variable.

Remark 4.2.7. Practical implementation of the theorem of change of variable:

The application of Theorem 4.2.6 for solving the primitive of a given function $f(x)$ defined on a non-degenerate interval I consists of four steps:

1) Look for a suitable change of variable

$$x = g(t), \quad t \in J$$

such that $g(J) = I$. Finding the right change $g(t)$ requires some expertise and the following chapters explain how to choose $g(t)$ for some standard situations.

2) The symbols x and dx are respectively replaced by $g(t)$ and by $dg(t) = g'(t)\, dt$ in the expression $\int f(x)\, dx$ to produce the transformation

$$\int f(x)\, dx, \quad x \in I \xrightarrow{\;x = g(t)\;} \int (f \circ g)(t) \cdot g'(t)\, dt, \quad t \in J \qquad (4.5)$$

which recovers the primitives that occur in the statement of Theorem 4.2.6.

The replacements $x \leftrightarrow g(t)$ and $dx \leftrightarrow g'(t)\, dt$ are easy to remember with the help of the following tip: a usual notation for the derivative of g at t is

$$g'(t) = \frac{dg(t)}{dt}; \qquad (4.6)$$

removing the denominators in (4.6) is the trick for obtaining the expression

$$g'(t)\, dt = dg(t) = dx \qquad (4.7)$$

required for the prescribed substitutions.[1]

---

[1]Given the change of variable $g \colon J \longrightarrow I$, let us keep the letters t and x to denote elements of J and I respectively. Also retaining the notation introduced at the beginning of Section 2.3, the derivative of g at t_0 is $g'(t_0) = \lim_{\triangle t \to 0} \triangle x / \triangle t$, where $\triangle x$ is the increment of the variable x when the variable t is incremented by a quantity $\triangle t$ at t_0. Thus, $g'(t_0)$ is very close to $\triangle g(t_0)/\triangle t$ when $\triangle x$ is very close to zero; in short, $g'(t_0) \simeq \triangle g(t_0)/\triangle t$ when $0 \neq \triangle x \simeq 0$, in which case, we can remove the denominator multiplying by $\triangle t$ so that $\triangle x = \triangle g(t_0) \simeq g'(t_0) \cdot \triangle t$. The notation

$$g'(t_0) = \lim_{\triangle x \to 0} \frac{\triangle g(t_0)}{\triangle x} = \frac{dg(t_0)}{dt}$$

is also very popular, but it is just notation. Of course removing denominators in (4.6) is unorthodox, but it accomplishes the purpose of (4.7).

3) Find a solution $H(t)$ for the primitive in the right side of (4.5).

4) The change $x = g(t)$ ($t \in J$, $x \in I$) is reversed in $H(t)$ in order to obtain the function $F(x) := H(g^{-1}(x))$, and in accordance with Theorem 4.2.6, $F(x)$ is a solution for $\int f(r) \, dr$

Example 4.2.8. Calculate $\int \cos(3x + 2) \, dx$.

Solution. Note that the domain of $\cos(3x+2)$ is $\mathbb{R}$ and that the change of variable $x \mapsto t = 3x + 2$ maps $\mathbb{R}$ onto $\mathbb{R}$ and $dt = 3 \, dx$. Thus, since $\sin t$ is an antiderivative of $\cos t$ on $\mathbb{R}$,

$$\int \cos(3x + 2) \, dx \xrightarrow{t=3x+2} \int \frac{1}{3} \cos t \, dt = \frac{1}{3} \sin t + C \xrightarrow{t=3x+2} \frac{1}{3} \sin(3x + 2) + C.$$

Therefore, the desired primitive is $\sin(3x + 2)/3 + C$, with $C \in \mathbb{R}$. $\qquad\square$

The fact that the inverse of a change of variable is also a change of variable leads to the following analysis:

Remark 4.2.9. Reversibility of the change of variable:

Following Proposition 4.2.4, the inverse of the function g is a change of variable. Hence, if we apply the procedure described in Remark 4.2.7 to the function $(f \circ g)(t) \cdot g'(t)$ with the change of variable $t = g^{-1}(x)$, we get

$$\int f \circ g(t) \cdot g'(t) \, dt \xrightarrow{t=g^{-1}(x)} \int f \circ g\big(g^{-1}(x)\big) g'\big(g^{-1}(x)\big) \cdot (g^{-1})'(x) \, dx = \int f(x) \, dx \quad (4.8)$$

which reverses (4.5). Thus, a complete picture of the whole process should include double-headed arrows to show the two ways in which a change of variable works, as indicated in the following diagram, which merges both (4.5) and (4.8):

$$
\begin{array}{ccc}
\displaystyle\int f(x) \, dx & \xleftarrow{\quad x = g(t) \quad} & \displaystyle\int (f \circ g)(t) \cdot g'(t) \, dt \\[2ex]
\big\| & & \big\| \\[2ex]
F(x) + C = H \circ g^{-1}(x) + C & \xleftarrow{\quad x = g(t) \quad} & H(t) + C
\end{array}
$$

$$(4.9)$$

Diagram (4.9) clearly shows that solving $\int f \circ g(t) g'(t) \, dt$ is in fact equivalent to solving $\int f(x) \, dx$.

Remark 4.2.10. Practical implementation of the reverse change of variable:

The implementation of the reverse change $t = g^{-1}(x)$ as explained in Remark 4.2.9 allows a useful shortcut on some occasions: assume we want to solve the primitive of a given function $h(t)$. Instead of replacing dt by $(g^{-1})'(x)\,dx$ in the expression $\int h(t)\,dt$, we depart from the identity $x = g(t)$ to obtain (4.7), and multiplying both sides of (4.7) by $1/g'(t)$, we get

$$dt = \frac{dx}{g'(t)}; \tag{4.10}$$

then, plugging $t = g^{-1}(x)$ into (4.10), we get

$$dt = \frac{dx}{g'\left(g^{-1}(x)\right)} \tag{4.11}$$

which determines the substitution to be performed in $\int h(t)\,dt$ in order to implement the change of variable $t = g^{-1}(x)$.

In practice, the advantage of the shortcut described in Remark 4.2.10 is the fact that it is not necessary to differentiate g^{-1}, but g. Sometimes, it is not even necessary to deal with any explicit expression for g^{-1}. An example will suffice to explain this:

Example 4.2.11. Calculate the primitive of the function $f(x) = \sin^4 x \, \cos^3 x$ on the interval $(0, \pi/2)$.

Solution. A standard change of variable for solving the required primitive is $t = \sin x$ with $x \in J := (0, \pi/2)$ (see Chapter 11). However, an orthodox application of Theorem 4.2.6 means solving x as a function of t, that is, $x = \arcsin t$ with $t \in I := (0, 1)$ and then proceeding with the replacement $dx = \arcsin' t \, dt$.

But the reversibility of the change of variable $t = \sin x$ enables the application of the shortcut described in Remark 4.2.10:

$$t = \sin x, \ x \in J \Rightarrow dt = \cos x \, dx \Rightarrow \frac{dt}{\cos x(t)} = dx$$

where $x(t)$ is the inverse of the change $t = \sin x$, $x \in J$ (note that we are not giving any specific expression for $x(t)$).

Thus, plugging the identities $t = \sin x$ (with $x \in J$ and $t \in I$) and $dt/\cos x(t) = dx$ into $\int \sin^4 x \, \cos^3 x \, dx$, we get

$$\int \sin^4 x \, \cos^3 x \, dx, \ x \in J \xleftarrow{\ \sin x = t\ } \int t^4 \cos^3 x(t) \frac{dt}{\cos x(t)}$$

$$= \int t^4 \cos^2 x(t) \, dt, \ t \in I,$$

and as $\cos^2 x(t) = 1 - \sin^2 x(t) = 1 - t^2$, it follows that

$$\int t^4 \cos^2 x(t) \, dt = \int t^4 (1 - t^2) \, dt = \int t^4 - t^6 \, dt = \frac{t^5}{5} - \frac{t^7}{7} + K, \quad t \in I.$$

Reversing the change of variable, we obtain

$$\frac{t^5}{5} - \frac{t^7}{7} + K, \ t \in I \xleftarrow{\sin x = t} \frac{\sin^5 x}{5} - \frac{\sin^7 x}{7} + K, \quad x \in J$$

which is the primitive of $\sin^4 x \cos^3 x$ on $(0, \pi/2)$. $\qquad\square$

The change of variable in definite integrals is not required to be injective as shown in Corollary 2.4.9; but this special property must be used with caution: the following example shows a misuse of the above mentioned corollary.

Example 4.2.12. Plugging the change of variable $x^2 = t$ ($\Rightarrow 2x \, dx = dt$) into the integral of x^2 on $[-1, 1]$, we get the following contradiction:

$$0 < \int_{-1}^{1} x^2 \, dx = \int_{(-1)^2}^{1^2} \frac{\sqrt{t}}{2} \, dt = \int_{1}^{1} \frac{\sqrt{t}}{2} \, dt = 0.$$

What went wrong?

The reader is encouraged to solve this question independently, to realize that this sort of mistake can hardly happen in performing a change of variable when calculating a primitive.

4.3 Important results concerning a change of variable

When solving the primitive of a function $f(x)$ whose domain is an interval I by means of a change of variable $x = g(t)$, it may occur that the domains of f and g do not adjust well to each other. In many of these cases, the mere application of the change $x = g(t)$ as in Theorem 4.2.6 only yields a collection of differentiable functions $F_1, F_2, ..., F_n$ for which $F_i'(x) = f(x)$ on I_i, where $I_1, I_2, ..., I_n$ is a collection of pairwise disjoint intervals such that $I = \cup_{i=1}^{n} I_i$. The following results explain how to construct an antiderivative of f on I by means of these functions F_i.

Proposition 4.3.1. *Let $f \colon [a, b] \longrightarrow \mathbb{R}$ be a function such that $-\infty < a < b < \infty$:*

(i) *If f is continuous at b and there exists a function $F(x)$ defined on $[a, b]$, differentiable on $[a, b)$, continuous at b and $F'(x) = f(x)$ for all $x \in [a, b)$, then F is an antiderivative of f on $[a, b]$.*

(ii) *If f is continuous at a and there exists a function $F(x)$ defined on $[a, b]$, differentiable on $(a, b]$, continuous at a and $F'(x) = f(x)$ for all $x \in (a, b]$, then F is an antiderivative of f on $[a, b]$.*

Proof. (i) We only need to prove that F is differentiable at b and that $F'(b) = f(b)$. In order to do this, the continuity of F and f at b allows us to apply L'Hôpital's rule to get

$$\frac{F(x) - F(b)}{x - b} \xrightarrow{x \to b} \lim_{x \to b} f(x) = f(b)$$

which proves that F is differentiable at b and $F'(b) = f(b)$.

(ii) The proof of this part is similar to that of (i). $\qquad\square$

Strengthening the hypotheses of Proposition 4.3.1 leads to the following result:

Proposition 4.3.2. *Let $f(x)$ be a continuous function on the interval $[a, b]$ where $-\infty < a < b < \infty$:*

(i) *If $G(x)$ is a differentiable function on $[a, b)$ such that $G'(x) = f(x)$ for all $x \in [a, b)$, then $G(x)$ can be extended to a continuous function $\tilde{G}(x)$ on $[a, b]$ which is an antiderivative of f on $[a, b]$.*

(ii) *If $G(x)$ is a differentiable function on $(a, b]$ such that $G'(x) = f(x)$ for all $x \in (a, b]$, then $G(x)$ can be extended to a continuous function $\tilde{G}(x)$ on $[a, b]$ which is an antiderivative of f on $[a, b]$.*

Proof. (i) Since $f(x)$ is continuous on $[a, b]$, its integral function $F(x) = \int_a^x f(t)\, dt$ is an antiderivative of $f(x)$ on $[a, b]$, and therefore, on $[a, b)$ too. Thus, by Proposition 4.1.3, there exists a constant C such that $G(x) = F(x) + C$ for all $x \in [a, b)$, and as $F(x)$ is continuous at b, it follows that

$$G(x) = F(x) + C \xrightarrow{\ x \to b\ } F(b) + C.$$

Thus, defining $\tilde{G}(x) := G(x)$ for $x \in [a, b)$ and $\tilde{G}(b) := F(b) + C$, we have that $\tilde{G}(x)$ is a continuous extension of $G(x)$ on $[a, b]$, and by virtue of Proposition 4.3.1, $\tilde{G}(x)$ is an antiderivative of $f(x)$ on $[a, b]$.

(ii) The proof of this statement is similar to that of (i). $\qquad\square$

Given a function $f(x)$ on a regular domain D and a non-degenerate interval $I \subset D$, a function $F : I \longrightarrow \mathbb{R}$ is said to be a *local antiderivative of $f(x)$ on I* if $F(x)$ is differentiable and $F'(x) = f(x)$ for all $x \in I$.

If f is continuous on D and a collection of local antiderivatives whose domains cover D is known, then Proposition 4.3.2 tells us that these local antiderivatives can be assembled to construct a global antiderivative of $f(x)$ on D, and it is here where the constants of integration play a central role and have to be carefully chosen. This procedure is detailed in the following result:

Proposition 4.3.3. *Let f be a continuous function on (a, b) and let $a < \xi < b$. Let $G_1 : (a, \xi) \longrightarrow \mathbb{R}$ and $G_2 : (\xi, b) \longrightarrow \mathbb{R}$ be a pair of differentiable functions such that $G_1'(x) = f(x)$ for all $x \in (a, \xi)$ and $G_2'(x) = f(x)$ for all $x \in (\xi, b)$. Let $\lambda_1 := G_1(\xi^-)$ and $\lambda_2 := G_2(\xi^+)$. Then the function*

$$F(x) = \begin{cases} G_1(x) & \text{if } a < x < \xi \\ \lambda_1 & \text{if } x = \xi \\ G_2(x) + \lambda_1 - \lambda_2 & \text{if } \xi < x < b \end{cases}$$

is an antiderivative of $f(x)$ on (a, b).

Proof. First of all, note that the existence of the one-sided limits λ_1 and λ_2 as real numbers is ensured by Proposition 4.3.2, so the function $F(x)$ is well defined and continuous on (a, b). Moreover, for every $x \in (a, \xi) \cup (\xi, b)$, there exists $F'(x) = f(x)$, and as $F(x)$ is the only continuous extension of $F|_{(a,\xi) \cup (\xi,b)}(x)$ to (a, b), applying Proposition 4.3.2 on (a, ξ) and on (ξ, b) gives

$$F'(\xi^-) = f(\xi) = F'(\xi^+)$$

which in turn shows that $F(x)$ is differentiable at ξ and $F'(\xi) = f(\xi)$. This proves that $F(x)$ is an antiderivative of $f(x)$ on (a, b). □

A common mistake in antidifferentiation calculus is to assume that the function

$$G(x) = \begin{cases} G_1(x) & \text{if } a \leqslant x < \xi \\ G_2(x) & \text{if } \xi < x < b \end{cases}$$

is an antiderivative of $f(x)$ on (a, b) (we are maintaining the same notation given in Proposition 4.3.3). Indeed, nothing ensures G can be extended to a differentiable function at ξ: if $\lambda_1 \neq \lambda_2$, any extension of G to ξ has a jump discontinuity at ξ.

An example is worth a thousand words:

Example 4.3.4. Calculate $\int f(x)\,dx$ where $f(x) := \sin x$ for $x \in [0, \pi/2]$ and $f(x) = 1 - \cos x$ for $x \in (\pi/2, 1]$.

Solution. Since

$$f(\pi/2^-) = \lim_{x \to \pi/2^-} \sin x = 1$$
$$f(\pi/2^+) = \lim_{x \to \pi/2^+} 1 - \cos x = 1$$

it follows that $f(x)$ is continuous on the interval $[0, \pi]$, so it possesses an antiderivative $F(x)$ and its primitive is $F(x) + K$ for all $K \in \mathbb{R}$.

In order to find an antiderivative F for f, as f is piecewise defined, we first get local antiderivatives G_1 and G_2 of f on the respective intervals $[0, \pi/2]$ and on $(\pi/2, \pi]$:

$$G_1(x) = -\cos x, \quad x \in [0, \pi/2]$$
$$G_2(x) = x - \sin x, \quad x \in (\pi/2, \pi].$$

At this point, it might be tempting to consider the piecewise defined function

$$G(x) = \begin{cases} G_1(x), & 0 \leqslant x \leqslant \pi/2 \\ G_2(x), & \pi/2 < x \leqslant \pi \end{cases}$$

as a genuine antiderivative for f. In fact, G is differentiable at each $x \in [0, \pi] \setminus \{\pi/2\}$ and $G'(x) = f(x)$, but G has a jump discontinuity at $\pi/2$ because

$$G(\pi/2^-) = 0 < \frac{\pi}{2} - 1 = G(\pi/2^+), \tag{4.12}$$

hence G is not differentiable at $\pi/2$.

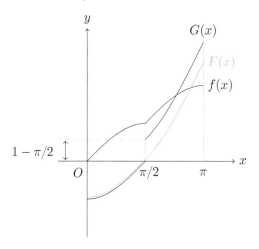

$$(4.13)$$

However, following the construction described in Proposition 4.3.3, the jump discontinuity at $\pi/2$ is fixed by moving the graph of $G_2(x)$ downwards to the right height to give the function

$$F(x) = \begin{cases} G_1(x), & 0 \leqslant x \leqslant \pi/2 \\ G_2(x) - \frac{\pi}{2} + 1, & \pi/2 < x \leqslant \pi \end{cases}$$

which is a true antiderivative of $f(x)$ on $[0, \pi]$.   □

Proposition 4.3.1 yields the following generalization of the theorem of the change of variable.

Corollary 4.3.5. *Let J be a non-degenerate interval, let $t_0 \in J$ and let $g \colon J \longrightarrow \mathbb{R}$ be a strictly monotone, continuous function on J, differentiable on $J\backslash\{t_0\}$ and such that $g'(x) \neq 0$ for all $x \in J\backslash\{t_0\}$. Let $f \colon g(J) \longrightarrow \mathbb{R}$ be a continuous function at $g(t_0)$.*
If $H \colon J \longrightarrow \mathbb{R}$ is an antiderivative of $(f \circ g)(t) \cdot g'(t)$ on $J\backslash\{t_0\}$ and is continuous on J, then $H \circ g^{-1}(x)$ is an antiderivative of $f(x)$ on $g(J)$.

Proof. Let $x_0 := g(t_0)$ and $I := g(J)$. By virtue of Theorem 4.2.6, $H \circ g^{-1}(x)$ is an antiderivative of $f(x)$ on $I\backslash\{x_0\}$. Moreover, by Theorem 2.2.11, g^{-1} is continuous on I. Hence, $H \circ g^{-1}(x)$ is also continuous on I, and in turn, it is an antiderivative of $f(x)$ on the entire interval I by virtue of Proposition 4.3.1.   □

Note that the set $I\backslash\{x_0\}$ occurring in the proof of Corollary 4.3.5 is either a single non-degenerate interval or the union of a pair of disjoint, non-degenerate intervals.

The meaning of Corollary 4.3.5 is that any function g which satisfies its hypotheses works as if it were an authentic change of variable. The use of this fact is left to the discretion of the reader.

Example 4.3.6. Calculate the primitive of $\sqrt[3]{x}$ using the function $g(t) = t^3$ as a change of variable.

Solution. Without the mediation of any change of variable, it is immediate that the primitive of $\sqrt[3]{x}$ is

$$\int \sqrt[3]{x}\, dx = \frac{3}{4} x^{4/3} + C, \quad x \in [0, \infty) \tag{4.14}$$

where C is any real constant.

Clearly, the function $g: [0, \infty) \longrightarrow [0, \infty)$ given by $g(t) = t^3$ just misses being a change of variable by the single point $t = 0$, where $g'(0) = 0$. But both $f(x)$ and $g(t)$ meet all the conditions of Corollary 4.3.5 and therefore, $g(t)$ serves as an authentic change of variable for solving $\int \sqrt[3]{x}\, dx$. Thus, following the mentioned corollary, we remove the point 0 from the domain and plug the identities $x = t^3$ and $dx = 3t^2\, dt$ into the expression $\int \sqrt[3]{x}\, dx$ so that

$$\int \sqrt[3]{x}\, dx, \ x \in (0, \infty) \xleftarrow{\ x=t^3\ } \int \sqrt[3]{t^3}3t^2\, dt = \int 3t^3\, dt = \frac{3}{4}t^4 + C, \ t \in (0, \infty),$$

where C is any real constant.

Clearly, $3t^4/4$ is an antiderivative of $\sqrt[3]{t^3}3t^2$ on $[0, \infty)$, so by virtue of Corollary 4.3.5, the reversion of the change $x = \sqrt[3]{x}$ in this function gives the solution for $\int \sqrt[3]{x}\, dx$ as

$$\frac{3}{4}t^4 + C, \ t \in [0, \infty) \xleftarrow{\ t=x^{1/3}\ } \frac{3}{4}x^{4/3} + C, \ x \in [0, \infty)$$

which is exactly the same solution as that obtained in (4.14). □

It is not very difficult to realize that Corollary 4.3.5 also works for any function $g: J \longrightarrow \mathbb{R}$ continuous on J, strictly monotone and such that

$$S_g := \{x \in J: \text{either } \nexists g'(x) \text{ or } g'(x) = 0\}$$

is a finite set. Note that g^{-1} is also continuous on the interval $g(J)$, is strictly monotone, and by virtue of Proposition 2.3.7, the set $S_{g^{-1}}$ is finite too.

Under the hypotheses of Corollary 4.3.5, the function $(f \circ g) \cdot g'$ is well defined on $J \backslash S_g$ but need not have a continuous extension at each $t_0 \in S_g$. For instance, consider the functions $f: [0, \infty) \longrightarrow \mathbb{R}$ given by $f(x) = 3$ for all x and $g: [0, \infty) \longrightarrow \mathbb{R}$ given by $g(t) = \sqrt[3]{t}$. Clearly, $(f \circ g)(t) \cdot g'(t) = 1/\sqrt[3]{t^2}$ is well defined on $(0, \infty)$ but it does not admit a continuous extension at $t = 0$ because $\lim_{t \to 0^+} 1/\sqrt[3]{t^2} = \infty$.

4.4　A practical method to perform a change of variable

When solving the primitive of a continuous function $f \colon D \longrightarrow \mathbb{R}$ by means of a change of variable, it sometimes happens that the function $t = g(x)$ chosen as change of variable is not an authentic change of variable because it misses some of the required conditions given in Definition 4.2.3. Usually, this situation is accompanied by a bad adjustment between the domains of the given function $f(x)$ and the chosen function $g(x)$. In these cases, Proposition 4.3.3 and Corollary 4.3.5 must be carefully applied and it is recommended to adopt the informally-described procedure that follows.

Suppose that the domain of $f(x)$ is $D = \cup_{k=1}^{5} J_i$ where

$$
\begin{aligned}
J_1 &:= (-\pi/2 - \pi, -\pi) \\
J_2 &:= (-\pi, 0) \\
J_3 &:= (0, 2\pi) \\
J_4 &:= (2\pi, \pi/2 + 2\pi) \\
J_5 &:= (\pi/2 + 2\pi, \pi/2 + 3\pi]
\end{aligned}
$$

and assume that the choice for $g(x)$ is $\sin x$. However, $\sin x$ is not a change of variable on its natural domain because it is not injective and its derivative is null at certain points. This inconvenience is fixed by taking the following steps:

1) Divide the natural domain of $g(x)$ into pairwise disjoint open intervals so that $g(x)$ is a change of variable on each of these intervals. Of course, this action may leave out some points of the natural domain of $g(x)$. For the considered case $g(x) = \sin x$, let

$$
I_n := \left(-\frac{\pi}{2} + n\pi, \frac{\pi}{2} + n\pi \right), \quad n \in \mathbb{Z}.
$$

Each restriction

$$
g_n(x) := g|_{I_n}(x), \quad n \in \mathbb{Z} \tag{4.15}
$$

is a change of variable in accordance with Definition 4.2.3. In particular, g_n is strictly increasing if n is even, and it is strictly decreasing if n is odd. Observe that the union of the domains of all g_n does not cover $\mathbb{R}$ completely. Indeed, $\mathbb{R} \backslash D_g := \{\pi/2 + n\pi \colon n \in \mathbb{Z}\}$ is not empty.

2) Draw three parallel horizontal lines. Represent the domain of $f(x)$ on the upper line and the domains of the functions g_n on the middle line. The lower line is reserved for the assembly of the local antiderivatives of $f(x)$ (see Diagram 4.16).

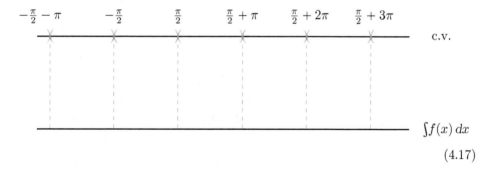

$$\int f(x)\,dx \tag{4.16}$$

3) Draw a dashed vertical line from each point of $\mathbb{R}\backslash D_g$ in the middle line to the line assigned to $\int f(x)\,dx$ (Diagram 4.17).

$$\int f(x)\,dx \tag{4.17}$$

4) Draw a continuous vertical line from each point of $\mathbb{R}\backslash D$ in the line assigned to $f(x)$ to the line assigned to $\int f(x)\,dx$ (in the particular case considered here, the intervals $(-\infty, -\pi/2 - \pi)$ and $(\pi/2 + 3\pi, \infty)$ have been removed from the first and third horizonal lines in Diagram 4.18 instead of drawing the corresponding continuous vertical lines in order to produce a clear picture).

Do not worry if some dashed vertical line is covered by a continuous vertical line. In fact, this is not an unwanted effect.

The action carried out in the fourth step moves each piece J_i of D into the third line. The combined action of the dashed lines and the continuous lines determines the open intervals

$$L_1^1 := (-\pi/2 - \pi, -\pi)$$
$$L_2^1 := (-\pi, -\pi/2), \ L_2^2 := (-\pi/2, 0)$$
$$L_3^1 := (0, \pi/2), \ L_3^2 := (\pi/2, \pi/2 + \pi), \ L_3^3 := (\pi/2 + \pi, 2\pi)$$
$$L_4^1 := (2\pi, \pi/2 + 2\pi)$$
$$L_5^1 := (\pi/2 + 2\pi, 3\pi)$$

(see Diagram 4.18). Note that each interval L_i^j is contained in J_i.

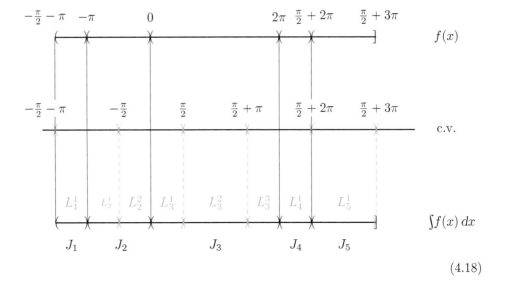

$$(4.18)$$

5) Assume that a local antiderivative $H_i^j(t)$ on L_i^j of $(f \circ g)(t) \cdot g'(t)$ is known for every i and j. Then, since each restriction $g|_{L_i^j}$ is a change of variable, the theorem of change of variable (Theorem 4.2.6) shows that $F_i^j(x) := H_i^j \circ g|_{L_i^j}(x)$ is a local antiderivative of $f(x)$ on L_i^j.

6) A local antiderivative $F_i(x)$ of $f(x)$ on each interval J_i is obtained from the local antiderivatives $F_i^j(x)$ by means of Propositions 4.3.2, 4.3.3 and Corollary 4.3.5. For this stage, it is very important to pay attention to those points in the lower line which are reached only by a dashed line because this determines whether or not Proposition 4.3.3 or Proposition 4.3.2 (or even Corollary 4.3.5) must be applied.

In the case considered here, the points reached by a dashed line in the lower line are

$$-\pi/2 \text{ in } J_2$$
$$\pi/2 \text{ and } \pi/2 + \pi \text{ in } J_3$$
$$\pi/2 + 3\pi \text{ in } J_5.$$

Thus, the local antiderivatives of $f(x)$ on J_1 and J_4 can be respectively chosen as $F_1^1(x)$ and $F_4^1(x)$.

For the local antiderivative of $f(x)$ on J_2, Corollary 4.3.5 shows that the continuous extension of $F_2^1(x)$ to $x = -\pi/2$ is an antiderivative of $f(x)$ on the interval $L_2^1 \cup \{-\pi/2\}$. Let us denote this extension as $F_2^1(x)$ too. An application of Proposition 4.3.3 allows to proceed as in Example 4.3.4 to find a suitable constant C so that

$$F_2(x) := \begin{cases} F_2^1(x), & x \in L_2^1 \cup \{-\pi/2\} \\ F_2^2(x) + C, & x \in L_2^2 \end{cases}$$

is the desired local antiderivative of $f(x)$ on J_2.

A similar reasoning based upon Proposition 4.3.3, Proposition 4.3.2 and Corollary 4.3.5 yields a local antiderivative $F_3(x)$ from $F_3^1(x)$, $F_3^2(x)$ and $F_3^3(x)$.

And following Corollary 4.3.5, $F_5(x)$ is the continuous extension of $F_5^1(x)$ to $x = \pi/2 + 3\pi$.

7) Once all local antiderivatives $F_i(x)$ have been found, the primitive of $f(x)$ on D is

$$\int f(x)\, dx = F_i(x) + K_i, \ x \in J_i, \ i = 1, 2, 3, 4, 5$$

where $\{K_i\}_{i=1}^5$ is any collection of real constants.

Chapters 10 and 11 contain plenty of examples where this procedure is applied.

4.5 Continuous functions with no elementary antiderivative

Although every continuous function $f(x)$ on an interval possesses an antiderivative, it may happen that it does not admit any elementary antiderivative, even if $f(x)$ is elementary. Nevertheless, such functions are often found in many natural contexts. For instance, the function e^{x^2} is probably the most important function in Probability Theory: $e^{x^2/2}/\sqrt{2\pi}$ is the density function of a normal distribution; the function $\sin x/x$ occurs in Fourier Analysis, the functions $\sqrt{a + \sin x}$ occur in Geometry, etc.

The following list of such functions is not exhaustive but contains the most frequent ones. The student must bear them in mind because their primitives cannot be solved by means of the methods explained in this book, which only deals with elementary solutions.

Table 4.5.1. **Functions with no elementary antiderivative:**

$e^{(x^2)}$	$e^{(x^3)}$	$e^{1/x}$	e^{1/x^2}
e^{-1/x^2}	$e^{(e^x)}$	$\dfrac{e^x}{x}$	$\dfrac{e^x}{x^2}$
$\dfrac{x}{x+e^x}$	$\dfrac{x}{1+e^x}$	$\dfrac{1}{x+e^x}$	
$e^x \log x$	$\dfrac{e^x}{\log x}$	$\dfrac{\log x}{e^x}$	$e^{\sqrt{\log x}}$
$e^{\sin x}$	$e^{\cos x}$	$e^{\tan x}$	$e^{\arcsin(\log x)}$
$e^{\arccos(\log x)}$	$e^{\arctan x}$		
$\sqrt{1+\log x}$	$\sqrt{x+\log x}$	$\sqrt{x+e^x}$	$\sqrt{\cos x}$
$\sqrt{a+\sin x}\ (a\neq1)$	$\sqrt{a+\cos x}\ (a\neq1)$	$e^x\sqrt{x}$	
$\sqrt{x}\sin x$	$\sqrt{x}\cos x$	$\sqrt{\sin x}$	$\sqrt{\cos x}$
$\log(1\pm e^x)$	$\log(\sin x)$	$\log(\cos x)$	$\log(\tan x)$
$\dfrac{1}{\log x}$	$\dfrac{x}{\log x}$	$\dfrac{1}{x+\log x}$	$\log(\log x)$
$\dfrac{\log x}{x+1}$	$\dfrac{\log x}{x-1}$		
$\sin(x^2)$	$\cos(x^2)$	$\tan(x^2)$	$\tan(\sqrt{x})$
$\sin(1/x)$	$\cos(1/x)$	$\tan(1/x)$	$x\tan x$
$\dfrac{\sin x}{x}$	$\dfrac{\cos x}{x}$	$\dfrac{\tan x}{x}$	
$\dfrac{x}{\sin x}$	$\dfrac{x}{\cos x}$	$\dfrac{x}{\tan x}$	
$\dfrac{1}{x+\sin x}$	$\dfrac{1}{x+\cos x}$	$\dfrac{1}{x+\tan x}$	

$\sin(\sin x)$	$\sin(\cos x)$	$\sin(\tan x)$
$\cos(\sin x)$	$\cos(\cos x)$	$\cos(\tan x)$
$\tan(\sin x)$	$\tan(\cos x)$	$\tan(\tan x)$
$\sin(e^x)$	$\cos(e^x)$	$\tan(e^x)$
$\dfrac{1}{\arcsin x}$	$\dfrac{1}{\arccos x}$	$\dfrac{1}{\arctan x}$
$\dfrac{\arcsin x}{x}$	$\dfrac{\arccos x}{x}$	$\dfrac{\arctan x}{x}$
$\dfrac{x}{\arcsin x}$	$\dfrac{x}{\arccos x}$	$\dfrac{x}{\arctan x}$
$\arcsin(e^x)$	$\arccos(e^x)$	$\arctan(e^x)$
$\arcsin(\log x)$	$\arccos(\log x)$	$\arctan(\log x)$
$\arcsin(x^2)$	$\arccos(x^2)$	

An explanation of why these functions do not admit an elementary antiderivative is beyond the scope of this book. The reader interested in this subject should consult references [Risch (1969)] and [Risch (1970)].

4.6 Basic primitives

When solving primitives, having a table of basic antiderivatives at hand is recommended, but in order to construct one, we must first tackle the primitives of $1/x$ and $1/\sqrt{x^2 - 1}$ because they are defined on a regular domain which does not consist of a single interval:

(a) The natural domain of $g(x) = 1/x$ is $(-\infty, 0) \cup (0, \infty)$. Since the derivative of $\log x$ is $g(x)|_{(0,\infty)}$, the chain rule gives that $\log(-x)$ is the derivative of $g(x)|_{(-\infty,0)}$, and in view of the regular decomposition of the domain of $g(x)$, Proposition 4.1.13 yields

$$\int \frac{1}{x}\, dx = \log|x| + C(x)$$

where $C(x) = C_1$ if $x \in (-\infty, 0)$, $C(x) = C_2$ if $x \in (0, \infty)$, and C_1 and C_2 are any pair of constants.

(b) The natural domain of $h(x) = 1/\sqrt{x^2 - 1}$ is $(-\infty, -1) \cup (1, \infty)$. Since the derivative of $\arg\cosh x$ is $h(x)|_{(1,\infty)}$, the chain rule gives that $-\arg\cosh(-x)$ is the derivative of $g(x)|_{(-\infty,-1)}$, and in view of the regular decomposition of the domain of $g(x)$, Proposition 4.1.13 yields

$$\int \frac{1}{\sqrt{x^2 - 1}} \, dx = \begin{cases} \arg\cosh(x) + C_1 & \text{if } x \in (1, \infty) \\ -\arg\cosh(-x) + C_2 & \text{if } x \in (-\infty, -1) \end{cases}$$

where C_1 and C_2 are arbitrary real constants.

In the following table, all the functions are defined on their natural domains.

Table 4.6.1. **Basic primitives:**

$$(1) \quad \int x^\beta \, dx = \frac{1}{\beta + 1} x^{\beta + 1} + C, \ \beta \in [0, \infty)$$

$$(2) \quad \int x^\beta \, dx = \frac{1}{\beta + 1} x^{\beta + 1} + \begin{cases} C_1 & \text{if } x > 0 \\ C_2 & \text{if } x < 0 \end{cases}, \ \beta \in (-\infty, 0) \backslash \{-1\}$$

$$(3) \quad \int \frac{1}{x} \, dx = \log |x| + \begin{cases} C_1 & \text{if } x > 0 \\ C_2 & \text{if } x < 0 \end{cases}$$

$$(4) \quad \int \frac{1}{1 + x^2} \, dx = \arctan x + C$$

$$(5) \quad \int e^x \, dx = e^x + C$$

$$(6) \quad \int \sin x \, dx = -\cos x + C$$

$$(7) \quad \int \cos x \, dx = \sin x + C$$

$$(8) \quad \int \frac{1}{\sqrt{1 - x^2}} \, dx = \arcsin x + C$$

$$(9) \quad \int \frac{1}{\sqrt{1 + x^2}} \, dx = \arg\sinh x + C$$

$$(10) \quad \int \frac{1}{\sqrt{x^2 - 1}} \, dx = \begin{cases} \arg\cosh(x) + C_1 & \text{if } x \in (1, \infty) \\ -\arg\cosh(-x) + C_2 & \text{if } x \in (-\infty, -1) \end{cases}$$

$$(11) \quad \int \sinh x \, dx = \cosh x + C$$

$$(12) \quad \int \cosh x \, dx = \sinh x + C$$

$$(13) \quad \int \tanh x \, dx = \log(\cosh x) + C$$

$$(14) \quad \int \frac{1}{\tanh x} \, dx = \log(|\sinh x|) + \begin{cases} C_1 & \text{if } x > 0 \\ C_2 & \text{if } x < 0 \end{cases}$$

$$(15) \quad \int \arg\sinh x \, dx = x \arg\sinh x - \sqrt{x^2 + 1} + C$$

$$(16) \quad \int \arg\cosh x \, dx = x \arg\cosh x - \sqrt{x^2 - 1} + C$$

$$(17) \quad \int \arg \tanh x \, dx = x \arg \tanh x + \frac{\log(1 - x^2)}{2} + C$$

The following two special cases must be added to the list above:

$$(18) \quad \int \frac{f'(x)}{f(x)} \, dx = \log |f(x)| + C(x)$$

$$(19) \quad \int \frac{ax + (b/2)}{\sqrt{ax^2 + bx + c}} \, dx = \sqrt{ax^2 + bx + c} + K(x)$$

where $f(x)$ is a differentiable function on a regular domain D, $C(x)$ is a locally constant function on D, $a \neq 0$, the two roots of $ax^2 + bx + c$ are different and $K(x)$ is locally constant on $D \cap \Delta$, where $\Delta := \{x \colon ax^2 + bx + c > 0\}$.

If x_0 and x_1 are the two solutions to $ax^2 + bx + c = 0$, then Δ is regular if and only if one of the following statements holds true:

(i) $a > 0$, both x_0 and x_1 are real and $x_0 \leqslant x_1$, in which case $\Delta = (-\infty, x_0) \cup (x_1, \infty)$;

(ii) $a < 0$, both x_0 and x_1 are real and $x_0 < x_1$, which gives $\Delta = (x_0, x_1)$:

(iii) $a > 0$ and neither x_0 nor x_1 are real, in which case $\Delta = (-\infty, \infty)$.

The cases (9) and (10) can be merged by means of (3.50) and (3.51) as

$$(20) \quad \int \frac{dx}{\sqrt{x^2 \pm 1}} = \log(x + \sqrt{x^2 \pm 1}) + C(x)$$

where $C(x)$ is locally constant on the domain of $1/\sqrt{x^2 \pm 1}$.

Chapter 5

Antidifferentiation by parts

In Chapter 4 two techniques of antidifferentiation were considered from a mere theoretical point of view: antidifferentiation by parts and change of variable. This chapter is oriented towards the practical application of antidifferentiation by parts, while the technique of change of variable will be applied directly to specific types of functions in later chapters.

5.1 Antidifferentiating by parts

The method of antidifferentiation by parts is a direct application of Theorem 4.2.2, which states that

$$\int f(x) \cdot g'(x)\, dx = f(x) \cdot g(x) - \int f'(x) \cdot g(x)\, dx \tag{5.1}$$

for a pair of differentiable functions $f(x)$ and $g(x)$ on a regular domain. As anyone might guess from (5.1), this method yields the primitive of $f(x)g'(x)$ provided the primitive of $f'(x)g(x)$ is known. For instance, it is particularly useful when $f(x)$ is a polynomial and $g(x)$ is a circular function or the exponential e^x. Sometimes, a reiterative application of this method is necessary in order to obtain the desired primitive.

It is customary to write (5.1) in a shorter form as

$$\int u\, dv = uv - \int v\, du \tag{5.2}$$

where $u = f(x)$, $v = g(x)$, $du = f'(x)\, dx$ and $dv = g'(x)\, dx$.

5.2 Examples

Example 5.2.1. Calculate $\int x \log x\, dx$.

Solution. Since $x \log x$ is defined and continuous on $(0, \infty)$, its antiderivatives are defined on the same interval.

Following the notation of (5.1), let $f(x) := \log x$ and $g'(x) := x$. Thus, $f'(x) = 1/x$, and $g(x) = x^2/2$. (Following the notation of (5.2), our choices for $f(x)$ and $g'(x)$ are $u = \log x$ and $dv = x\, dx$, so $du = dx$ and $v = x^2/2$.)

Applying (5.1) to the chosen functions $f(x)$ and $g(x)$,

$$\int x \log x \, dx = \frac{1}{2} x^2 \log x - \frac{1}{2} \int x \, dx = \frac{1}{2} x^2 \log x - \frac{1}{4} x^2 + C, \quad x \in (0, \infty).$$

$\square$

Remark 5.2.2. Any antiderivative $g(x)$ of $g'(x)$ is allowed in formula (5.1).

For instance, let us choose $g(x) = (x^2 + 1)/2$ in Example 5.2.1. Thus, the formula of antidifferentiation by parts yields

$$\int x \log x \, dx = \frac{1}{2} x^2 \log x + \frac{1}{2} \log x - \frac{1}{2} \int \frac{x^2 + 1}{x} \, dx$$

$$= \frac{1}{2} x^2 \log x + \frac{1}{2} \log x - \frac{1}{2} \int x + \frac{1}{x} \, dx$$

$$= \frac{1}{2} x^2 \log x + \frac{1}{2} \log x - \frac{1}{4} x^2 - \frac{1}{2} \log x + C, \quad x \in (0, \infty)$$

which matches the solution given in Example 5.2.1.

The following is a typical example where antidifferentiation by parts needs to be reiteratively applied.

Example 5.2.3. Calculate the primitive of $x^2 \sin x$.

Solution. Following the notation of (5.2), set $u = x^2$ and $dv = \sin x \, dx$, so $du = 2x \, dx$ and $v = -\cos x$. Then we get

$$\int x^2 \sin x \, dx = -x^2 \cos x + 2 \int x \cos x \, dx. \tag{5.3}$$

In order to solve $\int x \cos x \, dx$, we use the method of antidifferentiation by parts again. This time, we set $u = x$ and $dv = \cos x \, dx$, hence $du = dx$ and $v = \sin x$ which yields

$$\int x \cos x \, dx = x \sin x - \int \sin x \, dx = x \sin x + \cos x + C, \quad x \in \mathbb{R} \tag{5.4}$$

and plugging (5.4) into (5.3), we get

$$\int x^2 \sin x \, dx = -x^2 \cos x + 2x \sin x + 2 \cos x + C, \quad x \in \mathbb{R}.$$

$\square$

Note that the choice $u = \cos x$ and $dv = x \, dx$ for the second application of the antidifferentiation by parts in Example 5.2.3 has not been adopted, because it undoes the work done in (5.3). Indeed, setting $u = \cos x$ ($\Rightarrow du = -\sin x \, dx$) and $dv = x \, dx$ ($\Rightarrow v = \frac{1}{2} x^2$) to solve $\int x \cos x \, dx$ yields

$$\int x \cos x \, dx = \frac{1}{2} x^2 \cos x + \frac{1}{2} \int x^2 \sin x \, dx \tag{5.5}$$

and plugging (5.5) into (5.3) leads to the useless identity

$$\int x^2 \sin x \, dx = -x^2 \cos x + 2 \left[\frac{1}{2} x^2 \cos x + \frac{1}{2} \int x^2 \sin x \, dx \right].$$

Other times, the reiteration of the method of antidifferentiation by parts has to be applied in an indirect manner:

Example 5.2.4. Calculate $\int e^x \sin x \, dx$.

Solution. Let $u = \sin x$ and $dv = e^x \, dx$, so $du = \cos x \, dx$ and $v = e^x$. Thus (5.2) yields,

$$\int e^x \sin x \, dx = e^x \sin x - \int e^x \cos x \, dx.$$

Applying antidifferentiation by parts once again, but this time to $\int e^x \cos x \, dx$ with $u = \cos x$ and $dv = e^x \, dx$ (hence $du = -\sin x$ and $v = e^x$), we obtain

$$\int e^x \sin x \, dx = e^x \sin x - e^x \cos x - \int e^x \sin x \, dx$$

and solving for $\int e^x \sin x \, dx$, we get

$$\int e^x \sin x \, dx = \frac{1}{2} (e^x \sin x - e^x \cos x) + C, \quad x \in \mathbb{R}.$$

□

Example 5.2.5. Calculate $\int e^x \cos x \, dx$.

Solution. Let $u = \cos x$ and $dv = e^x \, dx$. Thus, $du = -\sin x \, dx$ and $v = e^x$. Then (5.2) gives

$$\int e^x \cos x \, dx = e^x \cos x + \int e^x \sin x \, dx,$$

and by means of Example 5.2.4, we conclude that

$$\int e^x \cos x \, dx = e^x \cos x + \frac{1}{2} (e^x \sin x - e^x \cos x) + C$$

$$= \frac{1}{2} (e^x \cos x + e^x \sin x) + C, \quad x \in \mathbb{R}.$$

□

In the following example, there is one only possible choice for u and dv:

Example 5.2.6. Calculate $\int \log x \, dx$.

Solution. Take $u = \log x$ and $dv = dx$ in (5.2), so $du = 1/x$ and $v = x$, and therefore,

$$\int \log x \, dx = x \log x - \int dx = x \log x - x + C, \quad x \in (0, \infty).$$

□

However, the next example allows multiple choices for u and dv:

Example 5.2.7. Calculate $\int xe^x \sin x\, dx$.

Solution. In order to antidifferentiate by parts, any one of the following choices works:

(1) $u = x$ and $dv = e^x \sin x\, dx$;
(2) $u = e^x x$ and $dv = \sin x\, dx$;
(3) $u = x \sin x$ and $dv = e^x\, dx$.

Choice (1) means $du = dx$ and $v \in \int e^x \sin x\, dx$; following Example 5.2.4, v can be chosen as $v = \frac{1}{2}(e^x \sin x - e^x \cos x)$. For these choices, (5.2) gives

$$\int xe^x \sin x\, dx = \frac{1}{2}x(e^x \sin x - e^x \cos x) - \frac{1}{2}\int e^x \sin x - e^x \cos x\, dx \qquad (5.6)$$

and since $\int e^x \sin x\, dx$ and $\int e^x \cos x\, dx$ have been solved in Examples 5.2.4 and 5.2.5, equation (5.6) gives

$$\int xe^x \sin x\, dx = \frac{1}{2}x(e^x \sin x - e^x \cos x)$$

$$-\frac{1}{4}(e^x \sin x - e^x \cos x) + \frac{1}{4}(e^x \cos x + e^x \sin x) + C$$

$$= \frac{1}{2}\big(xe^x(\sin x - \cos x) + e^x \cos x\big) + C, \quad x \in \mathbb{R}.$$

Choice (2) means $du = (e^x x + e^x)\, dx$ and $v = -\cos x$ so that

$$\int xe^x \sin x\, dx = -xe^x \cos x + \int xe^x \cos x + e^x \cos x\, dx. \qquad (5.7)$$

Since the primitive of $e^x \cos x$ has been found in Example 5.2.5, it only remains to solve $\int xe^x \cos x\, dx$. For this last primitive, set $u = xe^x$ and $dv = \cos x\, dx$, hence $du = (xe^x + e^x)\, dx$ and $v = \sin x$, and applying (5.2) gives

$$\int xe^x \cos x\, dx = -e^x x \sin x + \int e^x x \sin x + e^x \sin x\, dx.$$

Thus Example 5.2.4 yields

$$\int xe^x \cos x\, dx = xe^x \sin x - \int xe^x \sin x + e^x \sin x\, dx. \qquad (5.8)$$

Plugging (5.8) into (5.7), solving $\int xe^x \sin x$ in the resulting expression, and bearing in mind that $\int e^x \cos x\, dx = \frac{1}{2}(e^x \cos x + e^x \sin x) + C$, we obtain

$$\int xe^x \sin x\, dx = \frac{1}{2}(xe^x \sin x - xe^x \cos x + e^x \cos x) + C, \quad x \in \mathbb{R}$$

as desired.

Choice (3): $u = x \sin x \Rightarrow du = \sin x + x \cos x$ and $dv = e^x \, dx \Rightarrow v = e^x$; in this case, formula (5.2) yields

$$\int xe^x \sin x \, dx = xe^x \sin x - \int e^x \sin x \, dx - \int xe^x \cos x \, dx. \qquad (5.9)$$

The primitive $\int e^x \sin x \, dx$ has already been found in Example 5.2.4, so we only need to tackle $\int xe^x \cos x \, dx$. For this primitive, set $u = x \cos x \Rightarrow du = \cos x - x \sin x$ and $dv = e^x \, dx \Rightarrow v = e^x$. Then

$$\int xe^x \cos x \, dx = xe^x \cos x - \int e^x \cos x - xe^x \sin x \, dx. \qquad (5.10)$$

Plugging (5.10) into (5.9) and solving $\int xe^x \sin x \, dx$ we get

$$\int xe^x \sin x \, dx = \frac{1}{2} \left(xe^x \sin x - xe^x \cos x - \int e^x \sin x \, dx + \int e^x \cos x \, dx \right)$$

and application of Examples 5.2.4 and 5.2.5 leads to the desired solution. $\qquad \square$

5.3 Exercises

Find the domain and the primitive of each of the following functions.

Question 5.3.1

$f(x) = \sin(x)^2$

Question 5.3.2

$f(x) = \cos(x)^2$

Question 5.3.3

$f(x) = \log(x)^2$

Question 5.3.4

$f(x) = \log(x)^3$

Question 5.3.5

$f(x) = e^x x$

Question 5.3.6

$f(x) = e^x x^2$

Question 5.3.7

$f(x) = x \tan(x)^2$

Question 5.3.8

$f(x) = e^{5x} x$

Question 5.3.9

$f(x) = \sin(x)^2 e^x$

Question 5.3.10

$f(x) = x \log(x)^2$

Question 5.3.11

$f(x) = \log(x)^4 / x^2$

Question 5.3.12

$f(x) = x \cot(x)^2$

Answers to the Questions

Answer 5.3.1

$D_f = \mathbb{R}$ and $\int f(x)\,dx = (-\cos(x)\sin(x) + x)/2 + C$

Answer 5.3.2

$D_f = \mathbb{R}$ and $\int f(x)\,dx = (x + \cos(x)\sin(x))/2$

Answer 5.3.3

$D_f = (0, \infty)$ and $\int f(x)\,dx = x\log(x)^2 - 2x\log(x) + 2x + C$

Answer 5.3.4

$D_f = (0, \infty)$ and $\int f(x)\,dx = x\log(x)^3 - 3x\log(x)^2 + 6x\log(x) - 6x + C$

Answer 5.3.5

$D_f = \mathbb{R}$ and $\int f(x)\,dx = xe^x - e^x + C$

Answer 5.3.6

$D_f = \mathbb{R}$ and $\int f(x)\,dx = x^2 e^x - 2xe^x + 2e^x + C$

Answer 5.3.7

$D_f = \mathbb{R}\backslash\{\pi/2 + k\pi : k \in \mathbb{Z}\}$ and $\int f(x)\,dx = -\frac{1}{2}x^2 + x\tan(x) + \log(\cos x) + C(x)$

Answer 5.3.8

$D_f = \mathbb{R}$ and $\int f(x)\,dx = \frac{1}{25}[5xe^{5x} - e^{5x}] + C$

Answer 5.3.9

$D_f = \mathbb{R}$, $\int f(x)\,dx = \frac{1}{2}e^x - \frac{1}{10}[e^x\cos(2x) + 2e^x\sin(2x)] + C$

Answer 5.3.10

$D_f = (0, \infty)$, $\int f(x)\,dx = \frac{1}{2}x^2\log(x)^2 - \frac{1}{2}x^2\log(x) + \frac{1}{4}x^2 + C$

Answer 5.3.11

$D_f = (0, \infty)$,
$\int f(x)\, dx = -x^{-1} \log(x)^4 - 4x^{-1} \log(x)^3 - 12x^{-1} \log(x)^2 - 24x^{-1} \log(x) - 24x^{-1} + C$

Answer 5.3.12

$D_f = \mathbb{R} \backslash \{k\pi : k \in \mathbb{Z}\}$, $\int f(x)\, dx = -\frac{1}{2}x^2 - x \cot(x) + \log|\sin x| + C(x)$ where $C(x)$ is locally constant on D_f.

Chapter 6

Rational functions

A rational function or fraction of polynomials over $\mathbb{R}$ is a function that maps each $x \in \mathbb{R} \backslash R_Q$ to

$$\frac{P(x)}{Q(x)}$$

where $P(x)$ and $Q(x)$ are a pair of polynomials over $\mathbb{R}$, $Q(x)$ is not identically null and R_Q is the set of all real roots of $Q(x)$.

The *degree* of $P(x)/Q(x)$ is defined as the integer

$$\deg \frac{P(x)}{Q(x)} := \deg P(x) - \deg Q(x).$$

Since both $P(x)$ and $Q(x)$ are continuous on $\mathbb{R}$ and the only real values of x where $Q(x)$ vanishes belong to R_Q, it follows that $P(x)/Q(x)$ is continuous on $\mathbb{R} \backslash R_Q$ (Theorem 2.2.5). Moreover, R_Q is finite so $\mathbb{R} \backslash R_Q$ is a regular domain and therefore, by virtue of Theorem 4.1.9, there is an antiderivative $F(x)$ of $P(x)/Q(x)$ on $\mathbb{R} \backslash R_Q$. Consequently, its primitive is given by

$$\int \frac{P(x)}{Q(x)} \, dx = F(x) + C(x)$$

where $C(x)$ runs over all locally constant functions on $\mathbb{R} \backslash R_Q$. This chapter deals with how to calculate the antiderivative $F(x)$ explicitly. To carry out this task, the use of fractions of polynomials over $\mathbb{C}$ is indispensable.

A fraction of polynomials over $\mathbb{C}$ is a function that maps each $z \in \mathbb{C} \backslash C_Q$ to the complex number $P(z)/Q(z)$, where $P(x)$ and $Q(x)$ are polynomials over $\mathbb{C}$, $Q(x)$ is not identically null and C_Q is the set of all roots of $Q(x)$ in $\mathbb{C}$.

Let $\mathbb{K}$ denote either $\mathbb{R}$ or $\mathbb{C}$. A fraction $P(x)/Q(x)$ of polynomials over $\mathbb{K}$ is said to be *irreducible* if the polynomials $P(x)$ and $Q(x)$ are mutually irreducible over $\mathbb{K}$ and $\deg P(x) < \deg Q(x)$. If $P(x)$ and $Q(x)$ are a pair of polynomials over $\mathbb{R}$ and $P(x)/Q(x)$ is an irreducible fraction over $\mathbb{R}$, then it is also irreducible over $\mathbb{C}$. This is due to the fact that a number $z \in \mathbb{C}$ is a root of $Q(x)$ if and only if $\bar{z}$ is, in which case, both z and $\bar{z}$ have the same multiplicity, as may be deduced from Proposition 3.2.7 and the algorithm of division of polynomials (see Appendix A.7).

The primitives of fractions of polynomials over $\mathbb{R}$ are calculated in accordance with the classification provided in the following sections.

6.1 Fractions with denominator of degree one and numerator of degree zero

The fractions considered in this section can be expressed as
$$\frac{P(x)}{Q(x)} = \frac{A}{x-a}, \qquad x \in (-\infty, a) \cup (a, \infty)$$
where $A \in \mathbb{R}\backslash\{0\}$ and $a \in \mathbb{R}$. Clearly, $P(x)/Q(x)$ is irreducible. Its primitive is immediate and is included in Table 4.6.1 (3). Thus
$$\int \frac{A}{x-a}\, dx = A \log |x - a| + C(x), \quad \text{with} \quad C(x) = \begin{cases} C_1 & \text{if } x > a \\ C_2 & \text{if } x < a \end{cases} \qquad (6.1)$$
where C_1 and C_2 are any pair of real constants.

6.2 Fractions with irreducible denominator of degree two and numerator of degree zero

These fractions take the form:
$$\frac{P(x)}{Q(x)} = \frac{A}{ax^2 + bx + c}, \qquad x \in (-\infty, \infty)$$
where A, a, b and c are real numbers with $a > 0$ and $ax^2 + bx + c$ has no real root. Clearly, the fraction $P(x)/Q(x)$ is irreducible. In order to find its primitive, the denominator is reduced by completing squares (see Appendix A.9) so that
$$Q(x) = \gamma[(\alpha x + \beta)^2 + 1]$$
for certain real numbers $\gamma > 0$, $\alpha > 0$ and β. According to Table 4.6.1 (4), the primitive of $P(x)/Q(x)$ is obtained as follows:
$$\int \frac{A}{ax^2 + bx + c}\, dx = \int \frac{A/\gamma}{(\alpha x + \beta)^2 + 1}\, dx$$
$$= \frac{A}{\gamma\alpha} \arctan(\alpha x + \beta) + C \qquad (6.2)$$
where C is any real constant.

Example 6.2.1. Find the primitive of $P(x)/Q(x)$, where $P(x) = 1$ and $Q(x) = -x^2 + x - 2$.

Solution. The discriminant of $Q(x)$ is negative so $Q(x)$ has no real root, is irreducible and the domain of $P(x)/Q(x)$ is $\mathbb{R}$.

Since the leading coefficient of $Q(x)$ is negative, the fraction $P(x)/Q(x)$ is rewritten as
$$\frac{P(x)}{Q(x)} = \frac{-1}{x^2 - x + 2}.$$
By completing squares,
$$\frac{-1}{x^2 - x + 2} = \frac{-1}{\left(x - \frac{1}{2}\right)^2 + \frac{7}{4}} = \frac{-4/7}{\left(\frac{2}{\sqrt{7}}x - \frac{1}{\sqrt{7}}\right)^2 + 1}$$
which finally leads to the solution
$$\int \frac{-1}{x^2 - x + 2}\, dx = -\frac{2}{\sqrt{7}} \arctan\left(\frac{2}{\sqrt{7}}x - \frac{1}{\sqrt{7}}\right) + C, \quad C \in \mathbb{R}.$$

$\square$

6.3 Fractions with irreducible denominator of degree two and numerator of degree one

These fractions take the form
$$\frac{P(x)}{Q(x)} = \frac{Ax + B}{ax^2 + bx + c}, \quad x \in \mathbb{R}$$

where A, B, a, b and c are all real numbers with $a > 0$ and $ax^2 + bx + c$ has no real root, so $P(x)/Q(x)$ is irreducible. In this case, by adjusting coefficients, the fraction $P(x)/Q(x)$ can be rewritten as follows:
$$\frac{Ax + B}{ax^2 + bx + c} = \frac{A}{2a} \frac{2ax + b}{ax^2 + bx + c} + \frac{R}{ax^2 + bx + c}$$

where $R = B - \frac{A}{2a}b$. Thus, as $(ax^2 + bx + c)' = 2ax + b$, Table 4.6.1 (18) shows that the primitive of $P(x)/Q(x)$ is
$$\int \frac{Ax + B}{ax^2 + b + c}\, dx = \frac{A}{2a} \int \frac{2ax + b}{ax^2 + bx + c}\, dx + \int \frac{R}{ax^2 + bx + c}\, dx$$
$$= \frac{A}{2a} \log\left(ax^2 + bx + c\right) + f(x) + C \tag{6.3}$$

where C is any real constant and $f(x)$ is an antiderivative of $R/(ax^2 + bx + c)$ which can be calculated as explained in Section 6.2.

Example 6.3.1. Find the primitive of $P(x)/Q(x)$, where $P(x) = x + 1$ and $Q(x) = 3x^2 + x + 1$.

Solution. Since the discriminant of $Q(x)$ is negative, it follows that $Q(x)$ has no real root, so $Q(x)$ is irreducible and the domain of $P(x)/Q(x)$ is $\mathbb{R}$.

Bearing in mind that $6x + 1$ is the derivative of $3x^2 + x + 1$, the fraction $P(x)/Q(x)$ is decomposed as
$$\frac{x + 1}{3x^2 + x + 1} = \frac{1}{6} \cdot \frac{6x + 1}{3x^2 + x + 1} + \frac{5/6}{3x^2 + x + 1}, \tag{6.4}$$

and as the second fraction on the right side of (6.4) is irreducible, its numerator is a polynomial of degree zero and its denominator is an irreducible polynomial of degree two, we complete squares so that
$$\frac{5/6}{3x^2 + x + 1} = \frac{5/6}{\left(\sqrt{3}x + \frac{1}{2\sqrt{3}}\right)^2 + \frac{11}{12}} = \frac{2 \cdot 5/11}{\left(\frac{6}{\sqrt{11}}x + \frac{1}{\sqrt{11}}\right)^2 + 1}.$$

Thus
$$\int \frac{x + 1}{3x^2 + x + 1}\, dx$$
$$= \frac{1}{6} \int \frac{6x + 1}{3x^2 + x + 1}\, dx + \frac{2 \cdot 5}{11} \int \frac{1}{\left(\frac{6}{\sqrt{11}}x + \frac{1}{\sqrt{11}}\right)^2 + 1}\, dx$$
$$= \frac{1}{6} \log\left(3x^2 + x + 1\right) + \frac{5}{3\sqrt{11}} \arctan\left(\frac{6}{\sqrt{11}}x + \frac{1}{\sqrt{11}}\right) + C, \quad x \in \mathbb{R}.$$

□

6.4   Fractions of negative degree whose denominator only has simple real roots

Let $P(x)/Q(x)$ be a given fraction of polynomials over $\mathbb{R}$ where $\deg P(x) < \deg Q(x)$, $Q(x) = \prod_{i=1}^{n}(x - a_i)$, $a_i \in \mathbb{R}$ for all i and $a_1 < a_2 < \dots < a_n$. Thus $P(x)/Q(x)$ is continuous on its domain $D := \mathbb{R}\backslash\{a_i\}_{i=1}^{n}$.

By virtue of Theorem A.12.3, there exists a unique n-tuple $(A_i)_{i=1}^{n}$ of real coefficients such that

$$\frac{P(x)}{Q(x)} = \sum_{i=1}^{n} \frac{A_i}{x - a_i}. \tag{6.5}$$

Hence, the calculation of the primitive on D of $P(x)/Q(x)$ is reduced by linearity to the case of the irreducible fractions with denominator of degree one which were considered in Section 6.1. Indeed,

$$\int \frac{P(x)}{Q(x)}\, dx = \sum_{i=1}^{n} \int \frac{A_i}{x - a_i}\, dx = \sum_{i=1}^{n} A_i \log|x - a_i| + C(x), \quad x \in D \tag{6.6}$$

where $C(x)$ runs over all locally constant functions on D.

In order to calculate the coefficients A_i, multiply both sides of (6.5) by $Q(x)$ so that

$$P(x) = \sum_{i=1}^{n} A_i \prod_{j\neq i}(x - a_j), \text{ all } x \in \mathbb{R}\backslash\{a_i\}_{i=1}^{n}. \tag{6.7}$$

Since the set $\{a_i\}_{i=1}^{n}$ is finite and both sides of (6.7) are polynomials, an argument of continuity (see Proposition A.10.6) proves

$$P(x) = \sum_{i=1}^{n} A_i \prod_{j\neq i}(x - a_j), \quad x \in \mathbb{R}. \tag{6.8}$$

Evaluating both sides of (6.8) at $x = a_i$, all the addends of the right side but the i-th one disappear, leaving $A_i = P(a_i)/\prod_{j\neq i}(a_i - a_j)$ for each $1 \leqslant i \leqslant n$, and plugging the values of A_i in (6.6), the primitive is found.

Example 6.4.1. Calculate the primitive of $P(x)/Q(x)$, where $P(x) = 2x^2 - x + 1$ and $Q(x) = (x + 1)(x - 2)(x - 1)x$.

Solution. The domain of $P(x)/Q(x)$ is $D := \mathbb{R}\backslash\{-1, 0, 1, 2\}$ and all the roots of $Q(x)$ are real and simple, so we set the identity

$$\frac{2x^2 - x + 1}{Q(x)} = \frac{A_1}{x + 1} + \frac{A_2}{x} + \frac{A_3}{x - 1} + \frac{A_4}{x - 2}, \quad x \in \mathbb{R}\backslash\{-1, 0, 1, 2\} \tag{6.9}$$

where the numbers A_i are unknown. Multiplying both sides of (6.9) by $Q(x)$ yields the identity

$$2x^2 - x + 1 = A_1 x(x - 1)(x - 2) + A_2(x + 1)(x - 1)(x - 2)$$
$$+ A_3(x + 1)x(x - 2) + A_4(x + 1)x(x - 1), \quad x \in \mathbb{R}$$

and evaluating this identity at each root of $Q(x)$, we obtain the following system of equations, which readily yields the values of the unknowns A_i:

$$\begin{cases} x = -1: & 4 = -6A_1 \Rightarrow A_1 = -2/3 \\ x = 0, & 1 = 2A_2 \Rightarrow A_2 = 1/2 \\ x = 1: & 2 = -2A_3 \Rightarrow A_3 = -1 \\ x = 2: & 7 = 6A_4 \Rightarrow A_4 = 7/6. \end{cases}$$

Plugging the values of the the the coefficients A_i in (6.9), we get

$$\int \frac{2x^2 - x + 1}{(x+1)(x-2)(x-1)x} \, dx$$

$$= \int \frac{-2/3}{x+1} \, dx + \int \frac{1/2}{x} \, dx + \int \frac{-1}{x-1} \, dx + \int \frac{7/6}{x-2} \, dx$$

$$= -\frac{2}{3} \log |x+1| + \frac{1}{2} \log |x| - \log |x-1| + \frac{7}{6} \log |x-2| + C(x)$$

where $C(x)$ is any locally constant function on D. □

6.5 Fractions of negative degree whose denominator has only simple roots but at least one is not real

This section deals with fractions $P(x)/Q(x)$ where $\deg P(x) < \deg Q(x)$ and the irreducible factorization over $\mathbb{R}$ of $Q(x)$ is

$$Q(x) = \prod_{i=1}^{m} (x - a_i) \cdot \prod_{i=1}^{n} (x^2 + \mu_i x + \nu_i).$$

So the numbers a_i, μ_i and ν_i are real and labeled so that $a_1 < \ldots < a_n$ and $x^2 + \mu_i x + \nu_i = (x - c_i)(x - \overline{c_i})$ for all i, where the numbers c_i are pairwise different and $c_i \in \mathbb{C} \backslash \mathbb{R}$ for all i. Clearly, the domain of $P(x)/Q(x)$ is $D := \mathbb{R} \backslash \{a_i\}_{i=1}^{m}$.

Following Theorem A.12.3, there exist three unique tuples $(A_i)_{i=1}^{m}$, $(M_i)_{i=1}^{n}$ and $(N_i)_{i=1}^{n}$ of real numbers so that $P(x)/Q(x)$ is decomposed as

$$\frac{P(x)}{Q(x)} = \sum_{i=1}^{m} \frac{A_i}{x - a_i} + \sum_{i=1}^{n} \frac{M_i x + N_i}{x^2 + \mu_i x + \nu_i}. \tag{6.10}$$

Thus, if the coefficients A_i, M_i and N_i are known, the primitive of $P(x)/Q(x)$ on D can be found by linearity as follows:

$$\int \frac{P(x)}{Q(x)} \, dx = \sum_{i=1}^{m} \int \frac{A_i}{x - a_i} \, dx + \sum_{i=1}^{n} \int \frac{M_i x + N_i}{x^2 + \mu_i x + \nu_i} \, dx \tag{6.11}$$

where the primitives of the fractions $1/(x-a_i)$ and $(M_i x + N_i)/(x^2 + \mu_i x + \nu_i)$ can be calculated as indicated in Sections 6.1, 6.2 and 6.3. Thus the whole matter revolves around how to find the values of the above mentioned coefficients. To do this, two procedures are proposed, the first of which deals with real numbers only; however, it involves solving a linear system of several equations and several unknowns. The second procedure is formally identical to that of Section 6.4 with the only difference here that non-real roots involving complex numbers occur. The technical advantage of the second procedure with respect to the first procedure is that it involves solving a multiple system of equations each of which has only one unknown.

First procedure: First, both sides of (6.10) are multiplied by $Q(x)$ in order to remove the denominators, so

$$
\begin{aligned}
P(x) = &\sum_{i=1}^{m} A_i \prod_{j \neq i} (x - a_j) \cdot \prod_{j=1}^{n} (x^2 + \mu_j x + \nu_j) \\
&+ \sum_{i=1}^{m} (M_i x + N_i) \prod_{j=1}^{n} (x - a_j) \cdot \prod_{j \neq i} (x^2 + \mu_j x + \nu_j)
\end{aligned}
\tag{6.12}
$$

for all $x \in \mathbb{R}\backslash\{a_i\}_{i=1}^{m}$, and in turn, (6.12) holds for all $x \in \mathbb{R}$ by virtue of the continuity of polynomials on $\mathbb{R}$. Next, a system of $m + 2n$ equations where the unknowns are the coefficients A_i, M_i and N_i must be extracted from (6.12). To do this, we use any of the following methods:

(i) evaluation of (6.12) for at least at $m + 2n$ real numbers x_i, where $x_i = a_i$ for all $1 \leqslant i \leqslant m$;
(ii) expansion of the right-side of (6.12) and identification of the terms of the same degree in the polynomials of both sides.

The choice $x_i = a_i$ in method (i) yields the value A_i readily because all the addends but one on the right side of (6.12) vanish, so the problem is reduced to only finding the coefficients M_i and N_i; that means solving a smaller system of $2n$ equations and $2n$ unknowns.

A suitable combination of methods (i) and (ii) may reduce the difficulty of finding the unknowns M_i and N_i. In particular, method (ii) should be applied only after the calculation of the unknowns A_i.

It should be noted that the evaluation at $x = 0$ produces the same equation yielded by the identification of the terms of degree zero in (6.12).

Example 6.5.1. Find the primitive of $P(x)/Q(x)$ where $P(x) = 2x - 1$ and $Q(x) = (x + 1)(x^2 - 2x + 3)$.

Solution. The only real root of $Q(x)$ is -1 so $P(x)/Q(x)$ is defined and is continuous on $D := \mathbb{R}\backslash\{-1\}$. In order to find its primitive, first set

$$
\frac{2x - 1}{(x + 1)(x^2 - 2x + 3)} = \frac{A}{x + 1} + \frac{Mx + N}{x^2 - 2x + 3}
\tag{6.13}
$$

where A, M and N are unknown real numbers. Multiplying both sides of (6.13) by $(x + 1)(x^2 - 2x + 3)$ leads to

$$
2x - 1 = A(x^2 - 2x + 3) + (Mx + N)(x + 1), \quad x \in \mathbb{R}.
\tag{6.14}
$$

By evaluation at $x = -1$, (6.14) becomes $-3 = 6A$. Hence $A = -1/2$. In order to calculate the values of M and N, a glance at the second degree terms on both sides in (6.14) reveals that $0 = A + M$. In order to obtain one more equation, evaluation of (6.14) at $x = 0$ yields $-1 = 3A + N$. The system thus obtained is

$$\begin{cases} x^2 : & 0 = A + M = -\frac{1}{2} + M \\ x = 0 : & -1 = 3A + N = -\frac{3}{2} + N \end{cases}$$

so $M = 1/2$ and $N = 1/2$. Equation (6.13) now leads to

$$\int \frac{2x - 1}{(x+1)(x^2 - 2x + 3)}\, dx = -\int \frac{1/2}{x+1}\, dx + \frac{1}{2} \int \frac{x+1}{x^2 - 2x + 3}\, dx. \tag{6.15}$$

Next, the two primitives occurring at the right side of (6.15) are solved in accordance with the instructions given in Sections 6.1 to 6.3 so that

$$\int \frac{2x - 1}{(x+1)(x^2 - 2x + 3)}\, dx = -\frac{1}{2} \log |x + 1|$$

$$+ \frac{1}{4} \log (x^2 - 2x + 3) + \frac{1}{\sqrt{2}} \arctan \left(\frac{x-1}{\sqrt{2}} \right) + C(x)$$

where $C(x)$ is any locally constant function on $(-\infty, -1) \cup (-1, \infty)$. $\qquad\square$

Example 6.5.2. Find the primitive of $P(x)/Q(x)$ where

$$P(x) = 16x^4 - 17x^3 + 24x^2 - 15x + 4$$

$$Q(x) = x(x^2 + 1)(4x^2 - 4x + 2).$$

Solution. The only real root of $Q(x)$ is 0, so $P(x)/Q(x)$ is defined and continuous on $D := \mathbb{R}\backslash\{0\}$.

Set the identity

$$\begin{aligned} \frac{P(x)}{Q(x)} &= \frac{16x^4 - 17x^3 + 24x^2 - 15x + 4}{x(x^2 + 1)(4x^2 - 4x + 2)} \\ &= \frac{A}{x} + \frac{M_1 x + N_1}{x^2 + 1} + \frac{M_2 x + N_2}{4x^2 - 4x + 2} \end{aligned} \tag{6.16}$$

where A, M_1, N_1, M_2 and N_2 are unknown real numbers to be calculated. First, remove the denominators so that

$$\begin{aligned} P(x) &= 16x^4 - 17x^3 + 24x^2 - 15x + 4 \\ &= A(x^2 + 1)(4x^2 - 4x + 2) \\ &\quad + (M_1 x + N_1)x(4x^2 - 4x + 2) \\ &\quad + (M_2 x + N_2)x(x^2 + 1), \quad x \in \mathbb{R}. \end{aligned} \tag{6.17}$$

Evaluation of (6.17) at $x = 0$ and the identification of the terms of degree 4, 3, 2 and 1 on both sides of (6.17) lead to the system

$$\begin{cases} x = 0 : & P(0) = 4 = 2A \Rightarrow A = 2 \\ x^4 : & 16 = 4A + 4M_1 + M_2 = 8 + 4M_1 + M_2 \\ x^3 : & -17 = -4A - 4M_1 + 4N_1 + N_2 = -8 - 4M_1 + 4N_1 + N_2 \\ x^2 : & 24 = 6A + 2M_1 - 4N_1 + M_2 = 12 + 2M_1 - 4N_1 + M_2 \\ x^1 : & -15 = -4A + 2N_1 + N_2 = -8 + 2N_1 + N_2 \end{cases}$$

which yields $M_1 = 0$, $N_1 = -1$, $M_2 = 8$ and $N_2 = -5$. Plug these values into (6.16) so that

$$\frac{P(x)}{Q(x)} = \frac{2}{x} + \frac{-1}{x^2 + 1} + \frac{8x - 5}{4x^2 - 4x + 2} \tag{6.18}$$

$$= \frac{2}{x} + \frac{-1}{x^2 + 1} + \frac{8x - 4}{4x^2 - 4x + 2} - \frac{1}{(2x - 1)^2 + 1}.$$

Therefore,

$$\int \frac{P(x)}{Q(x)} \, dx = \int \frac{2}{x} \, dx - \int \frac{1}{x^2 + 1} \, dx + \int \frac{8x - 4}{4x^2 - 4x + 2} \, dx - \int \frac{1}{(2x - 1)^2 + 1} \, dx$$

$$= 2 \log |x| - \arctan x + \log(4x^2 - 4x + 2) - \frac{1}{2} \arctan (2x - 1) + C(x)$$

where $C(x)$ is any locally constant function on $(-\infty, 0) \cup (0, \infty)$. □

Second procedure: by virtue of Theorem A.12.3, for each $1 \leqslant i \leqslant n$ there exists a complex number C_i such that the fraction $(M_i x + N_i)/(x^2 + \mu_i x + \nu_i)$ can be decomposed as

$$\frac{M_i x + N_i}{x^2 + \mu_i x + \nu_i} = \frac{\overline{C_i}}{x - \overline{c_i}} + \frac{C_i}{x - c_i}, \quad x \in \mathbb{C} \backslash \{c_i, \overline{c_i}\} \tag{6.19}$$

where $(x - z_i)(x - \overline{z_i})$ is the irreducible decomposition of $x^2 + \mu_i x + \nu_i$ over $\mathbb{C}$. Thus

$$\frac{P(x)}{Q(x)} = \sum_{i=1}^{m} \frac{A_i}{x - a_i} + \sum_{i=1}^{n} \frac{\overline{C_i}}{x - \overline{c_i}} + \sum_{i=1}^{n} \frac{C_i}{x - c_i}, \quad x \in \mathbb{C} \backslash \{a_i\}_{i=1}^{m} \cup \{c_i\}_{i=1}^{n} \cup \{\overline{c_i}\}_{i=1}^{m}. \tag{6.20}$$

As in Section 6.4, both sides of (6.20) are multiplied by $Q(x)$ in order to remove denominators. This yields the identity

$$P(x) = \sum_{i=1}^{m} A_i \cdot \prod_{j \neq i} (x - a_j) \cdot \prod_{j=1}^{n} (x - \overline{c_j})(x - c_j)$$

$$+ \sum_{i=1}^{n} \overline{C_i} \cdot \prod_{j=1}^{m} (x - a_j) \cdot \prod_{j \neq i} (x - \overline{c_j}) \cdot \prod_{j=1}^{n} (x - c_j)$$

$$+ \sum_{i=1}^{n} C_i \cdot \prod_{j=1}^{m} (x - a_j) \cdot \prod_{j=1}^{n} (x - \overline{c_j}) \cdot \prod_{j \neq i} (x - c_j) \tag{6.21}$$

which holds for all $x \in \mathbb{C}$ by virtue of Proposition A.10.7. Thus, the value of A_i is directly obtained by the evaluation of $P(x)$ at a_i in (6.21), and the value of C_i is obtained by the evaluation of $P(x)$ at c_i in (6.21). Once the coefficients C_i are known, the identity (6.19) yields the values of the coefficients M_i and N_i.

Now let us solve Example 6.5.2 again but this time by means of the second procedure:

Example 6.5.3. Find the primitive of $P(x)/Q(x)$ where

$$P(x) = 16x^4 - 17x^3 + 24x^2 - 15x + 4$$

$$Q(x) = x(x^2 + 1)(4x^2 - 4x + 2)$$

Solution. Since the factorizations over $\mathbb{C}$ of $x^2 + 1$ and $4x^2 - 4x + 2$ are

$$x^2 + 1 = (x + i)(x - i)$$

$$4x^2 - 4x + 2 = 4\left(x - \frac{1}{2} - \frac{i}{2}\right)\left(x - \frac{1}{2} + \frac{i}{2}\right)$$

the fraction $P(x)/Q(x)$ over $\mathbb{C}$ is decomposed as

$$\frac{P(x)}{Q(x)} = \frac{4x^4 - \frac{17}{4}x^3 + 6x^2 - \frac{15}{4}x + 1}{x(x^2 + 1)(x^2 - x + \frac{1}{2})}$$

$$= \frac{A}{x} + \frac{C_1}{x - i} + \frac{\overline{C_1}}{x + i} + \frac{C_2}{x - \frac{1}{2} - \frac{1}{2}i} + \frac{\overline{C_2}}{x - \frac{1}{2} + \frac{1}{2}i} \qquad (6.22)$$

where A, C_1 and C_2 are unknown complex numbers. Removing denominators on both sides of (6.22), we get

$$P(x) = \frac{1}{4}(16x^4 - 17x^3 + 24x^2 - 15x + 4)$$

$$= A(x - i)(x + i)\left(x - 1/2 - i/2\right)\left(x - 1/2 + i/2\right)$$

$$+ C_1 x(x + i)\left(x - 1/2 - i/2\right)\left(x - 1/2 + i/2\right)$$

$$+ \overline{C_1} x(x - i)\left(x - 1/2 - i/2\right)\left(x - 1/2 + i/2\right)$$

$$+ C_2 x(x - i)(x + i)\left(x - 1/2 + i/2\right)$$

$$+ \overline{C_2} x(x - i)(x + i)\left(x - 1/2 - i/2\right), \qquad x \in \mathbb{C}. \qquad (6.23)$$

Evaluating $P(x)$ at the complex roots of $Q(x)$ on both sides of (6.23), the values A, C_1 and C_2 are readily calculated as

$$\begin{cases} x = 0: & P(0) = 1 = \frac{1}{2}A \Rightarrow A = 2 \\ x = i: & P(i) = -1 + \frac{1}{2}i = (1 + 2i)C_1 \Rightarrow C_1 = \frac{1}{2}i \\ x = \frac{1}{2} + \frac{i}{2}: & P\left(\frac{1}{2} + \frac{1}{2}i\right) = -\frac{13}{16} + \frac{1}{16}i = C_2\left(-\frac{3}{2} + \frac{1}{2}i\right) \Rightarrow C_2 = 1 + \frac{1}{4}i. \end{cases}$$

Hence

$$\frac{C_1}{x - i} + \frac{\overline{C_1}}{x + i} = \frac{-1}{x^2 + 1} \qquad (6.24)$$

$$\frac{C_2}{x - \frac{1}{2} - \frac{1}{2}i} + \frac{\overline{C_2}}{x - \frac{1}{2} + \frac{1}{2}i} = \frac{2x - \frac{5}{4}}{x^2 - 1 + \frac{1}{2}} \qquad (6.25)$$

and in combination with (6.22), the partial decomposition of (6.18) is recovered. From this point, the procedure is exactly the same as in Example 6.5.2. $\square$

It is also possible to mix the two procedures: decompose $P(x)/Q(x)$ as the sum of fractions of polynomials over $\mathbb{R}$ as in (6.10) and then, in order to find out the values of the real coefficients A_i, M_i and N_i, the equation (6.12) is evaluated at each root of $Q(x)$ including the non-real ones. See the following example.

Example 6.5.4. Find $\displaystyle\int \frac{x^2}{x^4 + 1}\, dx$.

Solution. The polynomial $x^4 + 1$ has no real root. Its four complex roots can be easily found by means of (A.14) in Appendix A.6: the modulus of all of these roots must be 1 and the argument of two of them must be $\pi/4$ and $\pi + \pi/4$, that is,

$$z_1 = \frac{1}{\sqrt{2}} + \frac{1}{\sqrt{2}}i$$

$$z_2 = -\frac{1}{\sqrt{2}} - \frac{1}{\sqrt{2}}i.$$

Thus, the other roots of $x^4 + 1$ are $\bar{z}_1$ and $\bar{z}_2$. Hence, the irreducible factorization of $x^4 + 1$ over the reals is

$$x^4 + 1 = \big[(x - z_1)(x - \bar{z}_1)\big]\big[(x - z_2)(x - \bar{z}_2)\big]$$
$$= (x^2 - \sqrt{2}x + 1)(x^2 + \sqrt{2}x + 1).$$

Thus

$$\frac{x^2}{x^4 + 1} = \frac{Ax + B}{x^2 - \sqrt{2}x + 1} + \frac{Cx + D}{x^2 + \sqrt{2}x + 1}, \quad x \in \mathbb{R}. \tag{6.26}$$

By multiplying both sides by $x^4 + 1$, we obtain an identity of polynomials over $\mathbb{R}$ that works for all $x \in \mathbb{R}$, and, in turn, for all $x \in \mathbb{C}$ by virtue of Proposition A.10.7:

$$x^2 = (Ax + B)(x^2 + \sqrt{2}x + 1) + (Cx + D)(x^2 - \sqrt{2}x + 1), \quad x \in \mathbb{C}. \tag{6.27}$$

Next, the identity (6.27) is evaluated at the roots $x = z_1$ and $x = z_2$, which readily gives the system

$$\begin{cases} t = z_1: & i = 2B + (2\sqrt{2}A + 2B)i \quad \begin{cases} 0 = 2B \\ 1 = 2\sqrt{2}A + 2B \end{cases} \\ t = z_2: & i = -2D + (-2\sqrt{2}C + 2D)i \quad \begin{cases} 0 = -2D \\ 1 = -2\sqrt{2}C + 2D \end{cases} \end{cases}$$

whose solutions are readily calculated as $A = 1/2\sqrt{2}$, $B = 0$, $C = -1/2\sqrt{2}$ and $D = 0$. Hence, (6.26) is

$$\frac{x^2}{x^4 + 1} = \frac{1}{2\sqrt{2}}\frac{x}{x^2 - \sqrt{2}x + 1} - \frac{1}{2\sqrt{2}}\frac{x}{x^2 + \sqrt{2}x + 1}, \quad x \in \mathbb{R}. \tag{6.28}$$

By adjusting coefficients and completing squares in the two fractions on the right side of (6.28),

$$\frac{x}{x^2 - \sqrt{2}x + 1} = \frac{1}{2}\frac{2x - \sqrt{2}}{x^2 - \sqrt{2}x + 1} + \frac{\sqrt{2}}{(\sqrt{2}x - 1)^2 + 1} \tag{6.29}$$

$$\frac{x}{x^2 + \sqrt{2}x + 1} = \frac{1}{2}\frac{2x + \sqrt{2}}{x^2 + \sqrt{2}x + 1} - \frac{\sqrt{2}}{(\sqrt{2}x + 1)^2 + 1} \tag{6.30}$$

and finding the primitives of the right sides of (6.29) and (6.30), it follows from (6.27) that

$$\int \frac{x^2}{x^4 + 1}\,dx = \frac{1}{4\sqrt{2}}\log(x^2 - \sqrt{2}x + 1) + \frac{1}{2\sqrt{2}}\arctan(\sqrt{2}x - 1)$$

$$= -\frac{1}{4\sqrt{2}}\log(x^2 + \sqrt{2}x + 1) + \frac{1}{2\sqrt{2}}\arctan(\sqrt{2}x + 1) + C$$

where C is any real constant.   □

6.6 Fractions of negative degree whose denominator has only real roots but at least one is multiple

This section deals with fractions $P(x)/Q(x)$

$$\frac{P(x)}{Q(x)} = \frac{P(x)}{\prod_{i=1}^{m}(x-a_i)^{m_i}}, \tag{6.31}$$

where $\deg P(x) < \deg Q(x)$, all numbers a_i are real and pairwise different, and $m_i > 1$ at least for one index i.

Clearly, the domain of $P(x)/Q(x)$ is $D := \mathbb{R}\backslash\{a_i\}_{i=1}^{m}$. In order to solve $\int P(x)/Q(x)\,dx$, two different methods are proposed:

First method (Hermite):

Let $Q_0(x) := \prod_{i=1}^{m}(x-a_i)^{m_i-1}$ and let $n := \deg Q_0(x) - 1$. By virtue of Theorem A.12.18, there exists a unique polynomial

$$P_0(x) = \alpha_n x^n + \ldots + \alpha_1 x + \alpha_0$$

and a unique collection $(A_i)_{i=1}^{m}$ of real numbers such that

$$\frac{P(x)}{Q(x)} = \left(\frac{P_0(x)}{Q_0(x)}\right)' + \sum_{i=1}^{m}\frac{A_i}{x-a_i}, \qquad x \in D. \tag{6.32}$$

Thus, as soon as the coefficients α_i and A_i are identified, it can be established by linearity that

$$\int \frac{P(x)}{Q(x)}\,dx = \frac{P_0(x)}{Q_0(x)} + \sum_{i=1}^{m} A_i \log|x-a_i| + C(x) \tag{6.33}$$

where $C(x)$ is any locally constant function on D.

In order to calculate the values of α_i and A_i, it is necessary to differentiate the fraction $P_0(x)/Q_0(x)$, which yields

$$\frac{P_0'(x)Q_0(x) - P_0(x)Q_0'(x)}{Q_0(x)^2}. \tag{6.34}$$

Plug (6.34) into (6.32) and, in order to remove denominators, multiply both sides of the resulting identity by $Q(x)$. Thus, since $Q(x) = Q_0(x) \cdot \prod_{i=1}^{m}(x-a_i)$, these last actions leave the following identity of polynomials:

$$P(x) = P_0'(x)\prod_{i=1}^{m}(x-a_i) - P_0(x)\left(\sum_{i=1}^{m}(m_i-1)\prod_{j\neq i}(x-a_j)\right)$$
$$+ \sum_{i=1}^{m} A_i(x-a_i)^{m_i-1}\prod_{j\neq i}(x-a_i)^{m_i} \tag{6.35}$$

which holds for all $x \in D$, and as the polynomials are continuous on $\mathbb{R}$, the identity (6.35) also holds for all $x \in \mathbb{R}$. Next, a system of $n+m+1$ equations whose unknowns are $\alpha_0, \ldots, \alpha_n; A_1, \ldots, A_m$ is obtained by means of any one of the following methods:

(i) evaluation of (6.35) at the roots a_i of $Q(x)$,
(ii) evaluation of (6.35) at certain values $x \in \mathbb{R}$ other than the numbers a_i,
(iii) identification of same degree terms of the two sides of (6.35).

Of course, the simpler the system of equations is, the easier its solution will be. For this purpose it is strongly recommended to apply (i) because many of the factors in (6.35) vanish, which yields m simple equations. Moreover, it is advisable not to expand all the factors occurring in (6.35); indeed, this does not obstruct the identification of terms of high and low degree and helps to figure out other points at which the evaluation of (6.35) might yield simple equations.

For instance, in the following example it is necessary to find out six unknowns labeled A to F, which requires obtaining a system of six equations from (6.35). However, a sensible choice of points at which (6.35) can be evaluated reduces that system to another of three equations, as can be seen in the solution offered.

Example 6.6.1. Find the primitive of $1/Q(x)$, where
$$Q(x) = (x-1)(x+1)^2(x-2)^3.$$

Solution. The domain of $1/(x-1)(x+1)^2(x-2)^3$ is $D := \mathbb{R}\backslash\{-1,1,2\}$.
Let us set the identity

$$\frac{1}{(x-1)(x+1)^2(x-2)^3} = \left(\frac{Ax^2+Bx+C}{(x+1)(x-2)^2}\right)' \qquad (6.36)$$
$$+ \frac{D}{x-1} + \frac{E}{x+1} + \frac{F}{x-2}$$

where A, B, C, D, E and F are six unknown real numbers to be found.

The derivative of the fraction $(Ax^2+Bx+C)/(x+1)(x-2)^2$ is
$$\frac{(2Ax+B)(x+1)(x-2)^2 - (Ax^2+Bx+C)\big((x-2)^2+2(x+1)(x-2)\big)}{(x+1)^2(x-2)^4}$$
$$= \frac{(2Ax+B)(x+1)(x-2) - (Ax^2+Bx+C)(3x)}{(x+1)^2(x-2)^3}.$$

Plugging the last expression into (6.36) and removing denominators, the following identity of polynomials is obtained:

$$1 = (2Ax+B)(x-1)(x+1)(x-2) - (Ax^2+Bx+C)3x(x-1)$$
$$+D(x+1)^2(x-2)^3$$
$$+E(x-1)(x+1)(x-2)^3$$
$$+F(x-1)(x+1)^2(x-2)^2, \quad x \in \mathbb{R}. \qquad (6.37)$$

Next, the identity of (6.37) is evaluated at the roots of $Q(x)$, which yields the following system:

$$\begin{cases} x=1: & 1=-4D \\ x=2: & 1=-(4A+2B+C)6 \\ x=-1: & 1=-(A-B+C)6; \end{cases} \qquad (6.38)$$

the first equation readily yields $D = -1/4$, and the other two equations yield $A = -B$. Three more equations are needed, and in order to manage them, taking advantage from the fact that the factor x occurs at the second addend on the right side of (6.37), we evaluate (6.37) at $x = 0$, which renders a simple equation. Moreover, the terms of fifth degree and fourth degree in (6.37) are easily identifiable, which yields the two additional required equations:

$$\begin{cases} x = 0: & 1 = 2B - 8D + 8E - 4F \\ x^5: & 0 = D + E + F \\ x^4: & 0 = -A - 4D - 6E - 3F. \end{cases} \tag{6.39}$$

Since $D = -1/4$, and bearing in mind that $A = -B$, both systems (6.38) and (6.39) readily give $A = 1/6$, $B = -1/6$, $C = -1/2$, $E = 1/36$ and $F = 2/9$. Thus, from (6.36), we obtain

$$\int \frac{1}{(x-1)(x+1)^2(x-2)^3}\,dx = \frac{1}{6}\frac{x^2 - x - 3}{(x+1)(x-2)^2}$$
$$- \frac{1}{4}\log|x-1| + \frac{1}{36}\log|x+1| + \frac{2}{9}\log|x-2| + C(x)$$

where $C(x)$ is locally constant on $\mathbb{R}\setminus\{-1,1,2\}$. □

Second method:

In accordance with Theorem A.12.18, the fraction given in (6.31) can be decomposed as

$$\frac{P(x)}{Q(x)} = \sum_{i=1}^{m}\sum_{l=1}^{m_i} \frac{A_{il}}{(x-a_i)^l}, \qquad x \in D \tag{6.40}$$

for a unique collection $(A_{il})_{l=1\,i=1}^{m_i\,m}$ of real numbers. In order to find out the values of the coefficients A_{ik}, the two sides of (6.40) are multiplied by $Q(x)$ which produces an identity of polynomials. Since D is finite and since all polynomials are continuous on $\mathbb{R}$, this identity holds for all $x \in \mathbb{R}$. This allows us to evaluate this identity at the roots $x = a_i$, which readily yields the coefficients A_{i,m_i} for $1 \leqslant i \leqslant m$. The remaining coefficients can be solved from a system of equations obtained by evaluating the same polynomial identity at other points x or by identifying terms of the same degree.

Example 6.6.2. Find $\displaystyle\int \frac{x+1}{(x-1)^2(x-2)}\,dx$.

Solution. Let $D := \mathbb{R}\setminus\{1,2\}$ and set

$$\frac{x+1}{(x-1)^2(x-2)} = \frac{A}{x-1} + \frac{B}{(x-1)^2} + \frac{C}{x-2}, \qquad x \in D. \tag{6.41}$$

Multiplying the two sides of (6.41) by $(x-1)^2(x-2)$ yields

$$x + 1 = A(x-1)(x-2) + B(x-2) + C(x-1)^2, \qquad x \in \mathbb{R}. \tag{6.42}$$

The evaluation of (6.42) at the roots $x = 1$ and $x = 2$, and the identification of the terms of second degree, gives the system

$$\begin{cases} x = 1: & 2 = -B \\ x = 2: & 3 = C \\ x^2: & 0 = A + C, \end{cases}$$

thus $A = -3$, and from (6.41), it can be concluded that

$$\int \frac{x+1}{(x-1)^2(x-2)}\, dx = \int \frac{-3}{x-1}\, dx + \int \frac{-2}{(x-1)^2}\, dx + \int \frac{3}{x-2}\, dx$$

$$= -3 \log|x-1| + \frac{2}{x-1} + 3\log|x-2| + C(x)$$

where $C(x)$ is locally constant on $\mathbb{R}\backslash\{1,2\}$.   □

This method is very simple when the polynomial of the denominator only has one root.

Example 6.6.3. Find $\displaystyle\int \frac{3x^2+1}{(x+1)^4}\, dx$.

Solution. Let $D := \mathbb{R}\backslash\{-1\}$ and set

$$\frac{3x^2+1}{(x+1)^4} = \frac{A}{x+1} + \frac{B}{(x+1)^2} + \frac{C}{(x+1)^3} + \frac{D}{(x+1)^4}, \quad x \in D. \tag{6.43}$$

Multiply the two sides of (6.43) by $(x+1)^4$ so that

$$3x^2 + 1 = A(x+1)^3 + B(x+1)^2 + C(x+1) + D, \quad x \in \mathbb{R}. \tag{6.44}$$

Evaluation of (6.44) at $x = -1$ readily yields $4 = D$, and plugging this value in (6.44) gives

$$3x^2 - 3 = A(x+1)^3 + B(x+1)^2 + C(x+1), \quad x \in \mathbb{R}. \tag{6.45}$$

Dividing both sides of (6.45) by $x + 1$, we get

$$3x - 3 = A(x+1)^2 + B(x+1) + C, \quad x \in \mathbb{R}. \tag{6.46}$$

Evaluation of (6.46) at $x = -1$ yields $C = -6$. Once C is known, (6.46) becomes $3x + 3 = A(x+1)^2 + B(x+1)$, and dividing both sides of this identity by $x + 1$ yields

$$3 = A(x+1) + B, \quad x \in \mathbb{R}. \tag{6.47}$$

Now it is clear that $A = 0$ and $B = 3$. Thus, from (6.43)

$$\int \frac{3x^2+1}{(x+1)^4}\, dx = \int \frac{3}{(x+1)^2}\, dx - \int \frac{6}{(x+1)^3}\, dx + \int \frac{4}{(x+1)^4}\, dx$$

$$= -\frac{3}{x+1} + \frac{3}{(x+1)^2} - \frac{4/3}{(x+1)^3} + C(x)$$

where $C(x)$ is a locally constant on D.   □

6.7 Fractions of negative degree whose denominator has at least one non-real root and some multiple root

This section deals with fractions of polynomials $P(x)/Q(x)$ where $\deg P(x) < \deg Q(x)$ and the irreducible factorisation over $\mathbb{R}$ of $Q(x)$ is

$$Q(x) = \prod_{i=1}^{m}(x - a_i)^{m_i} \cdot \prod_{i=1}^{n}(x^2 + \mu_i x + \nu_i)^{n_i},$$

where the non-real roots of $Q(x)$ have been paired in the second degree factors $x^2 + \mu_i x + \nu_i$. Thus, the factorization over $\mathbb{C}$ of $x^2 + \mu_i x + \nu_i$ is $(x - c_i)(x - \overline{c_i})$ and all the numbers a_i and c_i are pairwise different.

There are two main ways to solve the primitives of such fractions. One method is based upon the partial decompositions *(i)* and *(ii)* of Theorem A.12.15, but this is very laborious unless the reader is acquainted with Complex Analysis. The second method is based upon Theorem A.12.18, which allows us to work with functions of one real variable without much trouble.

As one might expect, the second method follows the first method described in Section 6.6 but taking into account the peculiarities induced by the non-real roots of $Q(x)$, which must be treated with the techniques explained in Sections 6.2, 6.3 and 6.5:

Since it is assumed that the reader is not familiar with Complex Analysis, we only explain the second method.

Hermite's method:

Let

$$Q_0(x) = \prod_{i=1}^{m}(x - a_i)^{m_i-1} \cdot \prod_{i=1}^{n}(x^2 + \mu_i x + \nu_i)^{n_i-1}$$

and let $k := \deg Q_0(x) - 1$. Then, by virtue of Theorem A.12.18, there exists a unique polynomial

$$P_0(x) = \alpha_k x^k + \ldots + \alpha_1 x + \alpha_0$$

and unique collections $(A_i)_{i=1}^{m}$, $(M_i)_{i=1}^{n}$ and $(N_i)_{i=1}^{n}$ of real numbers such that

$$\frac{P(x)}{Q(x)} = \left(\frac{P_0(x)}{Q_0(x)}\right)' + \sum_{i=1}^{m}\frac{A_i}{x - a_i} + \sum_{i=1}^{n}\frac{M_i x + N_i}{x^2 + \mu_i x + \nu_i}, \quad x \in D. \tag{6.48}$$

As soon as the coefficients α_i, A_i, M_i and N_i are known, we get by linearity that

$$\int \frac{P(x)}{Q(x)}\,dx = \frac{P_0(x)}{Q_0(x)} + \sum_{i=1}^{m} A_i \log|x - a_i|$$

$$+ \sum_{i=1}^{n} \int \frac{M_i x + N_i}{x^2 + \mu_i x + \nu_i}\,dx + C(x), \quad x \in D \tag{6.49}$$

where each $\int (M_i x + N_i)/(x^2 + \mu_i x + \nu_i)\,dx$ must be calculated according to Sections 6.2 and 6.3 and where $C(x)$ is any locally constant function on D.

In order to find out the values of α_i, A_i, M_i and N_i, the same steps of the first method described in Section 6.6 must be repeated: calculate the derivative of $P_0(x)/Q_0(x)$ and plug it into (6.48). The expression thus obtained is transformed into an identity of polynomials by multiplying its two sides by $Q(x)$. By virtue of the continuity of the polynomials on $\mathbb{R}$, this identity holds for all $x \in \mathbb{R}$, which allows us to apply any of the following techniques in order to obtain a system of $k + 1 + m + 2n$ equations:

(i) evaluation at $x = a_i$ for $1 \leqslant i \leqslant m$,
(ii) evaluation at $x = c_i$ for $1 \leqslant i \leqslant n$,
(iii) evaluation at any other values of x,
(iv) identification of terms of the same degree,

and all the unknowns α_i, A_i, M_i and N_i are solved from this system.

It should be pointed out that the application of (i) or (ii) reduces the difficulty of the mentioned system of equations. Also worth noting is that if one does not want to deal with complex numbers, then (ii) should not be applied and (iii) should be applied only for real values of x. If (ii) is applied, the evaluations at c_i provides the same information as the evaluation at $\overline{c_i}$; in compensation, the evaluation at each c_i yields two equations corresponding to the real part and to the imaginary part of this evaluation. Moreover, if the factors of $\big(P_0(x)/Q_0(x)\big)'$ are not expanded, or if they are only partially expanded, then it is easier to realize which factors cancel out in the numerator and denominator of the derivative of $P_0(x)/Q_0(x)$, and it also facilitates figuring out which points x are eligible to produce simple equations when (iii) is applied.

Example 6.7.1. Find the primitive of $P(x)/Q(x)$ where $P(x) = x^4 + 2x^2 + x$ and $Q(x) = (x-1)(x^2+1)^2$.

Solution. The domain of $P(x)/Q(x)$ is $D := \mathbb{R}\backslash\{1\}$. In order to calculate its primitive, let us set the identity

$$\frac{x^4 + 2x^2 + x}{(x-1)(x^2+1)^2} = \left(\frac{Ax + B}{x^2 + 1}\right)' + \frac{C}{x-1} + \frac{Dx + E}{x^2 + 1}$$
$$= \frac{A(x^2+1) - 2x(Ax+B)}{(x^2+1)^2} + \frac{C}{x-1} + \frac{Dx + E}{x^2 + 1}. \tag{6.50}$$

Multiplying both sides of (6.50) by $Q(x)$, we obtain the following identity of polynomials:

$$x^4 + 2x^2 + x = A(x^2+1)(x-1) - 2x(Ax+B)(x-1)$$
$$+ C(x^2+1)^2 + (Dx+E)(x-1)(x^2+1), \quad x \in \mathbb{R}. \tag{6.51}$$

Next, by evaluation of (6.51) at the roots $x = 1$ and $x = i$ of $Q(x)$, identification of its fourth degree terms and evaluation at $x = 0$ provide us with the system

$$\begin{cases} x = 1: & 4 = 4C \Rightarrow C = 1 \\ x = i: & -1 + i = -2A + 2B + (2A + 2B)i \Rightarrow \begin{cases} -1 & = -2A + 2B \\ 1 & = 2A + 2B \end{cases} \\ x^4. & 1 = C + D \to D = 0 \\ x = 0: & 0 = -A + C - E. \end{cases} \tag{6.52}$$

As the reader can see, the choice of the root $x = 1$ for the evaluation of (6.51) has readily yielded the value of C, and in turn, of D, reducing (6.52) to a system of three equations whose solution is $A = 1/2$, $B = 0$ and $E = 1/2$.

Moreover, the choice $x = 0$ as a point at which (6.51) is evaluated has been motivated by the fact that x is a factor of an addend on the right side of (6.51).

Thus, from (6.50),

$$\int \frac{x^4 + 2x^2 + x}{(x-1)(x^2+1)^2} \, dx = \frac{\frac{1}{2}x}{x^2+1} + \int \frac{1}{x-1} \, dx + \int \frac{1/2}{x^2+1} \, dx$$

$$= \frac{1}{2} \frac{x}{x^2+1} + \log|x-1| + \frac{1}{2} \arctan x + C(x)$$

where $C(x)$ is a locally constant function on $\mathbb{R} \backslash \{1\}$. $\qquad \square$

Example 6.7.2. Find the primitive of $P(x)/Q(x)$ where $P(x) = 4$ and $Q(x) = (x-1)^2(x^2+1)^2$.

Solution. The domain of $P(x)/Q(x)$ is $D := \mathbb{R} \backslash \{1\}$. In order to calculate its primitive, consider the identity

$$\frac{4}{(x-1)^2(x^2+1)^2} = \left(\frac{Ax^2 + Bx + C}{(x-1)(x^2+1)} \right)' + \frac{D}{x-1} + \frac{Ex+F}{x^2+1}$$

$$= \frac{(2Ax + B)(x-1)(x^2+1) - (Ax^2 + Bx + C)(3x^2 - 2x + 1)}{(x-1)^2(x^2+1)^2}$$

$$+ \frac{D}{x-1} + \frac{Ex+F}{x^2+1}. \tag{6.53}$$

Multiplying both sides of (6.53) by $Q(x)$, we obtain

$$4 = (2Ax + B)(x-1)(x^2+1) - (Ax^2 + Bx + C)(3x^2 - 2x + 1)$$
$$+ D(x-1)(x^2+1)^2 + (Ex+F)(x-1)^2(x^2+1), \quad x \in \mathbb{R}. \tag{6.54}$$

The evaluation of (6.54) at the roots $x = 1$ and $x = i$ of $Q(x)$ yields three equations. In order to get three more equations, the fifth, fourth and zero degree terms in both sides in (6.54) are identified, which produces the following system:

$$\begin{cases} x = 1: & 4 = -(A + B + C)2 \\ x = i: & 4 = -(-A + Bi + C)(-3 - 2i + 1) \Rightarrow \begin{cases} 4 & = -2A - 2B + 2C \\ 0 & = 2A - 2B - 2C \end{cases} \\ x^5: & 0 = D + E \\ x^4: & 0 = -A - D - 2E + F \\ x = 0: & 4 = -B - C - D + F \end{cases} \tag{6.55}$$

Note that the first three equations in (6.55) readily yield the values $A = -1$, $B = -1$ and $C = 0$. Plugging these three values into the last three equations of (6.55), we get a system of three equations whose unknowns are D, E and C, and the solutions to this last system are $D = -2$, $E = 2$ and $F = 1$.

Now, from (6.53), we have

$$\int \frac{4}{(x-1)^2(x^2+1)^2}\,dx = \frac{-x^2-x}{(x-1)(x^2+1)} - \int \frac{2}{x-1}\,dx + \int \frac{2x+1}{x^2+1}\,dx. \qquad (6.56)$$

Since

$$\frac{2x+1}{x^2+1} = \frac{2x}{x^2+1} + \frac{1}{x^2+1}$$

according to Sections 6.1 to 6.3 it follows that

$$\int \frac{4}{(x-1)^2(x^2+1)^2}\,dx$$

$$= -\frac{x^2+x}{(x-1)(x^2+1)} - 2\log|x-1| + \log(x^2+1) + \arctan x + C(x)$$

where $C(x)$ is a locally constant function on $\mathbb{R}\backslash\{1\}$.   □

Example 6.7.3. Find the primitive of $1/Q(x)$ where $Q(x) = (x^2+1)^3$.

Solution. The domain of $1/Q(x)$ is $\mathbb{R}$, so its antiderivatives are defined on $\mathbb{R}$ too. Let us set the identity

$$\frac{1}{(1+x^2)^3} = \left(\frac{Ax^3 + Bx^2 + Cx + D}{(x^2+1)^2} \right)' + \frac{Ex+F}{x^2+1}$$

$$= \frac{(3Ax^2 + 2Bx + C)(x^2+1)^2 - (Ax^3 + Bx^2 + Cx + D)4(x^2+1)x}{(x^2+1)^4}$$

$$+ \frac{Ex+F}{x^2+1}. \qquad (6.57)$$

Canceling x^2+1 in the numerator and the denominator of the first fraction of the second line of (6.57) and multiplying both sides of (6.57) by $Q(x)$, we obtain

$$1 = (3Ax^2 + 2Bx + C)(x^2+1)$$
$$- 4(Ax^3 + Bx^2 + Cx + D)x + (Ex+F)(x^2+1)^2, \quad x \in \mathbb{R}. \qquad (6.58)$$

The only root of $Q(x)$ is the non-real number $x = i$. Therefore, evaluating (6.58) at $x = i$ provides two equations. Since four more equations are needed, we identify the coefficients of degree 5, 4, 1 and 0 in (6.58). This provides the following system:

$$\begin{cases} x = i: & 1 = -4(-Ai - B + Ci + D)i \Rightarrow \begin{cases} 1 & = -4A + 4C \\ 0 & = B - D \end{cases} \\ x^5: & 0 = E \\ x^4: & 0 = -A + F \\ x^1: & 0 = 2B - 4D + E \\ x^0: & 1 = C + F. \end{cases}$$

The reader will realize that the second, third and fifth equations readily yield $B = 0$ and $D = 0$, and since $E = 0$, the system is reduced to one of three equations whose unknowns are A, C and F. The solution to this system is $A = 3/8$, $C = 5/8$ and $F = 3/8$, and from (6.57), we obtain

$$\int \frac{1}{(1 + x^2)^3}\, dx = \frac{\frac{3}{8}x^3 + \frac{5}{8}x}{(x^2 + 1)^2} + \int \frac{3/8}{x^2 + 1}\, dx$$

$$= \frac{\frac{3}{8}x^3 + \frac{5}{8}x}{(x^2 + 1)^2} + \frac{3}{8}\arctan x + C$$

where C is any real number. □

6.8 Fractions of non-negative degree

Given a fraction $P(x)/Q(x)$ of polynomials, the long division algorithm (Section A.7) shows that the fraction $P(x)/Q(x)$ can be rewritten as

$$\frac{P(x)}{Q(x)} = S(x) + \frac{R(x)}{Q(x)}$$

where $S(x)$ and $R(x)$ are polynomials over $\mathbb{R}$, $\deg R(x) < \deg Q(x)$ and $S(x)$ is identically null if and only if $\deg P(x) < \deg Q(x)$. Thus,

$$\int \frac{P(x)}{Q(x)}\, dx = \int S(x)\, dx + \int \frac{R(x)}{Q(x)}\, dx,$$

where the primitive of $S(x)$ is immediate and that of $R(x)/Q(x)$ must be calculated following the procedures of Sections 6.1 to 6.7.

Example 6.8.1. Calculate the primitive of $(2x^2 - 3)/(x^2 + 1)$.

Solution. Since $x^2 + 1$ has no real root, the domain of the fraction in the statement is $\mathbb{R}$. The long division algorithm yields the decomposition

$$\frac{2x^2 - 3}{x^2 + 1} = 2 - \frac{5}{x^2 + 1}.$$

Hence

$$\int \frac{2x^2 - 3}{x^2 + 1}\, dx = 2x - 5\arctan x + C$$

for any real number C. □

6.9 Exercises

Find the domain and primitive of each of the following functions.

Question 6.9.1

$$f(x) = -\frac{x^2}{1 - x^2}$$

Question 6.9.2

$$f(x) = \frac{1 - x^2}{1 + x^2}$$

Question 6.9.3

$$f(x) = \frac{1}{x^2(x^2 - 1)}$$

Question 6.9.4

$$f(x) = \frac{1}{(1 - x^2)^2}$$

Question 6.9.5

$$f(x) = \frac{x^2}{(x^2 - 1)^2}$$

Question 6.9.6

$$f(x) = \frac{x^2 - 2x - 1}{(x - 1)^2(1 + x^2)}$$

Question 6.9.7

$$f(x) = \frac{x}{1 + x^4}$$

Question 6.9.8

$$f(x) = \frac{1 + x^2}{x^2(1 + x + x^2)}$$

Question 6.9.9

$$f(x) = \frac{x^4}{(1 + x^2)^4}$$

Question 6.9.10

$$f(x) = \frac{1}{x^2 + 2x}$$

Question 6.9.11

$$f(x) = \frac{2}{x^2 + 2x - 1}$$

Question 6.9.12

$$f(x) = \frac{6 + 4x^2}{(x^2 + 2x - 1)(1 + x^2)}$$

Answers to the Questions

Answer 6.9.1

$D_f = \mathbb{R}\backslash\{-1, 1\}$, $\int f(x)\,dx = x + \frac{1}{2}\log\frac{1-x}{1+x} + C(x)$

Answer 6.9.2

$D_f = \mathbb{R}$, $\int f(x)\,dx = -x + 2\arctan(x) + C$

Answer 6.9.3

$D_f = \mathbb{R}\backslash\{-1, 0, 1\}$, $\int f(x)\,dx = x^{-1} + \frac{1}{2}\log|1 - x| - \frac{1}{2}\log|1 + x| + C(x)$

Answer 6.9.4

$D_f = \mathbb{R}\backslash\{-1, 1\}, \int f(x)\,dx = -\frac{1}{4}\log|x-1| - \frac{1}{4(x-1)} + \frac{1}{4}\log|x+1| - \frac{1}{4(x+1)} + C(x)$

Answer 6.9.5

$D_f = \mathbb{R}\backslash\{-1, 1\}, \int f(x)\,dx = \frac{1}{4}\log|1-x| - \frac{1}{4(x-1)} - \frac{1}{4}\log|1+x| - \frac{1}{4(x+1)} + C(x)$

Answer 6.9.6

$D_f = \mathbb{R}\backslash\{1\}, \int f(x)\,dx = \log|x-1| + \frac{1}{x-1} - \frac{1}{2}\log(1+x^2) + \arctan(x) + C(x)$

Answer 6.9.7

$D_f = \mathbb{R}, \int f(x)\,dx = \frac{1}{2}\arctan(x^2) + C$

Answer 6.9.8

$D_f = \mathbb{R}\backslash\{0\}, \int f(x)\,dx =$
$-\log|x| - \frac{1}{x} + \frac{1}{2}\log(x^2 + x + 1) + \frac{1}{\sqrt{3}}\arctan\left(\frac{2}{\sqrt{3}}x + \frac{1}{\sqrt{3}}\right) + C(x)$

Answer 6.9.9

$D_f = \mathbb{R}, \int f(x)\,dx = \frac{\frac{1}{16}x^5 - \frac{1}{6}x^3 - \frac{1}{16}x}{(1+x^2)^3} + \frac{1}{16}\arctan(x) + C$

Answer 6.9.10

$D_f = \mathbb{R}\backslash\{-\sqrt{2}, 0, \sqrt{2}\}, \int f(x)\,dx = \frac{1}{2}\log\left|\frac{x}{x+2}\right| + C(x)$

Answer 6.9.11

$D_f = \mathbb{R}\backslash\{-1-\sqrt{2}, -1+\sqrt{2}\}, \int f(x)\,dx = \frac{1}{\sqrt{2}}\log\left|\frac{x+1-\sqrt{2}}{x+1+\sqrt{2}}\right| + C(x)$

Answer 6.9.12

$D_f = \mathbb{R}\backslash\{-1-\sqrt{2}, -1+\sqrt{2}\},$

$$\int f(x)\,dx = \frac{1}{4}\log|x^2 + 2x - 1| + \frac{5\sqrt{2}}{4}\log\left|\frac{x-\sqrt{2}+1}{x+\sqrt{2}+1}\right| - \frac{1}{4}\log(1+x^2)$$

$$-\frac{1}{2}\arctan x + C(x)$$

Chapter 7

Fractions of polynomials over $\sqrt{ax^2 + bx + c}$

Throughout this chapter, $ax^2 + bx + c$ is a second degree polynomial over $\mathbb{R}$ whose factorization over $\mathbb{C}$ is

$$ax^2 + bx + c = a(x - x_0)(x - x_1)$$

and $\Delta := b^2 - 4ac$ denotes its discriminant. One of the following conditions is assumed:

(i) $a < 0$, both x_0 and x_1 are real and $x_0 < x_1$,
(ii) $a > 0$ and neither of x_0 nor x_1 are real,
(iii) $a > 0$, both x_0 and x_1 are real and $x_0 < x_1$.

Note that the cases (i) and (iii) happen if and only if $\Delta > 0$, and (ii) happens if and only if $a > 0$ and $\Delta < 0$.

Thus, given a non-null polynomial $P(x)$, the natural domain of the fraction

$$F(x) := \frac{P(x)}{\sqrt{ax^2 + bx + c}} \tag{7.1}$$

is $D := \{x \colon ax^2 + bx + c > 0\}$. In accordance with conditions (i), (ii) and (iii), D is respectively

(i') $D := (x_0, x_1)$,
(ii') $D := (-\infty, \infty)$,
(iii') $D := (-\infty, x_0) \cup (x_1, \infty)$,

so D is regular in any of these three cases. Moreover, $F(x)$ is continuous on D and therefore, it has a primitive which shall be calculated in the following three sections according to the degree of $P(x)$.[1]

---

[1] The following cases are discarded for natural reasons: (iv) $a < 0$ and $\Delta \leqslant 0$, in which case $ax^2 + bx + c \leqslant 0$ for all $x \in \mathbb{R}$ and therefore, $F(x)$ is not well defined on any real number; (v) $a > 0$ and $x_0 = x_1 \in \mathbb{R}$, in which case, $F(x) = P(x)/\sqrt{a}(x-x_0)$ for $x > x_0$ and $F(x) = -P(x)/\sqrt{a}(x-x_0)$ for $x < x_0$, in which case there exists the primitive of $F(x)$ on $\mathbb{R}\backslash\{x_0\}$ and it must be calculated as the primitive of a fraction of polynomials on $(-\infty, x_0)$ and on (x_0, ∞) separately.

7.1   Fractions $1/\sqrt{ax^2 + bx + c}$

A glance at Appendix A.9 will show that by completing squares, $1/\sqrt{ax^2 + bx + c}$ is respectively equal to one of the following fractions for suitable constants $\alpha > 0$, β and $\gamma > 0$:

(i'') $\dfrac{1}{\gamma\sqrt{-(\alpha x + \beta)^2 + 1}}$,

(ii'') $\dfrac{1}{\gamma\sqrt{(\alpha x + \beta)^2 + 1}}$,

(iii'') $\dfrac{1}{\gamma\sqrt{(\alpha x + \beta)^2 - 1}}$.

According to (8), (9) and (10) in Table 4.6.1, the primitive of each of the fractions in (i''), (ii'') and (iii'') is respectively

(i''') $\frac{1}{\alpha\gamma} \arcsin(\alpha x + \beta) + C$, $x \in (x_0, x_1)$,

(ii''') $\frac{1}{\alpha\gamma} \operatorname{arg\,sinh}(\alpha x + \beta) + C$, $x \in \mathbb{R}$,

(iii''') $\begin{cases} \frac{1}{\alpha\gamma} \operatorname{arg\,cosh}(\alpha x + \beta) + C_1 & x > x_1 \\ -\frac{1}{\alpha\gamma} \operatorname{arg\,cosh}(-\alpha x - \beta) + C_2 & x < x_0, \end{cases}$

where C, C_1 and C_2 are any real constants.

The solutions (i''') and (ii''') are straightforward, but the solution (iii''') may be a little bit tricky, so it will be explained in more detail next. First, note that the values $x = x_0$ and $x = x_1$ in (iii) satisfy the identity

$$(\alpha x + \beta)^2 - 1 = 0.$$

Hence,

$$x_0 = -\frac{1 + \beta}{\alpha} \quad \text{and} \quad x_1 = \frac{1 - \beta}{\alpha}.$$

Consider the intervals

$$I_0 = (-\infty, x_0), \quad I_1 = (x_1, \infty)$$

$$J_0 = (-\infty, -1), \quad J_1 = (1, \infty)$$

and the changes of variable $g_0 \colon I_0 \longrightarrow J_0$ and $g_1 \colon I_1 \longrightarrow J_1$, both defined by $g_i(x) := \alpha x + \beta$ for $x \in I_i$.

Each change $t = g_i(x)$ yields a local antiderivative of $1/\sqrt{(\alpha x + \beta)^2 - 1}$ on I_i as follows:

$$\int \frac{1}{\sqrt{(\alpha x + \beta)^2 - 1}}\, dx, \ x \in I_i \quad \xleftarrow{t=\alpha x + \beta} \quad \int \frac{1/\alpha}{\sqrt{t^2 - 1}}\, dt, \ t \in J_i$$

thus, by virtue of Table 4.6.1 (10) and reversing the changes of variable in J_0 and J_1,

$$\int \frac{1/\alpha}{\sqrt{t^2 - 1}}\, dt = \frac{\operatorname{arg\,cosh} t}{\alpha}, \ t \in J_1 \quad \xleftarrow{t=\alpha x + \beta} \quad \frac{1}{\alpha} \operatorname{arg\,cosh}(\alpha x + \beta), \ x \in I_1$$

$$\int \frac{1/\alpha}{\sqrt{t^2 - 1}}\, dt = -\frac{\operatorname{arg\,cosh}(-t)}{\alpha}, \ t \in J_0 \quad \xleftarrow{t=\alpha x + \beta} \quad -\frac{1}{\alpha} \operatorname{arg\,cosh}(-\alpha x - \beta), \ x \in I_0.$$

Hence, the complete solution is

$$\int \frac{1}{\sqrt{(\alpha x + \beta)^2 - 1}} \, dx = \begin{cases} \frac{1}{\alpha} \operatorname{arg\,cosh} (\alpha x + \beta) + C_1, & x \in I_1 \\ -\frac{1}{\alpha} \operatorname{arg\,cosh} (-\alpha x - \beta) + C_2, & x \in I_0. \end{cases}$$

Example 7.1.1. Calculate $\displaystyle\int \frac{7}{\sqrt{x^2 + x + 1}} \, dx$.

Solution. Since $x^2 + x + 1$ has no real root and its leading coefficient is positive, the domain of $f(x) = 7/\sqrt{x^2 + x + 1}$ is the entire real line $(-\infty, \infty)$, so its primitive must be calculated following (ii‴). Indeed, by completing squares,

$$\int \frac{7}{\sqrt{x^2 + x + 1}} \, dx = \int \frac{7}{\sqrt{\left[x + \frac{1}{2}\right]^2 + \frac{3}{4}}} \, dx$$

$$= \frac{2 \cdot 7}{\sqrt{3}} \int \frac{1}{\sqrt{\left[\frac{2}{\sqrt{3}} x + \frac{1}{\sqrt{3}}\right]^2 + 1}} \, dx = 7 \operatorname{arg\,sinh} \left(\frac{2}{\sqrt{3}} x + \frac{1}{\sqrt{3}} \right) + C$$

where C is any constant. □

Example 7.1.2. Calculate $\displaystyle\int \frac{1}{\sqrt{-x^2 + 7x - 10}} \, dx$.

Solution. Since $-x^2 + 7x - 10$ has two single real roots, 2 and 5, and its leading coefficient is negative, the domain of $f(x) = 1/\sqrt{-x^2 + 7x - 10}$ is the bounded interval $(2, 5)$, so its primitive must be calculated from (i‴). Indeed, by completing squares,

$$\int \frac{1}{\sqrt{-x^2 + 7x - 10}} \, dx = \int \frac{1}{\sqrt{-\left[x - \frac{7}{2}\right]^2 + \frac{9}{4}}} \, dx$$

$$= \frac{2}{3} \int \frac{1}{\sqrt{-\left[\frac{2}{3} x - \frac{7}{3}\right]^2 + 1}} \, dx = \arcsin \left(\frac{2}{3} x - \frac{7}{3} \right) + C$$

where C is any real constant. □

Example 7.1.3. Calculate $\displaystyle\int \frac{1}{\sqrt{x^2 - 3x + 2}} \, dx$.

Solution. Since the roots of $x^2 - 3x + 2$ are 1 and 2, both real and simple, and since its leading coefficient is positive, the domain of $f(x) = 1/\sqrt{x^2 - 3x + 2}$ is $(-\infty, 1) \cup (2, \infty)$, the union of two disjoint intervals, so its primitive must be calculated using (iii‴). Indeed, by completing squares,

$$\int \frac{1}{\sqrt{x^2 - 3x + 2}} \, dx = \int \frac{1}{\sqrt{\left[x - \frac{3}{2}\right]^2 - \frac{1}{4}}} \, dx$$

$$= 2 \int \frac{1}{\sqrt{(2x - 3)^2 - 1}} \, dx = \begin{cases} \operatorname{arg\,cosh} (2x - 3) + C_1, & x \in (2, \infty) \\ -\operatorname{arg\,cosh} (-2x + 3) + C_2, & x \in (-\infty, 1) \end{cases}$$

where C_1 and C_2 are any pair of real constants. □

7.2   Fractions $(Ax + B)/\sqrt{ax^2 + bx + c}$

Let A be a non-null real number. Bearing in mind that $2ax + b$ is the derivative of $ax^2 + bx + c$, the fraction $(Ax + B)/\sqrt{ax^2 + bx + c}$ can be rewritten by adjusting coefficients as

$$
\begin{aligned}
\frac{Ax + B}{\sqrt{ax^2 + bx + c}} &= \frac{A}{a} \frac{2ax + (b - b) + \frac{2a}{A}B}{2\sqrt{ax^2 + bx + c}} \\
&= \frac{A}{a} \frac{2ax + b}{2\sqrt{ax^2 + bx + c}} + \frac{A}{2a} \frac{-b + 2aB/A}{\sqrt{ax^2 + bx + c}};
\end{aligned}
\tag{7.2}
$$

it is immediate that $\frac{A}{a}\sqrt{ax^2 + bx + c}$ is an antiderivative of the first addend in the bottom line of (7.2). Thus, letting $R := \frac{A}{2a}(-b + 2aB/A)$, from (7.2),

$$
\int \frac{Ax + B}{\sqrt{ax^2 + bx + c}}\, dx = \frac{A}{a}\sqrt{ax^2 + bx + c} + \int \frac{R}{\sqrt{ax^2 + bx + c}}\, dx \tag{7.3}
$$

where $\int R/\sqrt{ax^2 + bx + c}\, dx$ must be calculated according to Section 7.1.

Example 7.2.1. Find $\displaystyle\int f(x)\, dx$ where $f(x) = \dfrac{8x + 11}{\sqrt{-x^2 + 4x - 3}}$.

Solution. The roots of $x^2 - 4x + 3$ are 1 and 3 and its leading coefficient is negative, so the domain of $f(x)$ is the interval $(1, 3)$. Since the derivative of $-x^2 + 4x - 3$ is $-2x + 4$, an adjustment of coefficients leads to

$$
\begin{aligned}
\int \frac{8x + 11}{\sqrt{-x^2 + 4x - 3}}\, dx &= -8\int \frac{-2x - \frac{11}{4}}{2\sqrt{-x^2 + 4x - 3}}\, dx \\
&= -8\int \frac{-2x + 4}{2\sqrt{-x^2 + 4x - 3}}\, dx - 8\int \frac{-4 - \frac{11}{4}}{2\sqrt{-x^2 + 4x - 3}}\, dx.
\end{aligned}
\tag{7.4}
$$

The first primitive in the bottom line of (7.4) is immediately calculated as

$$
\int \frac{-2x + 4}{2\sqrt{-x^2 + 4x - 3}}\, dx = \sqrt{-x^2 + 4x - 3}
$$

and the second one is calculated by completing squares as

$$
\int \frac{4 + \frac{11}{4}}{2\sqrt{-x^2 + 4x - 3}}\, dx = \frac{27}{8}\int \frac{1}{\sqrt{-(x - 2)^2 + 1}}\, dx = \frac{27}{8}\arcsin(x - 2).
$$

Thus, from (7.4),

$$
\int \frac{8x + 11}{\sqrt{-x^2 + 4x - 3}}\, dx = -8\sqrt{-x^2 + 4x - 3} + 27\arcsin(x - 2) + C, \quad x \in (1, 3).
$$

$\square$

Example 7.2.2. Find $\displaystyle\int f(x)\, dx$ where $f(x) = \dfrac{5x + 1}{\sqrt{x^2 + x + 1}}$.

Solution. The polynomial $x^2 + x + 1$ has no real root, so the domain of $f(x)$ is $(-\infty, \infty)$.

Bearing in mind that $2x + 1$ is the derivative of $x^2 + x + 1$, an adjustment of coefficients gives

$$\int \frac{5x + 1}{\sqrt{x^2 + x + 1}} \, dx = 5 \int \frac{x + \frac{1}{5}}{2\sqrt{x^2 + x + 1}} \, dx$$

$$= 5 \int \frac{2x + 1}{2\sqrt{x^2 + x + 1}} \, dx + 5 \int \frac{-1 + \frac{2}{5}}{2\sqrt{x^2 + x + 1}} \, dx.$$

(7.5)

The first primitive in the bottom line of (7.5) is readily calculated as

$$\int \frac{2x + 1}{2\sqrt{x^2 + x + 1}} \, dx = \sqrt{x^2 + x + 1}$$

and the second one is calculated by completing squares as

$$\int \frac{1}{\sqrt{x^2 + x + 1}} \, dx = \int \frac{2/\sqrt{3}}{\sqrt{\left(\frac{2}{\sqrt{3}}x + \frac{1}{\sqrt{3}}\right)^2 + 1}} \, dx = \operatorname{arg\,sinh}\left(\frac{2}{\sqrt{3}}x + \frac{1}{\sqrt{3}}\right).$$

Thus, from (7.5),

$$\int \frac{5x + 1}{\sqrt{x^2 + x + 1}} \, dx = 5\sqrt{x^2 + x + 1} - \frac{3}{2} \operatorname{arg\,sinh}\left(\frac{2}{\sqrt{3}}x + \frac{1}{\sqrt{3}}\right) + C, \ x \in (-\infty, \infty).$$

□

Example 7.2.3. Find the primitive of $f(x) = \dfrac{x + 6}{\sqrt{x^2 - 3x + 2}}$.

Solution. The roots of $x^2 - 3x + 2$ are 1 and 2 and its leading coefficient is positive, thus the domain of $f(x)$ is $(-\infty, 1) \cup (2, \infty)$.

Since $2x - 3$ is the derivative of $x^2 - 3x + 2$, an adjustment of coefficients yields

$$\int \frac{x + 6}{\sqrt{x^2 - 3x + 2}} \, dx = \int \frac{2x + 12}{2\sqrt{x^2 - 3x + 2}} \, dx$$

$$= \int \frac{2x - 3}{2\sqrt{x^2 - 3x + 2}} \, dx + \int \frac{3 + 12}{2\sqrt{x^2 - 3x + 2}} \, dx.$$

(7.6)

The first primitive in the bottom line of (7.6) is

$$\int \frac{2x - 3}{2\sqrt{x^2 - 3x + 2}} \, dx = \sqrt{x^2 - 3x + 2}$$

and the second one is calculated by completing squares as

$$\int \frac{3 + 12}{2\sqrt{x^2 - 3x + 2}} \, dx = \frac{15}{2} \int \frac{1}{\sqrt{\left(x - \frac{3}{2}\right)^2 - \frac{1}{4}}} \, dx$$

$$= 15 \int \frac{1}{\sqrt{(2x - 3)^2 - 1}} \, dx = \begin{cases} \frac{15}{2} \operatorname{arg\,cosh}(2x - 3), & x \in (2, \infty) \\ -\frac{15}{2} \operatorname{arg\,cosh}(-2x + 3), & x \in (-\infty, 1). \end{cases}$$

Thus, from (7.6),

$$\int \frac{x + 6}{\sqrt{x^2 - 3x + 2}} \, dx$$

$$= \sqrt{x^2 - 3x + 2} + \begin{cases} \frac{15}{2} \operatorname{arg\,cosh}(2x - 3) + C_1, & x \in (2, \infty) \\ -\frac{15}{2} \operatorname{arg\,cosh}(-2x + 3) + C_2, & x \in (-\infty, 1) \end{cases}$$

where C_1 and C_2 is any pair of real numbers.

□

7.3   Fractions $P(x)/\sqrt{ax^2 + bx + c}$ for any polynomial $P(x)$

The calculation of the primitives of this type of function is based upon the following result. Its proof uses the notion of direct sum of subspaces (see Section 1.1).

Proposition 7.3.1. *Given a polynomial $P_1(x)$ of degree $n \geqslant 2$ and a polynomial $ax^2 + bx + c$ with $a \neq 0$, there exists a unique polynomial $P_2(x)$ of degree at most $n - 1$ and a unique real number λ for which*

$$\frac{P_1(x)}{\sqrt{ax^2 + bx + c}} = \left(P_2(x)\sqrt{ax^2 + bx + c}\right)' + \frac{\lambda}{\sqrt{ax^2 + bx + c}}.$$

Proof. Let $\mathcal{P}_{n-1}$ and $\mathcal{P}_n$ respectively denote the sets of all polynomials over $\mathbb{R}$ of degree at most $n - 1$ and n, and let $\mathcal{P}_0$ denote the set of all constant polynomials over $\mathbb{R}$. Clearly, $\mathcal{P}_{n-1}$, $\mathcal{P}_n$, and $\mathcal{P}_0$ are vector spaces over $\mathbb{R}$ and their dimensions are respectively n, $n+1$ and 1. In particular, $\mathcal{P}_0 = \mathbb{R}$. Let us write $Q(x) := ax^2 + bx + c$ for short.

Since $[P_2 \cdot \sqrt{Q}]' = P_2' \cdot \sqrt{Q} + P_2 Q'/(2\sqrt{Q})$, it is clear that in order to prove the statement, it suffices to prove that for every $P_1(x) \in \mathcal{P}_n$, there exists a unique polynomial $P_2(x) \in \mathcal{P}_{n-1}$ and a unique constant $\lambda \in \mathbb{R}$ such that

$$P_1(x) = P_2'(x)Q + \frac{1}{2}P_2(x)Q'(x) + \lambda. \tag{7.7}$$

To do that, consider the linear mapping $\phi\colon \mathcal{P}_{n-1} \longrightarrow \mathcal{P}_n$ given by

$$\phi(P) := P' \cdot Q + \frac{1}{2}P \cdot Q'.$$

Let $P \in \mathcal{P}_{n-1}$ be any non-identically null polynomial and let $a_m x^m$ be its leading term (so, $0 \leqslant m \leqslant n - 1$). A few simple calculations yield that the leading term of $\phi(P)$ is

$$(m + 1)a a_m x^{m+1}. \tag{7.8}$$

Formula (7.8) shows that $\phi(P) \neq 0$, that is, ϕ is injective and in turn, $\dim \phi(\mathcal{P}_{n-1}) = n$. Besides, (7.8) also proves that $\phi(\mathcal{P}_{n-1}) \cap \mathcal{P}_0 = \{0\}$. Thus since $\dim \mathcal{P}_n = n+1 = \dim \phi(\mathcal{P}_{n-1}) + \dim \mathcal{P}_0$, we conclude that

$$\mathcal{P}_n = \phi(\mathcal{P}_{n-1}) \oplus \mathcal{P}_0$$

and this shows that (7.7) holds. $\square$

Proposition 7.3.1 yields the following procedure to solve the primitive of $P(x)/\sqrt{ax^2 + bx + c}$, where $P(x)$ is a polynomial of degree $n \geqslant 1$.

First, set the identity

$$\frac{P(x)}{\sqrt{ax^2 + bx + c}}$$
$$= \left((A_{n-1}x^{n-1} + A_{n-2}x^{n-2} + \ldots + A_1 x + A_0)\sqrt{ax^2 + bx + c}\right)'$$
$$+ \frac{\lambda}{\sqrt{ax^2 + bx + c}} \tag{7.9}$$

where the coefficients A_i and λ are unknown but their existence and uniqueness is claimed in Proposition 7.3.1. Next, calculate the derivative that occurs in (7.9) and multiply both sides of the resulting identity by $\sqrt{ax^2 + bx + c}$. This yields the following identity of polynomials:

$$
\begin{aligned}
P(x) \quad &\big((n-1)A_{n-1}x^n{}^2 + (n-2)A_{n-2}x^{n-3} \dots + A_1\big)(ax^2 + bx + c) \\
&+ \frac{1}{2}(A_{n-1}x^{n-1} + A_{n-2}x^{n-2} \dots + A_1 x + A_0)(2ax + b) + \lambda.
\end{aligned} \tag{7.10}
$$

Identifying terms of the same degree in (7.10), the values of the unknowns A_i and λ are easily found in decreasing order from A_{n-1} to A_1 and ending with λ. Thus, the primitive of $P(x)/\sqrt{ax^2 + bx + c}$ is easily calculated from (7.9) as

$$
\int \frac{P(x)}{\sqrt{ax^2 + bx + c}}\, dx
$$
$$
= (A_{n-1}x^{n-1} + \dots + A_1 x + A_0)\sqrt{ax^2 + bx + c} + \int \frac{\lambda}{\sqrt{ax^2 + bx + c}}\, dx
$$

where $\int \frac{\lambda}{\sqrt{ax^2+bx+c}}\, dx$ must be found as indicated in Section 7.1.

Example 7.3.2. Find $\displaystyle\int \frac{x^4 + x^2}{\sqrt{x^2 + 4}}\, dx$.

Solution. Since $x^2 + 4$ has no real root, the domain of $\frac{x^4+x^2}{\sqrt{x^2+4}}$ is $(-\infty, \infty)$. Set the identity

$$
\frac{x^4 + x^2}{\sqrt{x^2 + 4}} = \left((Ax^3 + Bx^2 + Cx + D)\sqrt{x^2 + 4}\right)' + \frac{E}{\sqrt{x^2 + 4}} \tag{7.11}
$$

where A, B, C, D and E are unknowns. Calculating the derivative that occurs in (7.11), this identity becomes

$$
\frac{x^4 + x^2}{\sqrt{x^2 + 4}} = (3Ax^2 + 2Bx + C)\sqrt{x^2 + 4}
$$
$$
+ (Ax^3 + Bx^2 + Cx + D)\frac{x}{\sqrt{x^2 + 4}} + \frac{E}{\sqrt{x^2 + 4}}.
$$

Multiplying both sides by $\sqrt{x^2 + 4}$ yields

$$
x^4 + x^2 = (3Ax^2 + 2Bx + C)(x^2 + 4) + (Ax^3 + Bx^2 + Cx + D)x + E
$$

and identifying terms of the same degree, the following system is obtained:

$$
\begin{cases}
x^4: & 1 = 4A \Rightarrow A = 1/4 \\
x^3: & 0 = 3B \Rightarrow B = 0 \\
x^2: & 1 = 12A + 2C = 3 + 2C \Rightarrow C = -1 \\
x^1: & 0 = 8B + D = D \Rightarrow D = 0 \\
x^0: & 0 = 4C + E \Rightarrow E = 4.
\end{cases}
$$

Plug the values of A to E in (7.11) so that

$$\int \frac{x^4 + x^2}{\sqrt{x^2 + 4}} \, dx = \left(\frac{1}{4}x^3 - x\right)\sqrt{x^2 + 4} + \int \frac{4}{\sqrt{x^2 + 4}} \, dx. \tag{7.12}$$

As

$$\int \frac{1}{\sqrt{x^2 + 4}} \, dx = \frac{1}{2}\int \frac{1}{\sqrt{(x/2)^2 + 1}} \, dx = 4 \arg \sinh(x/2) + C, \ C \in \mathbb{R} \tag{7.13}$$

plugging (7.13) into (7.12),

$$\int \frac{x^4 + x^2}{\sqrt{x^2 + 4}} \, dx = \left(\frac{1}{4}x^3 - x\right)\sqrt{x^2 + 4} + 16 \arg \sinh(x/2) + C$$

where C is any real number.   □

Example 7.3.3. Find $\displaystyle\int \frac{x^3 + 1}{\sqrt{x^2 - 1}} \, dx$.

Solution. First, note that the domain of $\frac{x^3}{\sqrt{x^2-1}}$ is $(-\infty, -1) \cup (1, \infty)$. Next, set the identity

$$\frac{x^3 + 1}{\sqrt{x^2 - 1}} = \left((Ax^2 + Bx + C)\sqrt{x^2 - 1}\right)' + \frac{D}{\sqrt{x^2 - 1}}. \tag{7.14}$$

Thus,

$$\frac{x^3 + 1}{\sqrt{x^2 - 1}} = (2Ax + B)\sqrt{x^2 - 1} + (Ax^2 + Bx + C)\frac{x}{\sqrt{x^2 - 1}} + \frac{D}{\sqrt{x^2 - 1}}$$

and multiplying both sides by $\sqrt{x^2 - 1}$ yields

$$x^3 + 1 = (2Ax + B)(x^2 - 1) + (Ax^2 + Bx + C)x + D. \tag{7.15}$$

The following system, whose unknowns are A, B, C and D, is obtained by identification of terms of the same degree in (7.15):

$$\begin{cases} x^3: & 1 = 3A \Rightarrow A = 1/3 \\ x^2: & 0 = 2B \Rightarrow B = 0 \\ x^1: & 0 = -2A + C \Rightarrow C = 2/3 \\ x^0: & 1 = -B + D \Rightarrow D = 1 \end{cases}$$

and once the coefficients A to D are known, from (7.14),

$$\int \frac{x^3 + 1}{\sqrt{x^2 - 1}} \, dx = \left(\frac{1}{3}x^2 + \frac{2}{3}\right)\sqrt{x^2 - 1} + \int \frac{1}{\sqrt{x^2 - 1}} \, dx$$

$$= \left(\frac{1}{3}x^2 + \frac{2}{3}\right)\sqrt{x^2 - 1} + \begin{cases} \arg \cosh x + C_1, \ x \in (1, \infty) \\ -\arg \cosh(-x) + C_2, \ x \in (-\infty, -1) \end{cases}$$

where C_1 and C_2 is any pair of real numbers.   □

7.4 Exercises

Find the domain and an antiderivative of each of the following functions.

Question 7.4.1

$$f(x) = \frac{x^4 + x + 1}{\sqrt{x^2 + 2x - 2}}$$

Question 7.4.2

$$f(x) = \frac{x^3 - 1}{\sqrt{1 - 4x^2}}$$

Question 7.4.3

$$f(x) = \frac{x^3 + x^2 + x + 1}{\sqrt{2x^2 - x}}$$

Question 7.4.4

$$f(x) = \frac{x^3 + x - 1}{\sqrt{-x^2 + 3x}}$$

Question 7.4.5

$$f(x) = \frac{3x + 6}{\sqrt{-x^2 + 3x - 2}}$$

Question 7.4.6

$$f(x) = \frac{4x + 5}{\sqrt{-x^2 + x + 2}}$$

Question 7.4.7

$$f(x) = \frac{x + 5}{\sqrt{x^2 - x + 1}}$$

Question 7.4.8

$$f(x) = \frac{5x + 1}{\sqrt{x^2 + x + 3}}$$

Question 7.4.9

$$f(x) = \frac{2x - 3}{\sqrt{(x - 1)(x + 2)}}$$

Question 7.4.10

$$f(x) = \frac{4x + 3}{\sqrt{(x + 3)(x + 1)}}$$

Answers to the Questions

Answer 7.4.1

$D_f = \mathbb{R}$,

$$\left(\frac{1}{4}x^3 - \frac{7}{12}x^2 + \frac{53}{24}x - \frac{191}{24} \right) \sqrt{x^2 + 2x - 2} + \frac{107}{8} \operatorname{arg\,sinh} \left(\frac{x + 1}{\sqrt{3}} \right)$$

Answer 7.4.2

$D_f = (-1/2, 1/2)$,

$$\left(-\frac{1}{12}x^2 - \frac{1}{24} \right) \sqrt{1 - 4x^2} - \frac{1}{2} \arcsin(2x)$$

Answer 7.4.3

$D_f = (-\infty, 0) \cup (1/2, \infty)$,

$$\left(\frac{1}{6}x^2 + \frac{17}{48}x + \frac{49}{64}\right)\sqrt{2x^2 - x} + \frac{177}{256} \cdot \begin{cases} \operatorname{arg\,cosh}(4x - 1) & x \in (1/2, \infty) \\ -\operatorname{arg\,cosh}(1 - 4x) & x \in (-\infty, 0) \end{cases}$$

Answer 7.4.4

$D_f = (0, 3)$,

$$\left[-\frac{1}{3}x^2 - \frac{5}{4}x - \frac{53}{8}\right]\sqrt{-x^2 + 3x} + \frac{143}{16}\arcsin(2x/3 - 1)$$

Answer 7.4.5

$D_f = (1, 2)$,

$$-3\sqrt{-x^2 + 3x - 2} + \frac{21}{2}\arcsin(2x - 3)$$

Answer 7.4.6

$D_f = (-1, 2)$,

$$-4\sqrt{-x^2 + x + 2} + 7\arcsin\left((2x - 1)/3\right)$$

Answer 7.4.7

$D_f = \mathbb{R}$,

$$\sqrt{x^2 - x + 1} + \frac{11}{2}\operatorname{arg\,sinh}\left((2x - 1)/\sqrt{3}\right)$$

Answer 7.4.8

$D_f = \mathbb{R}$,

$$5\sqrt{x^2 + x + 3} - \frac{3}{2}\operatorname{arg\,sinh}\left((2x + 1)/\sqrt{11}\right)$$

Answer 7.4.9

$D_f = (-\infty, -2) \cup (1, \infty)$,

$$2\sqrt{x^2 + x - 2} + 4 \cdot \begin{cases} -\operatorname{arg\,cosh}\left((2x + 1)/3\right), & x \in (1, \infty) \\ \operatorname{arg\,cosh}\left(-(2x + 1)/3\right), & x \in (-\infty, -2) \end{cases}$$

Answer 7.4.10

$D_f = (-\infty, -3) \cup (-1, \infty)$,

$$4\sqrt{x^2 + 4x + 3} + 5 \cdot \begin{cases} -\operatorname{arg\,cosh}(x + 2), & x \in (-1, -\infty) \\ \operatorname{arg\,cosh}(-x - 2), & x \in (-\infty, -3) \end{cases}$$

Chapter 8

Fractions $\dfrac{1}{(\alpha x+\beta)^p\sqrt{ax^2+bx+c}}$, $p \in \mathbb{N}$

Throughout this chapter, p is a natural number, $\alpha x + \beta$ is a first degree polynomial over $\mathbb{R}$ whose unique root is denoted by $y_0 := -\beta/\alpha$, and $ax^2 + bx + c$ is a second degree polynomial over $\mathbb{R}$ whose factorization over $\mathbb{C}$ is

$$ax^2 + bx + c = a(x - x_0)(x - x_1).$$

One of the following conditions is assumed:

(i) $a < 0$, both x_0 and x_1 are real and $x_0 < x_1$,
(ii) $a > 0$ and neither x_0 nor x_1 is real,
(iii) $a > 0$, both x_0 and x_1 are real and $x_0 < x_1$.

Let Δ denote the discriminant of $ax^2 + bx + c$. Note that cases (i) and (iii) happen if and only if $\Delta > 0$, and (ii) happens if and only if $a > 0$ and $\Delta < 0$.

For $p \in \mathbb{N}$, let

$$F(x) := \frac{1}{(\alpha x + \beta)^p\sqrt{ax^2 + bx + c}}. \tag{8.1}$$

The natural domain of $F(x)$ is

$$D := \{x \in \mathbb{R}: ax^2 + bx + c > 0, \alpha x + \beta \neq 0\}.$$

In accordance with conditions (i), (ii) and (iii), and depending upon the position of y_0 with respect to x_0 and to x_1, D is

(i') $D = (x_0, y_0) \cup (y_0, x_1)$ if $x_0 < y_0 < x_1$,
  $D = (x_0, x_1)$ if either $y_0 \leqslant x_0$ or $x_1 \leqslant y_0$,
(ii') $D = (-\infty, y_0) \cup (y_0, \infty)$,
(iii') $D = (-\infty, y_0) \cup (y_0, x_0) \cup (x_1, \infty)$ whenever $y_0 < x_0$,
  $D = (-\infty, x_0) \cup (x_1, \infty)$ whenever $x_0 \leqslant y_0 \leqslant x_1$,
  $D = (-\infty, x_0) \cup (x_1, y_0) \cup (y_0, \infty)$ whenever $x_1 < y_0$.

Clearly, D is regular, its regular decomposition consists of one, two or three open, pairwise-disjoint, non-empty intervals and $F(x)$ is continuous on D, hence it has a primitive on its domain, which will be calculated by means of the change of variable studied in the following section.

125

8.1   The change of variable $t = 1/(\alpha x + \beta)$

Consider a function $g: I \longrightarrow \mathbb{R}$ where

$$\bullet \ g \text{ maps each } x \in I \text{ to } t = 1/(\alpha x + \beta), \tag{8.2}$$

$$\bullet \ I \text{ is one of the open intervals of}$$
$$\text{the regular decomposition of } D, \tag{8.3}$$

that is, $I = (z_0, z_1)$ is one of the intervals occurring at (i'), (ii') or (iii'). Note that I is included either in $(-\infty, y_0)$ or in (y_0, ∞).
    The derivative of g is

$$g'(x) = -\frac{\alpha}{(\alpha x + \beta)^2}.$$

Thus, $g'(x) \neq 0$ for all $x \in I$ and therefore, g is a change of variable.
    The inverse change g^{-1} maps each $t \in g(I)$ to

$$x = \frac{1}{\alpha} \cdot \frac{1}{t} - \frac{\beta}{\alpha} \tag{8.4}$$

hence, by differentiation on (8.4), the general expression (4.7) for the change of variable given in (8.2) is

$$dx = -\frac{1}{\alpha} \frac{1}{t^2} \, dt. \tag{8.5}$$

If $\alpha > 0$, then $g'(x) < 0$ for all $x \in I$, hence g is decreasing. Moreover,

$$\lim_{x \to y_0^+} 1/(\alpha x + \beta) = \infty,$$
$$\lim_{x \to \infty} 1/(\alpha x + \beta) = 0,$$
$$\lim_{x \to y_0^-} 1/(\alpha x + \beta) = -\infty,$$
$$\lim_{x \to -\infty} 1/(\alpha x + \beta) = 0.$$

Hence, as $y_0 \notin I$, then

$$\bullet \text{ either } \sup I \leqslant y_0, \text{ in which case } g(I) \subset (-\infty, 0) \tag{8.6}$$

$$\bullet \text{ or } y_0 \leqslant \inf I, \text{ in which case } g(I) \subset (0, \infty); \tag{8.7}$$

and

$$g(I) = \Big(\lim_{x \to z_1^-} g(x), \ \lim_{x \to z_0^+} g(x) \Big). \tag{8.8}$$

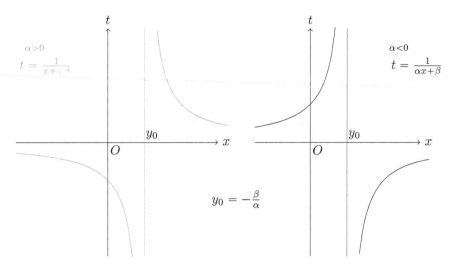

If $\alpha < 0$, then g is an increasing change of variable for which

- either $\sup I \leqslant y_0$, in which case $g(I) \subset (0, \infty)$ (8.9)
- or $y_0 \leqslant \inf I$, in which case $g(I) \subset (-\infty, 0)$ (8.10)

and

$$g(I) = \left(\lim_{x\to z_0^+} g(x),\ \lim_{x\to z_1^-} g(x) \right).$$

8.2   Primitives of $\dfrac{1}{(\alpha x+\beta)^p\sqrt{ax^2+bx+c}}$, $p\in\mathbb{N}$

The primitive of the fraction $F(x)$ given in (8.1) is locally solved on each interval I_j by means of the change of variable $g(x) = 1/(\alpha x + \beta)$ on I_j studied in Section 8.1, where

$$D = \bigcup_{1\leqslant j\leqslant k} I_j$$

is the regular decomposition of the domain of $F(x)$ according to (i$'$), (ii$'$) and (iii$'$) (thus k can be 1, 2 or 3). Thus, the change of variable given in (8.2) is implemented by means of the substitutions (8.4) and (8.5), so that

$$\int \frac{1}{(\alpha x + \beta)^p\sqrt{ax^2 + bx + c}}\, dx,\ x\in I_j \ \xleftrightarrow{\ x=\frac{1}{\alpha t}-\frac{\beta}{\alpha}\ }$$

$$\xleftarrow{\ x=\frac{1}{\alpha t}-\frac{\beta}{\alpha}\ } -\frac{1}{\alpha}\int \frac{t^p}{t^2\sqrt{a\left(\frac{1}{\alpha t}-\frac{\beta}{\alpha}\right)^2 + b\left(\frac{1}{\alpha t}-\frac{\beta}{\alpha}\right) + c}}\, dt$$

$$= -\frac{1}{\alpha}\int \frac{t^{p-1}}{t\sqrt{a'\frac{1}{t^2} + b'\frac{1}{t} + c'}}\, dt,\ t\in g(I_j) \qquad (8.11)$$

where

$$a' = \frac{a}{\alpha^2}, \quad b' = -\frac{2a\beta}{\alpha^2} + \frac{b}{\alpha}, \quad c' = \frac{a\beta^2}{\alpha^2} - \frac{b\beta}{\alpha} + c.$$

In cases (8.7) and (8.9), since the factor t in the denominator of the last fraction in (8.11) is positive, it is pushed into the square root so that

$$\int \frac{t^{p-1}\,dt}{t\sqrt{a'\frac{1}{t^2} + b'\frac{1}{t} + c'}} = \int \frac{t^{p-1}\,dt}{\sqrt{a' + b't + c't^2}}; \tag{8.12}$$

in cases (8.6) and (8.10) the same factor t in (8.11) is negative, hence

$$\int \frac{t^{p-1}\,dt}{t\sqrt{a'\frac{1}{t^2} + b'\frac{1}{t} + c'}} = -\int \frac{t^{p-1}\,dt}{\sqrt{a' + b't + c't^2}} \tag{8.13}$$

and both primitives given in (8.12) and in (8.13) can be solved as explained in Section 7.1 if $p = 1$, or as in Section 7.2 if $p \geqslant 2$. Next, after reversing the changes of variable, the local antiderivatives corresponding to (8.12) and (8.13) are assembled to make a global primitive of $1/(\alpha x + \beta)^p \sqrt{ax^2 + bx + c}$.

Example 8.2.1. Find $\displaystyle\int \frac{1}{x\sqrt{x^2 - 1}}\,dx.$

Solution. The roots of $x^2 - 1$ are -1 and 1, and $-1 < 0 < 1$. Thus, the domain of $f(x) = 1/x\sqrt{x^2 - 1}$ is $D = I_1 \cup I_2$, where $I_1 := (-\infty, -1)$ and $I_2 := (1, \infty)$.

The change of variable $g(x) = 1/x$ is implemented on every interval I_j by means of the substitutions $x = 1/t$ and $dx = -\,dt/t^2$, so that

$$\int \frac{1}{x\sqrt{x^2 - 1}}\,dx, \; x \in I_j \xleftarrow{x = \frac{1}{t}} -\int \frac{1}{t\sqrt{\frac{1}{t^2} - 1}}\,dt, \; t \in g(I_j). \tag{8.14}$$

In order to calculate the primitive of $f(x)$ on I_1, note that $g(I_1) = (-1, 0)$. Thus, the primitive on the right side of (8.14) for $j = 1$ is

$$-\int \frac{1}{t\sqrt{\frac{1}{t^2} - 1}}\,dt = \int \frac{1}{\sqrt{1 - t^2}}\,dt = \arcsin t, \; t \in (-1, 0);$$

reversing the change of variable,

$$\arcsin t, \; t \in (-1, 0) \xleftarrow{t = \frac{1}{x}} \arcsin(1/x), \; x \in (-\infty, -1). \tag{8.15}$$

And in order to calculate the primitive of $f(x)$ on I_2, note that $g(I_2) = (0, 1)$. Thus, the primitive on the right side of (8.14) for $j = 2$ is

$$-\int \frac{1}{t\sqrt{\frac{1}{t^2} - 1}}\,dt = -\int \frac{1}{\sqrt{1 - t^2}}\,dt = -\arcsin t, \; t \in (0, 1);$$

reversing the change of variable,

$$-\arcsin t, \; t \in (0, 1) \xleftarrow{t = \frac{1}{x}} -\arcsin(1/x), \; x \in (1, \infty). \tag{8.16}$$

Assembling the local antiderivatives given in (8.15) and in (8.16), we obtain the solution

$$\int \frac{1}{x\sqrt{x^4 - 1}}\, dx = \begin{cases} -\arcsin{(1/x)} + C_1, & x \in (1, \infty) \\ \arcsin{(1/x)} + C_2, & x \in (-\infty, -1) \end{cases}$$

where C_1 and C_2 are any pair of real constants.   □

Example 8.2.2. Find $\displaystyle\int \frac{1}{x\sqrt{1 - x^2}}\, dx$.

Solution. The roots of $x^2 - 1$ are -1 and 1, and $-1 < 0 < 1$. Thus, the domain of $f(x) = 1/x\sqrt{1 - x^2}$ is $D = I_1 \cup I_2$, where $I_1 := (-1, 0)$ and $I_2 := (0, 1)$.

The change of variable $g(x) = 1/x$ is implemented on every interval I_j by means of the substitutions $x = 1/t$ and $dx = -\, dt/t^2$, so that

$$\int \frac{dx}{x\sqrt{1 - x^2}}, \ x \in I_j \ \xleftarrow{\ x = \frac{1}{t}\ } \ -\int \frac{dt}{t\sqrt{1 - \frac{1}{t^2}}}, \ t \in g(I_j). \tag{8.17}$$

In order to calculate the primitive of $f(x)$ on I_2, note that $g(I_2) = (1, \infty)$. Thus, the primitive on the right side of (8.17) for $j = 2$ is

$$-\int \frac{1}{t\sqrt{1 - \frac{1}{t^2}}}\, dt = -\int \frac{1}{\sqrt{t^2 - 1}}\, dt = -\operatorname{arg\,cosh} t, \ t \in (1, \infty);$$

and reversing the change of variable,

$$-\operatorname{arg\,cosh} t, \ t \in (1, \infty) \ \xleftarrow{\ t = \frac{1}{x}\ } \ -\operatorname{arg\,cosh} (1/x), \ x \in (0, 1). \tag{8.18}$$

And in order to calculate the primitive of $f(x)$ on I_1, note that $g(I_1) = (-\infty, -1)$. Hence, the primitive on the right side of (8.17) for $j = 1$ is

$$-\int \frac{1}{t\sqrt{1 - \frac{1}{t^2}}}\, dt = \int \frac{1}{\sqrt{t^2 - 1}}\, dt = -\operatorname{arg\,cosh} (-t), \ t \in (-\infty, -1);$$

reversing the change of variable,

$$-\operatorname{arg\,cosh} (-t), \ t \in (-\infty, -1) \ \xleftarrow{\ t = \frac{1}{x}\ } \ -\operatorname{arg\,cosh} (-1/x), \ x \in (-1, 0). \tag{8.19}$$

Finally, assembling the local antiderivatives given in (8.18) and in (8.19), the global primitive is obtained as

$$\int \frac{dx}{x\sqrt{1 - x^2}} = \begin{cases} -\operatorname{arg\,cosh} (1/x) + C_1, & x \in (0, 1) \\ -\operatorname{arg\,cosh} (-1/x) + C_2, & x \in (-1, 0) \end{cases}$$

where C_1 and C_2 are any pair of real constants.   □

Example 8.2.3. Find $\displaystyle\int \frac{dx}{(x - 2)\sqrt{x^2 - 1}}\, dx$.

Solution. The roots of $x^2 - 1$ are -1 and 1, its leading coefficient is positive and the only root of $x - 2$ is 2, hence the domain of $f(x) = 1/(x-2)\sqrt{x^2 - 1}$ is $D = I_1 \cup I_2 \cup I_3$, where

$$I_3 := (2, \infty)$$
$$I_2 := (1, 2)$$
$$I_1 := (-\infty, -1).$$

The change of variable $t = 1/(x-2)$ is implemented on every interval I_j by means of the substitutions $x = 1/t + 2$ $dx = -dt/t^2$, so that

$$\int \frac{dx}{(x-2)\sqrt{x^2 - 1}}, \ x \in I_j \xleftarrow{x = \frac{1}{t} + 2} -\int \frac{dt}{t\sqrt{\left(\frac{1}{t} + 2\right)^2 - 1}}$$

$$= -\int \frac{dt}{t\sqrt{\frac{1}{t^2} + \frac{4}{t} + 3}}, \ t \in g(I_j). \tag{8.20}$$

Applying (8.20) to each interval $g(I_j)$, we get

$$-\int \frac{dt}{t\sqrt{\frac{1}{t^2} + \frac{4}{t} + 3}} = -\int \frac{dt}{\sqrt{1 + 4t + 3t^2}}, \ t \in g(I_3) = (0, \infty); \tag{8.21}$$

$$-\int \frac{dt}{t\sqrt{\frac{1}{t^2} + \frac{4}{t} + 3}} = \int \frac{dt}{\sqrt{1 + 4t + 3t^2}}, \ t \in g(I_2) = (-\infty, -1); \tag{8.22}$$

$$-\int \frac{dt}{t\sqrt{\frac{1}{t^2} + \frac{4}{t} + 3}} = \int \frac{dt}{\sqrt{1 + 4t + 3t^2}}, \ t \in g(I_1) = (-1/3, 0). \tag{8.23}$$

In order to find the three primitives of (8.21), (8.22) and (8.23), it is convenient to find the primitive of $h(t) := 1/\sqrt{1 + 4t + 3t^2}$ separately. Note that the natural domain of $h(t)$ is $(-\infty, -1) \cup (-1/3, \infty)$. By completing squares,

$$\int \frac{dt}{\sqrt{1 + 4t + 3t^2}} = \int \frac{dt}{\sqrt{\left(\sqrt{3}t + \frac{2}{\sqrt{3}}\right)^2 - \frac{1}{3}}} = \int \frac{\sqrt{3}\, dt}{\sqrt{(3t + 2)^2 - 1}},$$

and in accordance with (7.1),

$$\int \frac{\sqrt{3}\, dt}{\sqrt{(3t+2)^2 - 1}} = \begin{cases} \dfrac{1}{\sqrt{3}} \arg\cosh{(3t+2)} + C_1, \ t \in (-1/3, \infty) & (8.24) \\[2mm] -\dfrac{1}{\sqrt{3}} \arg\cosh{(-3t-2)} + C_2, \ t \in (-\infty, -1). & (8.25) \end{cases}$$

Thus, since $(0, \infty) \subset (-1/3, \infty)$, a local antiderivative of $f(x)$ on I_3 is obtained by a combination of (8.21) and (8.24) followed by a reversion of the change of variable, which yields

$$-\int \frac{dt}{t\sqrt{\frac{1}{t^2} + \frac{4}{t} + 3}} = -\frac{1}{\sqrt{3}} \arg\cosh{(3t+2)} + C_3, \ t \in (0, \infty)$$

$$\xleftarrow{t = \frac{1}{x-2}} -\frac{1}{\sqrt{3}} \arg\cosh\left(\frac{3}{x-2} + 2\right) + C_3, \ x \in I_3. \tag{8.26}$$

To calculate the primitive of $f(x)$ on I_2, since $g(I_2) = (-\infty, -1)$, a combination of (8.22) and (8.25) followed by a reversion of the change of variable gives

$$-\int \frac{dt}{t\sqrt{\frac{1}{t^2} + \frac{4}{t} + 3}} = -\frac{1}{\sqrt{3}} \arg\cosh\left(-3t - 2\right) + C_2, \ t \in (-\infty, -1)$$

$$\xleftarrow{t = \frac{1}{x-2}} -\frac{1}{\sqrt{3}} \arg\cosh\left(-\frac{3}{x-2} - 2\right) + C_2, \ x \in I_2. \qquad (8.27)$$

Analogously, for the primitive of $f(x)$ on I_1, since $(-1/3, 0) \subset (-1/3, \infty)$ a combination of (8.23) and (8.24) and the subsequent reversion of change of variable yields

$$-\int \frac{dt}{t\sqrt{\frac{1}{t^2} + \frac{4}{t} + 3}} = \frac{1}{\sqrt{3}} \arg\cosh\left(3t + 2\right) + C_1, \ t \in (-\infty, -1)$$

$$\xleftarrow{t = \frac{1}{x-2}} \frac{1}{\sqrt{3}} \arg\cosh\left(\frac{3}{x-2} + 2\right) + C_1, \ x \in I_1. \qquad (8.28)$$

Assembling the local solutions given in (8.26), (8.27) and (8.28), the desired primitive is

$$\int \frac{dx}{(x-2)\sqrt{x^2-1}} = \begin{cases} -\frac{1}{\sqrt{3}} \arg\cosh\left(\frac{3}{x-2} + 2\right) + C_3, \ x \in I_3 \\ -\frac{1}{\sqrt{3}} \arg\cosh\left(-\frac{3}{x-2} - 2\right) + C_2, \ x \in I_2 \\ \frac{1}{\sqrt{3}} \arg\cosh\left(\frac{3}{x-2} + 2\right) + C_1, \ x \in I_1 \end{cases}$$

where C_1, C_2 and C_3 are any real constants. $\qquad\qquad \square$

Example 8.2.4. Find $\displaystyle\int \frac{dx}{(x+1)^3\sqrt{x^2+2x}}$.

Solution. Since the roots of $x^2 + 2x$ are -2 and 0, its leading coefficient is positive and the root of $x + 1$ is $-1 \notin (-\infty, -2) \cup (0, \infty)$, it follows that the domain of $f(x) = 1/(x+1)^3\sqrt{x^2+2x}$ is $I_1 \cup I_2$ where $I_1 := (-\infty, -2)$ and $I_2 := (0, \infty)$.

The change of variable $g(x) = 1/(x+1)$ is implemented on every interval I_j by means of the substitutions $x = 1/t - 1$ and $dx = -dt/t^2$, so that

$$\int \frac{dx}{(x+1)^3\sqrt{x^2+2x}}, \ x \in I_j \xleftarrow{x = \frac{1}{t} - 1} -\int \frac{t^2\, dt}{t\sqrt{\left(\frac{1}{t} - 1\right)^2 + 2\left(\frac{1}{t} - 1\right)}}$$

$$= -\int \frac{t^2\, dt}{t\sqrt{\frac{1}{t^2} - 1}} \, dt, \ t \in g(I_j). \qquad (8.29)$$

Applying the solution above to each interval $g(I_j)$, we get

$$-\int \frac{t^2\, dt}{t\sqrt{\frac{1}{t^2} - 1}} = -\int \frac{t^2\, dt}{\sqrt{1 - t^2}}, \ t \in g(I_2) = (0, 1) \qquad (8.30)$$

$$-\int \frac{t^2\, dt}{t\sqrt{\frac{1}{t^2} - 1}} = \int \frac{t^2\, dt}{\sqrt{1 - t^2}}, \ t \in g(I_1) = (-1, 0). \qquad (8.31)$$

As identities (8.30) and (8.31) suggest, the primitive of $h(x) := t^2/\sqrt{1-t^2}$ needs to be found. To do that, and bearing in mind that the natural domain of $h(t)$ is $(-1,1)$, we set the following identity in accordance with Section 7.3:

$$\frac{t^2}{\sqrt{1-t^2}} = \left[(At+B)\sqrt{1-t^2}\right]' + \frac{C}{\sqrt{1-t^2}}$$

$$= A\sqrt{1-t^2} + (At+B)\frac{-t}{\sqrt{1-t^2}} + \frac{C}{\sqrt{1-t^2}}. \tag{8.32}$$

Multiply both sides by $\sqrt{1-t^2}$ so that

$$t^2 = A(1-t^2) - (At+B)t + C. \tag{8.33}$$

Equating coefficients of the same degree in (8.33) yields the following system:

$$\begin{cases} t^2: & 1 = -2A \Rightarrow A = -1/2 \\ t^1: & 0 = -B \Rightarrow B = 0 \\ t=0: & 0 = A + C \Rightarrow C = 1/2. \end{cases}$$

Next, plug the values of A, B and C into (8.32) so that

$$\int \frac{t^2\,dt}{\sqrt{1-t^2}} = -\frac{1}{2}t\sqrt{1-t^2} + \int \frac{1/2}{\sqrt{1-t^2}}\,dt$$

$$= -\frac{1}{2}t\sqrt{1-t^2} + \frac{1}{2}\arcsin t + C, \quad t \in (-1,1). \tag{8.34}$$

Thus, the primitive of $f(x)$ on I_2 is obtained from a combination of (8.30) and (8.34), and a subsequent reversion of change of variable:

$$-\int \frac{t^2\,dt}{t\sqrt{\frac{1}{t^2}-1}} = \frac{1}{2}t\sqrt{1-t^2} - \frac{1}{2}\arcsin t + C_2, \; t \in (0,1) \tag{8.35}$$

$$\xrightarrow{t=\frac{1}{1+x}} \frac{1}{2(1+x)}\sqrt{1 - \frac{1}{(1+x)^2}} - \frac{1}{2}\arcsin\left(\frac{1}{1+x}\right) + C_2, \; x \in I_2.$$

Analogously, the primitive of $f(x)$ on I_1 is calculated by means of the following combination of (8.31) and (8.34) and the subsequent reversion of change of variable:

$$-\int \frac{t^2\,dt}{t\sqrt{\frac{1}{t^2}-1}}\,dt = -\frac{1}{2}t\sqrt{1-t^2} + \frac{1}{2}\arcsin t + C_1, \; t \in (-1,0) \tag{8.36}$$

$$\xrightarrow{t=\frac{1}{1+x}} \frac{-1}{2(1+x)}\sqrt{1 - \frac{1}{(1+x)^2}} + \frac{1}{2}\arcsin\left(\frac{1}{1+x}\right) + C_1, \; x \in I_1.$$

Putting together the local solutions given in (8.35) and in (8.36), the complete solution is

$$\int \frac{dx}{(x+1)^3\sqrt{x^2+2x}} = \begin{cases} \frac{\sqrt{x^2+2x}}{2(1+x)^2} - \frac{1}{2}\arcsin\left(\frac{1}{1+x}\right) + C_2, & x \in I_2 \\[2mm] \frac{\sqrt{x^2+2x}}{2(1+x)^2} + \frac{1}{2}\arcsin\left(\frac{1}{1+x}\right) + C_1, & x \in I_1 \end{cases}$$

where C_1 and C_2 is any pair of real constants. $\qquad\Box$

Example 8.2.5. Find $\displaystyle\int \frac{dx}{(x-3)^2\sqrt{x^2-x-2}}$.

Solution. Since the roots of $x^2 - x - 2$ are -1 and 2, its leading coefficient is positive and the root of $x - 3$ is $3 \in (2, \infty)$, it follows that the domain of $f(x) = 1/((x-3)^2\sqrt{x^2-x-2})$ is $\ldots$ is $D := I_1 \cup I_2 \cup I_3$ where

$$I_3 = (3, \infty)$$
$$I_2 = (2, 3)$$
$$I_1 := (-\infty, -1).$$

The change of variable $g(x) = 1/(x-3)$ is implemented on every interval I_j by means of the substitutions $x = 1/t + 3$ and $dx = -dt/t^2$, so that

$$\int \frac{dx}{(x-3)^2\sqrt{x^2-x-2}}, \; x \in I_j \xleftarrow{x=\frac{1}{t}+3} -\int \frac{t^2\,dt}{t^2\sqrt{\left(\frac{1}{t}+3\right)^2 - \left(\frac{1}{t}+3\right) - 2}}$$

$$= -\int \frac{t\,dt}{t\sqrt{4 + \frac{5}{t} + \frac{1}{t^2}}}, \; t \in g(I_j) \tag{8.37}$$

so that

$$-\int \frac{t\,dt}{t\sqrt{4 + \frac{5}{t} + \frac{1}{t^2}}} = -\int \frac{t\,dt}{\sqrt{4t^2 + 5t + 1}}, \; t \in g(I_3) = (0, \infty) \tag{8.38}$$

$$-\int \frac{t\,dt}{t\sqrt{4 + \frac{5}{t} + \frac{1}{t^2}}} = \int \frac{t\,dt}{\sqrt{4t^2 + 5t + 1}}, \; t \in g(I_2) = (-\infty, -1) \tag{8.39}$$

$$-\int \frac{t\,dt}{t\sqrt{4 + \frac{5}{t} + \frac{1}{t^2}}} = \int \frac{t\,dt}{\sqrt{4t^2 + 5t + 1}}, \; t \in g(I_1) = (-1/4, 0). \tag{8.40}$$

As the identities (8.38), (8.39) and (8.40) suggest, the primitive of $h(t) := t/\sqrt{4t^2 + 5t + 1}$ needs to be found. The natural domain of $h(t)$ is $(-\infty, -1) \cup (-1/4, \infty)$. Thus, in accordance with Section 7.2, $h(t)$ is split into two fractions so that

$$\int \frac{t\,dt}{\sqrt{4t^2 + 5t + 1}} = \frac{1}{4}\int \frac{8t + (5 - 5)}{2\sqrt{4t^2 + 5t + 1}}\,dt$$

$$= \frac{1}{4}\int \frac{8t + 5}{2\sqrt{4t^2 + 5t + 1}}\,dt - \frac{5}{8}\int \frac{dt}{\sqrt{4t^2 + 5t + 1}}\,dt. \tag{8.41}$$

The first primitive of the bottom line in (8.41) is

$$\int \frac{8t + 5}{2\sqrt{4t^2 + 5t + 1}}\,dt = \sqrt{4t^2 + 5t + 1}$$

and the second is found by completing squares as follows:

$$\int \frac{dt}{\sqrt{4t^2 + 5t + 1}} = \frac{4}{3}\int \frac{dt}{\sqrt{\left(\frac{8}{3}t + \frac{5}{3}\right)^2 - 1}}$$

$$= \begin{cases} \frac{1}{2}\operatorname{arg\,cosh}\left(\frac{8}{3}t + \frac{5}{3}\right) + C_1, \; t > -\frac{1}{4} \\[2ex] -\frac{1}{2}\operatorname{arg\,cosh}\left(-\frac{8}{3}t - \frac{5}{3}\right) + C_2, \; t < -1. \end{cases}$$

Thus, the complete primitive of $h(t)$ is

$$\int \frac{t\,dt}{\sqrt{4t^2 + 5t + 1}} = \frac{1}{4}\sqrt{4t^2 + 5t + 1} +$$

$$+ \begin{cases} -\dfrac{5}{16}\,\text{arg cosh}\left(\dfrac{8}{3}t + \dfrac{5}{3}\right) + C_1, \ t > -\dfrac{1}{4} & (8.42) \\[4mm] \dfrac{5}{16}\,\text{arg cosh}\left(-\dfrac{8}{3}t - \dfrac{5}{3}\right) + C_2, \ t < -1 & (8.43) \end{cases}$$

The primitive of $f(x)$ on I_3 can be now calculated by combining (8.38) with (8.42) and reversing the change of variable as follows:

$$-\frac{1}{4}\sqrt{4t^2 + 5t + 1} + \frac{5}{16}\,\text{arg cosh}\left(\frac{8}{3}t + \frac{5}{3}\right) + C_3, \ t \in (0, \infty)$$

$$\xrightarrow{t=\frac{1}{x-3}} -\frac{1}{4}\sqrt{\frac{4}{(x-3)^2} + \frac{5}{x-3} + 1}$$

$$+ \frac{5}{16}\,\text{arg cosh}\left(\frac{8}{3}\frac{1}{x-3} + \frac{5}{3}\right) + C_3, \ x \in I_3. \qquad (8.44)$$

The primitive of $f(x)$ on I_2 is obtained by a combination of (8.39) and (8.43), and reversing the change of variable so that

$$\frac{1}{4}\sqrt{4t^2 + 5t + 1} + \frac{5}{16}\,\text{arg cosh}\left(-\frac{8}{3}t - \frac{5}{3}\right) + C_2, \ t \in (-\infty, -1)$$

$$\xrightarrow{t=\frac{1}{x-3}} \frac{1}{4}\sqrt{\frac{4}{(x-3)^2} + \frac{5}{x-3} + 1}$$

$$+ \frac{5}{16}\,\text{arg cosh}\left(-\frac{8}{3}\frac{1}{x-3} - \frac{5}{3}\right) + C_2, \ x \in I_2. \qquad (8.45)$$

And the primitive of $f(x)$ on I_1 is calculated by a combination of (8.40) with (8.42) and reversing the change of variable as follows:

$$\frac{1}{4}\sqrt{4t^2 + 5t + 1} - \frac{5}{16}\,\text{arg cosh}\left(\frac{8}{3}t + \frac{5}{3}\right) + C_1, \ t \in (-1/4, 0)$$

$$\xrightarrow{t=\frac{1}{x-3}} \frac{1}{4}\sqrt{\frac{4}{(x-3)^2} + \frac{5}{x-3} + 1}$$

$$- \frac{5}{16}\,\text{arg cosh}\left(\frac{8}{3}\frac{1}{x-3} + \frac{5}{3}\right) + C_1, \ x \in I_1. \qquad (8.46)$$

Putting together the local solutions given in (8.44), (8.45) and (8.46) the complete solution is

$$\int \frac{dx}{(x-3)^2\sqrt{x^2 - x - 2}} =$$

$$\begin{cases} -\dfrac{1}{4}\sqrt{\dfrac{4}{(x-3)^2}+\dfrac{5}{x-3}+1}+\dfrac{5}{16}\operatorname{arg\,cosh}\left(\dfrac{8/3}{x-3}+\dfrac{5}{3}\right)+C_3, & x \in I_3 \\[3mm] \dfrac{1}{4}\sqrt{\dfrac{4}{(x-3)^2}+\dfrac{5}{x-3}+1}+\dfrac{5}{16}\operatorname{arg\,cosh}\left(-\dfrac{8/3}{x-3}-\dfrac{5}{3}\right)+C_2, & x \in I_2 \\[3mm] \dfrac{1}{4}\sqrt{\dfrac{4}{(x-3)^2}+\dfrac{5}{x-3}+1}-\dfrac{5}{16}\operatorname{arg\,cosh}\left(\dfrac{8/3}{x-3}+\dfrac{5}{3}\right)+C_1, & x \in I_1 \end{cases}$$

where C_1, C_2 and C_3 are any real constants. ☐

8.3 Exercises

Find the domain and an antiderivative for each of the following functions.

Question 8.3.1

$$f(x) = \dfrac{1}{x^2\sqrt{4-x^2}}$$

Question 8.3.2

$$f(x) = \dfrac{3}{(x-2)\sqrt{x^2-4}}$$

Question 8.3.3

$$f(x) = \dfrac{1}{x\sqrt{2x^2+1}}$$

Question 8.3.4

$$f(x) = \dfrac{2}{(x-2)^2\sqrt{x^2+x+1}}$$

Question 8.3.5

$$f(x) = \dfrac{1}{(x-2)\sqrt{(x-1)(x+2)}}$$

Question 8.3.6

$$f(x) = \dfrac{1}{(x-2)\sqrt{-x^2+x}}$$

Question 8.3.7

$$f(x) = \dfrac{1}{x^3\sqrt{x^2-1}}$$

Question 8.3.8

$$f(x) = \dfrac{1}{(x+1)^2\sqrt{x^2+4}}$$

Question 8.3.9

$$f(x) = \dfrac{1}{(x+3)\sqrt{x^2-1}}$$

Answers to the Questions

Answer 8.3.1

$D_f = (-2,0) \cup (0,2)$,

$$\begin{cases} -\dfrac{1}{8x}\sqrt{\dfrac{4}{x^2}-1}-\dfrac{1}{16}\operatorname{arg\,cosh}(2/x), & x \in (0,2) \\[3mm] \dfrac{1}{8x}\sqrt{\dfrac{4}{x^2}-1}-\dfrac{1}{16}\operatorname{arg\,cosh}(-2/x), & x \in (-2,0) \end{cases}$$

Answer 8.3.2

$D_f = (-\infty, -2) \cup (2, \infty),$

$$\begin{cases} -\frac{3}{2}\sqrt{1 + \frac{4}{x-2}}, & x \in (2, \infty) \\ \frac{3}{2}\sqrt{1 + \frac{4}{x-2}}, & x \in (-\infty, 2) \end{cases}$$

Answer 8.3.3

$D_f = (-\infty, 0) \cup (0, \infty),$

$$\begin{cases} -\arg\sinh(1/\sqrt{2}x), & x \in (0, \infty) \\ \arg\sinh(1/\sqrt{2}x), & x \in (-\infty, 0) \end{cases}$$

Answer 8.3.4

$D_f = (-\infty, 2) \cup (2, \infty),$

$$\left[-\frac{2}{7}\sqrt{\frac{7}{(x-2)^2} + \frac{5}{x-2} + 1} + \frac{5}{7\sqrt{7}}\arg\sinh\left(\frac{14}{\sqrt{3}(x-2)} + \frac{5}{\sqrt{3}}\right) \right] \cdot \begin{cases} 1, & x \in (2, \infty) \\ -1, & x \in (-\infty, 2) \end{cases}$$

Answer 8.3.5

$D_f = (-\infty, -2) \cup (1, 2) \cup (2, \infty),$

$$\begin{cases} -\frac{1}{2}\arg\cosh\left(\frac{8}{3}\frac{1}{x-2} + \frac{5}{3}\right), & x \in (2, \infty) \\ -\frac{1}{2}\arg\cosh\left(-\frac{8}{3}\frac{1}{x-2} - \frac{5}{3}\right), & x \in (1, 2) \\ \frac{1}{2}\arg\cosh\left(\frac{8}{3}\frac{1}{x-2} + \frac{5}{3}\right), & x \in (-\infty, -2) \end{cases}$$

Answer 8.3.6

$D_f = (0, 1),$

$$\frac{1}{\sqrt{2}}\arcsin\left(\frac{4}{x-2} + 3\right)$$

Answer 8.3.7

$D_f = (-\infty, -1) \cup (1, \infty),$

$$\frac{1}{2x}\sqrt{1 - \frac{1}{x^2}} - \frac{1}{2}\arcsin(1/x) \cdot \begin{cases} 1, & x \in (1, \infty) \\ -1, & x \in (-\infty, -1) \end{cases}$$

Answer 8.3.8

$D_f = (-\infty, -1) \cup (-1, \infty),$

$-\dfrac{1}{5}\sqrt{\dfrac{5}{(r+1)^2} + -\dfrac{2}{r+1} + 1} - \dfrac{1}{5\sqrt{5}}\,\text{arg sinh}\left(\dfrac{5}{5(r+1)} - \dfrac{1}{5}\right)\cdot\begin{cases} 1, & x\in(-1,\infty) \\ 1, & x\in(-\infty,-1) \end{cases}$

Answer 8.3.9

$D_f = (-\infty, -3) \cup (-3, -1) \cup (1, \infty),$

$\begin{cases} \dfrac{\sqrt{2}}{4}\,\text{arg cosh}\left(-\dfrac{8}{x+3} + 3\right), & x\in(1,\infty) \\[2mm] -\dfrac{\sqrt{2}}{4}\,\text{arg cosh}\left(-\dfrac{8}{x+3} + 3\right), & x\in(-\infty,-3) \\[2mm] -\dfrac{\sqrt{2}}{4}\,\text{arg cosh}\left(\dfrac{8}{x+3} - 3\right), & x\in(-3,-1) \end{cases}$

Chapter 9

Rational functions of x and of rational powers of $\frac{ax+b}{cx+d}$

Throughout this chapter, a, b, c, d denote fixed real numbers which satisfy $ad \neq bc$ and one and only one of the following conditions:

- $a \neq 0$ and $c > 0$ (9.1)
- $a = 0$ and $c > 0$ ($\Rightarrow b \neq 0$)   (9.2)
- $a > 0$ and $c = 0$ ($\Rightarrow d \neq 0$).   (9.3)

Consider the restricted fraction

$$F(x) := \frac{ax+b}{cx+d}, \quad x \in E = \left\{ x : \frac{ax+b}{cx+d} \geq 0 \right\}. \tag{9.4}$$

Under the conditions (9.1)–(9.3), there exists the inverse of $t = F(x)$ which can be readily obtained as

$$x = \frac{b - dt}{ct - a}, \quad t \in F(E). \tag{9.5}$$

The aim of this chapter is to find the primitive of the function

$$\varrho(x) := R\left(x, F(x)^{\frac{p_1}{q_1}}, F(x)^{\frac{p_2}{q_2}}, \ldots, F(x)^{\frac{p_n}{q_n}} \right) \tag{9.6}$$

where $p_i \in \mathbb{Z}\backslash\{0\}$, $q_i \in \mathbb{N}$ and L.C.M.$(|p_i|, q_i) = 1$ for all i, $q_i > 1$ for some i,[1] and $R(x, x_1, \ldots, x_n)$ is a rational function of its arguments from $\mathbb{R}^{n+1}$ into $\mathbb{R}$, that is, a function given by a finite combination of sums, products and quotients of scalars, the variable x and the functions $F(x)^{p_i/q_i}$.

Note that E is the natural domain of $F(x)^{p_i/q_i}$ whenever $q_i > 1$ (see Appendix A.1 for more information).

9.1   The domain of $R\left(x, \left(\frac{ax+b}{cx+d}\right)^{p_1/q_1}, \ldots, \left(\frac{ax+b}{cx+d}\right)^{p_n/q_n}\right)$

A function of the form (9.6) can be expressed as a composition

$$\varrho(x) = Q \circ g(x), \quad x \in D_\varrho \tag{9.7}$$

---

[1] The conditions on p_i and q_i prevent the trivial case where $\varrho(x)$ is a rational function of one variable, in which case its primitive can be found following the techniques of Chapter 6. Also note that a fraction $(ax + b)/(cx + d)$ where $a = 0$, $c < 0$ and $b \neq 0$ is implicitly included in 9.2. Analogously, the case $a < 0$, $c = 0$ and $d \neq 0$ is included in 9.3.

where $Q(t) = P_1(t)/P_2(t)$ for a pair of polynomials $P_1(t)$ and $P_2(t)$ over $\mathbb{R}$ and

$$g(x) = \sqrt[q]{F(x)}, \ x \in E \tag{9.8}$$

where $q = \text{L.C.M.}(q_1, q_2, \ldots, q_n)$.

Thus, $\varrho(x)$ is a composition of continuous functions, and in combination with the following result, the existence of its primitive is sustained by Theorem 4.1.9.

The function $g(x)$ in (9.7) plays a crucial role because some of its restrictions provide the changes of variable to be used in order to solve the primitive of $\varrho(x)$.

Proposition 9.1.1. *The natural domain of $\varrho(x)$ is a union of finitely many disjoint intervals. Hence, it is a regular domain.*

Proof. Following the same notation as in (9.7), and bearing in mind that the natural domain of g is E, it follows that

$$D_\varrho = g^{-1}(D_Q) = E\backslash g^{-1}(Z) \tag{9.9}$$

where Z is the finite set of all roots of $P_2(t)$. Thus, since g is injective, $g^{-1}(Z)$ is also finite, and therefore, D_ϱ will be proved to be regular as soon as E is proved to be regular too. For this, a few computations show that in case (9.1),

$$E = \left(\left[-\frac{b}{a}, \infty\right) \cap \left(-\frac{d}{c}, \infty\right)\right) \cup \left(\left(-\infty, -\frac{b}{a}\right] \cap \left(-\infty, -\frac{d}{c}\right)\right) \text{ if } a > 0 \tag{9.10}$$

$$E = \left(-\frac{d}{c}, -\frac{b}{a}\right] \cup \left[-\frac{b}{a}, -\frac{d}{c}\right) \text{ if } a < 0. \tag{9.11}$$

In case (9.2),

$$E = \left(-\frac{d}{c}, \infty\right) \text{ if } b > 0 \tag{9.12}$$

$$E = \left(-\infty, -\frac{d}{c}\right) \text{ if } b < 0. \tag{9.13}$$

And in case (9.3),

$$E = \left[-\frac{b}{a}, \infty\right) \text{ if } d > 0 \tag{9.14}$$

$$E = \left(-\infty, -\frac{b}{a}\right] \text{ if } d < 0. \tag{9.15}$$

This proves that E is a regular domain, which concludes the proof. $\square$

Solving the primitive of $\varrho(x)$ requires the changes of variables induced by the function $g(x)$ defined in (9.8). These changes of variable are studied in the following section.

9.2 The change of variable $g(x) = \left(\frac{ax+b}{cx+d}\right)^{1/q}$, $1 < q \in \mathbb{N}$

Let $D_\varrho = \bigcup_{j=1}^k I_j$ be the regular decomposition of the natural domain of $\varrho(x)$ and for each j, consider the function $g_j : I_j^\circ \to \mathbb{R}$ given by

$$g_j(x) = \left(\frac{ax+b}{cx+d}\right)^{\frac{1}{q}}, \qquad x \in I_j^\circ \tag{9.16}$$

where as said before, q is the least common multiple of $\{q_i\}_{i=1}^n$, so that $g_j = g|_{I_j^\circ}$. The derivative of $g_j(x)$ is

$$g_j'(x) = \frac{1}{q}\left(\frac{ax+b}{cx+d}\right)^{\frac{1-q}{q}} \frac{ad-cb}{(cx+d)^2}, \qquad x \in I_j^\circ. \tag{9.17}$$

Thus, as g may vanish only at the endpoints of the intervals I_j, $g_j'(x) \neq 0$ for all $x \in I_j^\circ$. Hence, g_j is a change of variable.

The inverse change g_j^{-1} maps each $t \in g_j(I_j^\circ)$ to

$$x = \frac{b-dt^q}{ct^q-a}, \qquad t \in g_j(I_j^\circ) \tag{9.18}$$

and by differentiation on (9.18), the substitution formula (4.7) for this change of variable is

$$dx = q(ad-cb)\frac{t^{q-1}}{(ct^q-a)^2}\,dt. \tag{9.19}$$

Moreover, if $ad - cb > 0$, then the changes $g_j(x)$ are increasing; and if $ad - cb < 0$, then all changes $g_j(x)$ are decreasing.

9.3 The primitive of $R\left(x, \left(\frac{ax+b}{cx+d}\right)^{p_1/q_1}, \ldots, \left(\frac{ax+b}{cx+d}\right)^{p_n/q_n}\right)$

The implementation of the changes of variable g_j to find the primitive of the function $\varrho(x)$ given in (9.6) takes the following steps:

1: *Find D_ϱ and its regular decomposition.* If the regular decomposition of D_ϱ cannot be figured out at a glance, then apply the following procedure: find the fraction $Q(t)$ in explicit form as

$$Q(t) = \varrho \circ g_j^{-1}(t) \tag{9.20}$$

by plugging (9.18) into $\varrho(x)$.

Note that the formula of $Q(t)$ is the same for all indices j considered in (9.20). This is a consequence of the fact that all changes g_j are given by the same formula irrespectively of their domains (see (9.16)). Once $Q(t)$ is known, the natural domain of $\varrho(x)$ and its regular decomposition $D_\varrho = \bigcup_{j=1}^k I_j$ are easily found from (9.9).

2: *Find a local antiderivative on each piece I_j of D_ϱ.* Given any index $1 \leqslant j \leqslant k$, the change of variable g_j is implemented on I_j° by means of the substitutions (9.18) and (9.19), so that

$$\int \varrho(x)\,dx, \ x \in I_j^\circ \xleftarrow{\ x=g_j^{-1}(t)\ } q(ad-cb)\int Q(t)\frac{t^{q-1}}{(ct^q-a)^2}\,dt, \ t \in g(I_j^\circ).$$

Since $Q(t)t^{q-1}/(ct^q - a)^2$ is a rational function of one variable, its primitive can be solved in accordance with Chapter 6. Next, the reversion of the change $t = g_j(x)$ yields a local antiderivative $f_j(x)$ of $\varrho(x)$ on I_j°.

In accordance with Proposition 4.3.3, the continuous extension $\tilde{f}_j(x)$ of $f_j(x)$ to the entire interval I_j (only in the case that one or both endpoints of I_j were removed from I_j to get I_j°) is an antiderivative of $\varrho(x)$ on I_j. Note that all these functions $\tilde{f}_j(x)$ are defined on different domains but are given by the same mathematical expression.

3: *Construct the global primitive of $\varrho(x)$.* The primitive $\int \varrho(x)\, dx$ is given by

$$\tilde{f}_j(x) + C_j, \ x \in I_j, \ 1 \leqslant j \leqslant k$$

where $\{C_j\}_{j=1}^k$ is any collection of real constants. Thus, $C(x) := C_j(x)$ for $x \in I_j$ and $1 \leqslant j \leqslant k$ defines a locally constant function on D_ϱ. Note that the function $C(x)$ can be explicitly given only if the regular decomposition of D_ϱ is known.

Example 9.3.1. Find $\displaystyle\int \frac{\sqrt{x+1}+2}{(x+1)^2 - \sqrt{x+1}}\, dx.$

Solution. Let

$$\varrho(x) := \frac{\sqrt{x+1}+2}{(x+1)^2 - \sqrt{x+1}}, \ x \in D_\varrho \tag{9.21}$$

and consider the injective function $g\colon D_g \longrightarrow \mathbb{R}$ defined by the expression

$$t = \sqrt{x+1}, \ x \in D_g = [-1,\infty). \tag{9.22}$$

The inverse g^{-1} is determined by solving for x in (9.22), which yields

$$x = t^2 - 1, \ t \in g(D_g). \tag{9.23}$$

In order to determine D_ϱ, plug (9.22) into (9.21). This yields the composition

$$Q(t) := \varrho \circ g^{-1}(t) = \frac{t+2}{t^4 - t}.$$

As the real roots of $t^4 - t$ are 0 and 1, the regular domain of $\varrho(x)$ is

$$D_\varrho = [-1,\infty)\backslash g^{-1}(\{0,1\}) = [-1,\infty)\backslash\{-1,0\}$$

and the regular decomposition of D_ϱ is $I_1 \cup I_2$ where $I_1 := (-1,0)$ and $I_2 := (0,\infty)$. Thus, the changes of variable to be applied are $g_1 := g|_{I_1}$ and $g_2 := g|_{I_2}$. Both changes are increasing and their ranges are $g_1(I_1) = (0,1)$ and $g_2(I_2) = (1,\infty)$.

By differentiation on (9.23), we obtain the expression

$$dx = 2t\, dt \tag{9.24}$$

which is valid for both changes g_1 and g_2.

The changes of variable g_j can now be implemented by means of the substitutions (9.22) and (9.24), so that

$$\int \frac{\sqrt{x+1}+2}{(x+1)^2 - \sqrt{x+1}}\, dx, \quad x \in I_j \xrightarrow{t=g_j(x)} 2 \int \frac{t+2}{t^4 - t} t\, dt$$

$$= 2 \int \frac{t+2}{t^3 - 1}\, dt, \quad t \in g_j(I_j). \tag{9.25}$$

In order to solve this last primitive, the irreducible decomposition $t^3 - 1 = (t-1)(t^2 + t + 1)$ allows us to set the identity

$$\frac{t+2}{(t-1)(t^2 + t + 1)} = \frac{A}{t-1} + \frac{Bt + C}{t^2 + t + 1}. \tag{9.26}$$

In turn, (9.26) leads to the identity

$$t + 2 = A(t^2 + t + 1) + (Bt + C)(t - 1), \quad t \in \mathbb{R} \tag{9.27}$$

and the application of the usual techniques on (9.27) yields the system:

$$\begin{cases} t = 1: & 3 = 3A \Rightarrow A = 1 \\ t^0: & 2 = A - C \Rightarrow C = -1 \\ t^2: & 0 = A + B \Rightarrow B = -1. \end{cases}$$

Therefore, (9.26) becomes

$$\frac{t+2}{(t-1)(t^2 + t + 1)} = \frac{1}{t-1} - \frac{t+1}{t^2 + t + 1}. \tag{9.28}$$

By adjustment of coefficients and completing squares as explained in Section 6.3, the last fraction in (9.28) can be rewritten as

$$\frac{t+1}{t^2 + t + 1} = \frac{1}{2} \frac{2t+1}{t^2 + t + 1} + \frac{2}{3} \frac{1}{\left(\frac{2}{\sqrt{3}}t + \frac{1}{\sqrt{3}}\right)^2 + 1}. \tag{9.29}$$

Thus, from (9.28) and (9.29), it follows

$$\int \frac{t+2}{(t-1)(t^2 + t + 1)}\, dt = \log|t - 1| \tag{9.30}$$

$$- \frac{1}{2} \log\left(t^2 + t + 1\right) - \frac{1}{\sqrt{3}} \arctan\left(\frac{2}{\sqrt{3}}t + \frac{1}{\sqrt{3}}\right) + C, \ t \in g_j(I_j), \ j = 1, 2.$$

Reverse the changes of variables g_1 and g_2 by plugging (9.22) into (9.30). This yields two local antiderivatives $f_j(x)$ of $\varrho(x)$ on I_j for $j = 1, 2$ as

$$f_j(x) = 2 \log|\sqrt{x+1} - 1| - \log\left(|x+1| + \sqrt{x+1} + 1\right) \tag{9.31}$$

$$- \frac{2}{\sqrt{3}} \arctan\left(\frac{2}{\sqrt{3}}\sqrt{x+1} + \frac{1}{\sqrt{3}}\right), \quad x \in I_j.$$

Thus, the required primitive is obtained from (9.31) as

$$\int \frac{\sqrt{x+1}+2}{(x+1)^2 - \sqrt{x+1}}\, dx = F(x) + C(x), \ x \in (-1, 0) \cup (0, \infty)$$

where $F(x) := f_j(x)$ if $x \in I_j$ and $C(x)$ is locally constant on $(-1, 0) \cup (0, \infty)$. $\quad\square$

Example 9.3.2. Find $\int \sqrt{\dfrac{x+1}{x-1}}\, dx$.

Solution. The natural domain of $\varrho(x) := \sqrt{\dfrac{x+1}{x-1}}$ is formed by the real numbers x for which

$$\frac{x+1}{x-1} \geqslant 0,$$

namely, the real numbers x for which either $x+1 \geqslant 0$ and $x-1 > 0$, or $x+1 \leqslant 0$ and $x-1 < 0$. Hence,

$$D_\varrho = (-\infty, -1] \cup (1, \infty).$$

Let $I_1 := (-\infty, -1]$ and $I_2 = I_2^\circ := (1, \infty)$, and consider the decreasing changes of variable $g_j \colon I_j^\circ \longrightarrow \mathbb{R}$ defined by the expression

$$t = \sqrt{\frac{x+1}{x-1}}, \quad x \in I_j^\circ, \quad j = 1, 2 \tag{9.32}$$

so that

$$g_1(I_1^\circ) = \left(\lim_{x \to -1^-} g_1(x), \ \lim_{x \to -\infty} g_1(x) \right) = (0, 1)$$

$$g_2(I_2^\circ) = \left(\lim_{x \to \infty} g_2(x), \ \lim_{x \to 1^+} g_2(x) \right) = (1, \infty).$$

Solving for x in (9.32) gives

$$x = \frac{1+t^2}{t^2 - 1}, \quad t \in g_j(I_j^\circ), \tag{9.33}$$

which determines the inverse changes g_j^{-1}.

Differentiation on (9.33) provides the following substitution formula, valid for both changes g_j:

$$dx = -\frac{4t}{(t^2 - 1)^2}\, dt. \tag{9.34}$$

The chosen changes of variable can now be performed by means of the substitutions (9.32) and (9.34), so that

$$\int \sqrt{\frac{x+1}{x-1}}\, dx, \quad x \in I_j^\circ \xleftarrow{\;t = g_j(x)\;} -4 \int \frac{t^2}{(t^2-1)^2}\, dt, \quad t \in g_j(I_j^\circ). \tag{9.35}$$

In accordance with Section 6.6, since the irreducible factorization of the denominator of the last fraction is $(t^2 - 1)^2 = (t-1)^2(t+1)^2$, the fraction $t^2/(t^2-1)^2$ can be expressed as

$$
\begin{aligned}
\frac{t^2}{(t-1)^2(t+1)^2} &= \left(\frac{At + B}{(t-1)(t+1)} \right)' + \frac{C}{t-1} + \frac{D}{t+1} \\
&= \frac{A(t-1)(t+1) - 2t(At + B)}{(t-1)^2(t+1)^2} + \frac{C}{t-1} + \frac{D}{t+1}
\end{aligned}
\tag{9.36}
$$

and, eliminating denominators in (9.36), the continuity of the polynomials over $\mathbb{R}$ produces the identity

$$t^2 = A(t-1)(t+1) - 2t(At+B) + C(t-1)(t+1)^2 + D(t+1)(t-1)^2. \quad t \in \mathbb{R}.$$

The usual techniques give the following system:

$$\begin{cases} t = 1: & 1 = -2A - 2B \\ t = -1: & 1 = -2A + 2B \\ t^3: & 0 = C + D \\ t^0: & 0 = -A - C + D \end{cases}$$

and solving first for A and B from the first two equations, it readily yields $A = -1/2$, $B = 0$, $C = 1/4$ and $D = -1/4$.

Therefore, (9.36) becomes

$$\frac{t^2}{(t^2-1)^2} = \left(\frac{-\frac{1}{2}t}{(t-1)(t+1)} \right)' + \frac{1/4}{t-1} - \frac{1/4}{t+1} \tag{9.37}$$

and the primitive of $t^2/(t^2-1)^2$ on $g_j(I_j^\circ)$ is

$$\frac{-\frac{1}{2}t}{(t-1)(t+1)} + \frac{1}{4}\log|t-1| - \frac{1}{4}\log(t+1) + C, \quad t \in g_j(I_j^\circ). \tag{9.38}$$

Plugging (9.32) into (9.38) to reverse the changes g_j, and multiplying the result by -4 as indicated in (9.35), we obtain a pair of local antiderivatives f_1 and f_2 of $\varrho(x)$ on I_1° and I_2° respectively as

$$f_j(x) = \frac{2\sqrt{\frac{x+1}{x-1}}}{\frac{x+1}{x-1} - 1} - \log\left| \sqrt{\frac{x+1}{x-1}} - 1 \right| + \log\left(\sqrt{\frac{x+1}{x-1}} + 1 \right), \quad x \in I_j^\circ. \tag{9.39}$$

The expression for $f_1(x)$ in (9.39) also defines a continuous function on the entire interval I_1. Thus, labelling this extension to I_1 as f_1 too, it follows by virtue of Proposition 4.3.2 that $f_1(x)$ is a local antiderivative of $\varrho(x)$ on I_1 and therefore

$$\int \sqrt{\frac{x+1}{x-1}}\, dx = F(x) + C(x), \quad x \in I_1 \cup I_2$$

where $F(x) = f_j(x)$ if $x \in I_j$ and $C(x)$ is any locally constant function on $(-\infty, -1] \cup (1, \infty)$. $\square$

Example 9.3.3. Find $\displaystyle\int \frac{dx}{x(1 + 2\sqrt{x} + \sqrt[3]{x})}$.

Solution. Let $\varrho(x) := 1/x(1 + 2\sqrt{x} + \sqrt[3]{x})$. A glance reveals that the natural domain of $\varrho(x)$ is

$$D_\varrho = (0, \infty).$$

Since $6 = \text{L.C.M.}\{2, 3\}$, consider the increasing change of variable $g\colon (0, \infty) \longrightarrow \mathbb{R}$ given by

$$t = \sqrt[6]{x}, \quad x \in (0, \infty) \tag{9.40}$$

so that $g\big((0,\infty)\big) = (0,\infty)$. The inverse change g^{-1} is obtained by solving for x in (9.40), so that

$$x = t^6, \quad t \in (0,\infty) \tag{9.41}$$

and differentiation on (9.41) provides the substitution formula

$$dx = 6t^5 \, dt. \tag{9.42}$$

The change of variable $t = \sqrt[6]{x}$ can be now implemented by means of the substitutions (9.41) and (9.42) so that

$$\int \frac{dx}{x(1 + 2\sqrt{x} + \sqrt[3]{x})}, \quad x \in (0,\infty) \xleftarrow{\ x=t^6\ } \int \frac{6t^5 \, dt}{t^6(1 + 2t^3 + t^2)} \tag{9.43}$$

$$= 6 \int \frac{dt}{t(t+1)(2t^2 - t + 1)}, \quad t \in (0,\infty)$$

where the denominator of the fraction in the last primitive is given in irreducible form.

Applying the techniques of Section 6.5, the fraction $1/t(t+1)(2t^2 - t + 1)$ can be expressed as

$$\frac{1}{t(t+1)(2t^2 - t + 1)} = \frac{A}{t} + \frac{B}{t+1} + \frac{Ct + D}{2t^2 - t + 1}. \tag{9.44}$$

Multiplying the two sides of (9.44) by $t(t+1)(2t^2 - t + 1)$ and the continuity of the polynomials over $\mathbb{R}$ lead to the identity

$$1 = A(t+1)(2t^2 - t + 1) + Bt(2t^2 - t + 1) + (Ct + D)t(t+1), \quad t \in (0,\infty) \tag{9.45}$$

and in turn, (9.45) leads to the following system, which is readily solved:

$$\begin{cases} t = -1: & 1 = -4B \Rightarrow B = -1/4 \\ t = 0: & 1 = A \\ t^3: & 0 = 2A + 2B + C \Rightarrow C = -3/2 \\ t^1: & 0 = B + D \Rightarrow D = 1/4. \end{cases}$$

Therefore, (9.44) becomes

$$\frac{1}{t(t+1)(2t^2 - t + 1)} = \frac{1}{t} - \frac{1/4}{t+1} - \frac{\frac{3}{2}t - \frac{1}{4}}{2t^2 - t + 1}. \tag{9.46}$$

By adjusting coefficients and completing squares, the third fraction of the right side in (9.46) can be rewritten as

$$\frac{\frac{3}{2}t - \frac{1}{4}}{2t^2 - t + 1} = \frac{3}{8} \frac{4t - 1}{2t^2 - t + 1} + \frac{1}{7} \frac{1}{\left(\frac{4}{\sqrt{7}}t - \frac{1}{\sqrt{7}}\right)^2 + 1}.$$

Thus, the primitive of $\frac{6}{t(t+1)(2t^2-t+1)}$ on $(0,\infty)$ is

$$\int \frac{6\,dt}{t(t+1)(2t^2-t+1)} \tag{9.47}$$

$$= \int \frac{6\,dt}{t} - \int \frac{\frac{9}{2}\,dt}{t+1} - \int \frac{9}{4}\frac{4t-1}{2t^2-t+1} - \frac{0}{7}\int \frac{dt}{\left(\frac{4}{\sqrt{7}}t - \frac{1}{\sqrt{7}}\right)^2 + 1}$$

$$= 6\log t - \frac{3}{2}\log(t+1) - \frac{9}{4}\log(2t^2-t+1)$$

$$- \frac{3}{2\sqrt{7}}\arctan\left(\frac{4}{\sqrt{7}}t - \frac{1}{\sqrt{7}}\right), \quad t \in (0,\infty)$$

and reversing the change of variable by plugging (9.40) into (9.47), we obtain

$$\int \frac{1}{x(1 + 2\sqrt{x} + \sqrt[3]{x})}\,dx = \log(x) - \frac{3}{2}\log(\sqrt[6]{x}+1)$$

$$- \frac{9}{4}\log(2\sqrt[3]{x} - \sqrt[6]{x} + 1) - \frac{3}{2\sqrt{7}}\arctan\left(\frac{4}{\sqrt{7}}\sqrt[6]{x} - \frac{1}{\sqrt{7}}\right) + C, \quad x \in (0,\infty).$$

$\square$

Example 9.3.4. Find $\displaystyle\int \frac{1}{(2x+1)\left(2 - \sqrt{\frac{1-x}{1+2x}}\right)}\,dx$.

Solution. Let

$$\varrho(x) := \frac{1}{(2x+1)\left(2 - \sqrt{\frac{1-x}{1+2x}}\right)}, \quad x \in D_\varrho \tag{9.48}$$

and the injective function $g\colon D_g \longrightarrow \mathbb{R}$ be defined by the expression

$$t = \sqrt{\frac{1-x}{1+2x}}, \quad x \in D_g. \tag{9.49}$$

Solving for x in (9.49), we obtain

$$x = \frac{1-t^2}{2t^2+1}, \quad t \in g(D_g) \tag{9.50}$$

which determines the inverse $g^{-1}\colon g(D_g) \longrightarrow \mathbb{R}$.

A few simple computations show that the natural domain of g is

$$D_g = \{x\colon 1 - x \geqslant 0 \text{ and } 1 + 2x > 0\} = (-1/2, 1]$$

and

$$g((-1/2, 1]) = \left[g(1), \lim_{x \to -1/2} g(x)\right) = [0, \infty).$$

In order to determine D_ϱ, plug (9.50) into (9.48). This yields the composition

$$Q(t) := \varrho \circ g^{-1}(t) = \frac{1}{(2\frac{1-t^2}{2t^2+1} + 1)(2 - t)} = \frac{2t^2+1}{3(2-t)}. \tag{9.51}$$

Since the denominator in the right side of (9.51) only vanishes at $t = 2$, the natural domain of $\varrho(x)$ and its regular decomposition are

$$D_\varrho = (-1/2, 1] \backslash g^{-1}(\{2\}) = (-1/2, 1] \backslash \{-1/3\} = \left(-\frac{1}{2}, -\frac{1}{3}\right) \cup \left(-\frac{1}{3}, 1\right].$$

Denote $I_1 := (-1/2, -1/3)$ and $I_2 := (-1/3, 1]$. Since the regular decomposition of D_ϱ consists of the intervals I_1 and I_2, the changes of variable to be used are

$$g_j := g|_{I_j^\circ}, \quad j = 1, 2, \tag{9.52}$$

and their ranges are $g_1(I_1^\circ) = (2, \infty)$ and $g_2(I_2^\circ) = (0, 2)$.

Differentiation on (9.50) yields the substitution formula

$$dx = -\frac{6t}{(2t^2 + 1)^2}\, dt \tag{9.53}$$

which is valid for both changes g_1 and g_2.

The changes of variable g_1 and g_2 can now be performed by means of the substitutions (9.49)–(9.50) and (9.53) so that

$$\int \frac{dx}{(2x + 1)\left(2 - \sqrt{\frac{1-x}{1+2x}}\right)}, \quad x \in I_j^\circ \xleftarrow{t = g_j(x)} \int \frac{2t\, dt}{(t - 2)(2t^2 + 1)}, \quad t \in g_j(I_j^\circ). \tag{9.54}$$

In order to find the primitive of the right side of (9.54), set

$$\frac{2t}{(t - 2)(2t^2 + 1)} = \frac{A}{t - 2} + \frac{Bt + C}{2t^2 + 1}. \tag{9.55}$$

Thus

$$2t = A(2t^2 + 1) + (Bt + C)(t - 2), \quad t \in \mathbb{R}$$

which yields the following system:

$$\begin{cases} t = 2: & 4 = 9A \Rightarrow A = \frac{4}{9} \\ t = 0: & 0 = A - 2C = \frac{4}{9} - 2C \Rightarrow C = \frac{2}{9} \\ t^2: & 0 = 2A + B = \frac{8}{9} + B \Rightarrow B = -\frac{8}{9}. \end{cases}$$

Plugging the values of A, B and C into (9.55) and following Section 6.5, some adjustment of coefficients and a completion of squares gives

$$\frac{2t}{(t - 2)(2t^2 + 1)} = \frac{4/9}{t - 2} + \frac{-\frac{8}{9}t + \frac{2}{9}}{2t^2 + 1} = \frac{4/9}{t - 2} - \frac{2}{9}\frac{4t}{2t^2 + 1} + \frac{2/9}{(\sqrt{2}t)^2 + 1}.$$

Hence,

$$\int \frac{2t}{(t - 2)(2t^2 + 1)}\, dt \tag{9.56}$$

$$= \frac{4}{9}\int \frac{dt}{t - 2} - \frac{2}{9}\int \frac{4t\, dt}{2t^2 + 1} + \frac{2}{9}\int \frac{dt}{(\sqrt{2}t)^2 + 1}$$

$$= \frac{4}{9}\log|t - 2| - \frac{2}{9}\log(2t^2 + 1) + \frac{\sqrt{2}}{9}\arctan\sqrt{2}t + C_j, \quad t \in g_j(I_j^\circ), \; j = 1, 2.$$

Plugging (9.49) into (9.56) to reverse the changes $t = g_1(x)$ and $t = g_2(x)$, we obtain local antiderivatives $f_j(x)$ of $\varrho(x)$ on I_j° for $j = 1, 2$ as

$$f_j(x) = \frac{4}{9} \log \left| \sqrt{\frac{1-x}{1+2x}} - 2 \right| \tag{9.57}$$

$$- \frac{2}{9} \log \left(2\frac{1-x}{1+2x} + 1 \right) + \frac{\sqrt{2}}{9} \arctan \sqrt{2\frac{1-x}{1+2x}}, x \in I_j^\circ, \; j = 1, 2.$$

The expression for $f_1(x)$ in (9.57) also defines a continuous function on the entire interval I_1. Thus, labelling this extension as $f_1(x)$ too, Proposition 4.3.2 proves that $f_1(x)$ is a local antiderivative of $\varrho(x)$ on I_1.

Hence, the required primitive is

$$\int \frac{dx}{(2x+1)\left(2 - \sqrt{\frac{1-x}{1+2x}}\right)} = F(x) + C(x) \; x \in (-1/2, -1/3) \cup (-1/3, 1]$$

where $F(x) := f_j(x)$ if $x \in I_j$ $(j = 1, 2)$ and $C(x)$ is locally constant on $(-1/2, -1/3) \cup (-1/3, 1]$.   □

9.4 Exercises

Find the domain and an antiderivative for each of the following functions.

Question 9.4.1

$$f(x) = \frac{(2 + \sqrt{x})^2}{1 + \sqrt{x}}$$

Question 9.4.2

$$f(x) = \frac{1 - x}{1 + \sqrt{x}}$$

Question 9.4.3

$$f(x) = \frac{1}{x+1}\sqrt{\frac{3+x}{x-1}}$$

Question 9.4.4

$$f(x) = \frac{6}{1 + 3(x+4)^{2/3}}$$

Question 9.4.5

$$f(x) = \frac{(1 + \sqrt{x})^3}{x^{1/3}}$$

Question 9.4.6

$$f(x) = \frac{1}{x\sqrt{2x+1}}$$

Question 9.4.7

$$f(x) = \frac{x - \sqrt{x+1}}{x + \sqrt{x+1}}$$

Question 9.4.8

$$f(x) = \sqrt{\frac{x-3}{x+5}}$$

Question 9.4.9

$$f(x) = \frac{1}{\sqrt{3x-1}}$$

Answers to the Questions

Answer 9.4.1

$D_f = [0, \infty)$,

$$-2\log\left(1 + \sqrt{x}\right) + \frac{2}{3}\sqrt{x}^3 + 3x + 2\sqrt{x}, \ x \in [0, \infty)$$

Answer 9.4.2

$D_f = [0, \infty)$,

$$-\frac{2}{3}\sqrt{x}^3 + x, \ x \in [0, \infty)$$

Answer 9.4.3

$D_f = (-\infty, -3] \cup (1, \infty)$,

$$\frac{1}{2}\log\left|\frac{1 + \sqrt{(3+x)/(x-1)}}{1 - \sqrt{(3+x)/(x-1)}}\right| - \arctan\sqrt{\frac{3+x}{x-1}}$$

Answer 9.4.4

$D_f = [-4, \infty)$,

$$6\sqrt[3]{x+4} - 2\sqrt{3}\arctan\left(\sqrt{3}\sqrt[3]{x+4}\right)$$

Answer 9.4.5

$D_f = (0, \infty)$,

$$\frac{6}{13}x^{13/6} + \frac{9}{5}x^{5/3} + \frac{18}{7}x^{7/6} + \frac{3}{2}x^{2/3}$$

Answer 9.4.6

$D_f = (-1/2, 0) \cup (0, \infty)$,

$$\log\left|\frac{\sqrt{2x+1} - 1}{\sqrt{2x+1} + 1}\right|$$

Answer 9.4.7

$D_f = [-1, (1 - \sqrt{5})/2) \cup ((1 - \sqrt{5})/2, \infty)$

$$-\frac{2+2\sqrt{5}}{\sqrt{5}}\log\left|\sqrt{x+1} + \frac{1-\sqrt{5}}{2}\right| + \frac{2-2\sqrt{5}}{\sqrt{5}}\log\left|\sqrt{x+1} + \frac{1+\sqrt{5}}{2}\right| + x + 1 - 4\sqrt{x+1}$$

Answer 9.4.8

$D_f = (-\infty, -5) \cup [3, \infty),$

$$\log\left|\sqrt{\frac{x-3}{x+5}} - 1\right| - \frac{1}{\sqrt{\frac{x-3}{x+5}} - 1} - \log\left|\sqrt{\frac{x-3}{x+5}} + 1\right| - \frac{1}{\sqrt{\frac{x-3}{x+5}} + 1}$$

Answer 9.4.9

$D_f = (1/3, \infty),$

$$\frac{2}{3}\sqrt{3x-1}$$

Chapter 10

Binomial Differentials

A *binomial differential* is a function that takes one of the following forms:
$$f(x) := (a + bx^s)^{\frac{p}{q}}, \quad x \in D_f \tag{10.1}$$
or
$$f(x) := x^r (a + bx^s)^{\frac{p}{q}}, \quad x \in D_f \tag{10.2}$$
where $\{a, b\} \subset \mathbb{R}$, $\{r, s\} \subset \mathbb{Q}\backslash\{0\}$, $p \in \mathbb{Z}$ and $q \in \mathbb{N}$.

In order to avoid trivial cases, in this chapter we will consider only the binomial differentials whose parameters a, b, r, s, p and q satisfy all the following conditions:

- $\{a, b\} \subset \mathbb{R}\backslash\{0\}$ (10.3)
- $p \in \mathbb{Z}\backslash\{0\}$, $1 < q \in \mathbb{N}$ and G.C.D.$(p, q) = 1$ (10.4)
- if s is not an odd integer, then at least

        one of the numbers a, b is greater than zero.   (10.5)

Condition (10.5) prevents the case where D_f is either empty or a singleton. Each of the cases $a = 0$, $b = 0$ or $p = 0$ are excluded because they leave functions of the form $f(x) = x^t$ with t a rational number, whose primitive is immediate. The case $q = 1$ is also eliminated because the primitive of the corresponding binomial differentials can be found as described in Chapter 9.

10.1 The domain of a binomial differential

Throughout this section we denote $\mathbb{N}^- := \mathbb{Z} \cap (-\infty, 0)$, $\mathbb{Q}^+ := \mathbb{Q} \cap (0, \infty)$ and $\mathbb{Q}^- := \mathbb{Q} \cap (-\infty, 0)$.

Following the discussion in Section 3.4 and Appendix A.1, the natural domain and range of $y(x) := x^s$, where $s \in \mathbb{Q}\backslash\{0\}$, is

(p1) $s \in \mathbb{N}$ and odd $\Rightarrow D_y = (-\infty, \infty)$ and $y(D_y) = (-\infty, \infty)$;
(p2) $s \in \mathbb{N}$ and even $\Rightarrow D_y = (-\infty, \infty)$ and $y(D_y) = [0, \infty)$;
(p3) $s \in \mathbb{N}^-$ and odd $\Rightarrow D_y = (-\infty, 0) \cup (0, \infty)$ and $y(D_y) = (-\infty, 0) \cup (0, \infty)$;
(p4) $s \in \mathbb{N}^-$ and even $\Rightarrow D_y = (-\infty, 0) \cup (0, \infty)$ and $y(D_y) = (0, \infty)$;
(p5) $s \in \mathbb{Q}^+\backslash\mathbb{N} \Rightarrow D_y = [0, \infty)$ and $y(D_y) = [0, \infty)$;
(p6) $s \in \mathbb{Q}^-\backslash\mathbb{N}^- \Rightarrow D_y = (0, \infty)$ and $y(D_y) = (0, \infty)$.

Since $q > 1$ and G.C.D.$(p, q) = 1$, the natural domain of a binomial differential $f(x) = (a + bx^s)^{p/q}$, with $p \geqslant 1$, is the set $D_f = \{x: a + bx^s \geqslant 0\}$. Statements (p1) to (p6) and a few calculations readily yield Table 10.1.1, where D_f is classified depending upon the parameters s, a, b and p.

Table 10.1.1. Natural domain of $f(x) = (a + bx^s)^{p/q}$, where $p \in \mathbb{N}$ and $\tau = |a/b|^{1/|s|} > 0$:

s	$a<0, b>0$	$a>0, b<0$	$a>0, b>0$	$a<0, b<0$
$s \in \mathbb{N}$ even	$(-\infty,-\tau] \cup [\tau,\infty)$	$[-\tau,\tau]$	$(-\infty,\infty)$	$-$
$s \in \mathbb{N}$ odd	$[\tau,\infty)$	$(-\infty,\tau]$	$[-\tau,\infty)$	$(-\infty,-\tau]$
$s \in \mathbb{N}^-$ even	$[-\tau^{-1},0) \cup (0,\tau^{-1}]$	$(-\infty,-\tau^{-1}] \cup [\tau^{-1},\infty)$	$(-\infty,0) \cup (0,\infty)$	$-$
$s \in \mathbb{N}^-$ odd	$(0,\tau^{-1}]$	$(-\infty,0) \cup [\tau^{-1},\infty)$	$(-\infty,-\tau^{-1}] \cup (0,\infty)$	$[-\tau^{-1},0)$
$s \in \mathbb{Q}^+ \backslash \mathbb{N}$	$[\tau,\infty)$	$[0,\tau]$	$[0,\infty)$	$-$
$s \in \mathbb{Q}^- \backslash \mathbb{N}^-$	$(0,\tau^{-1}]$	$[\tau^{-1},\infty)$	$(0,\infty)$	$-$

If $p \leqslant -1$, the natural domain of $f(x) = (a + bx^s)^{p/q}$ is $D_f = \{x: a + bx^s > 0\}$. It can be classified in terms of the parameters s, a, b and p as shown in Table 10.1.2.

Table 10.1.2. Natural domain of $f(x) = (a + bx^s)^{p/q}$, where $p \in \mathbb{N}^-$ and $\tau = |a/b|^{1/|s|} > 0$:

s	$a<0, b>0$	$a>0, b<0$	$a>0, b>0$	$a<0, b<0$
$s \in \mathbb{N}$ even	$(-\infty,-\tau) \cup (\tau,\infty)$	$(-\tau,\tau)$	$(-\infty,\infty)$	$-$
$s \in \mathbb{N}$ odd	(τ,∞)	$(-\infty,\tau)$	$(-\tau,\infty)$	$(-\infty,-\tau)$
$s \in \mathbb{N}^-$ even	$(-\tau^{-1},0) \cup (0,\tau^{-1})$	$(-\infty,\tau^{-1}) \cup (\tau^{-1},\infty)$	$(-\infty,0) \cup (0,\infty)$	$-$
$s \in \mathbb{N}^-$ odd	$(0,\tau^{-1})$	$(-\infty,0) \cup (\tau^{-1},\infty)$	$(-\infty,-\tau^{-1}) \cup (0,\infty)$	$(-\tau^{-1},0)$
$s \in \mathbb{Q}^+ \backslash \mathbb{N}$	(τ,∞)	$[0,\tau)$	$[0,\infty)$	$-$
$s \in \mathbb{Q}^- \backslash \mathbb{N}^-$	$(0,\tau^{-1})$	(τ^{-1},∞)	$(0,\infty)$	$-$

It can be observed from Table 10.1.1 that for any binomial differential $f(x)$ of type (10.1), D_f is a regular domain consisting of one or two disjoint intervals. In particular, if D_f is not an interval, then $0 \notin D_f$ and, in that case, the regular decomposition of D_f consists of the pair of intervals $D_f \cap (-\infty, 0)$ and $D_f \cap (0, \infty)$.

A binomial differential of type (10.2) is of the form $f_0(x) = y(x)f(x)$ where $y(x) = x^r$ and $f(x) = (a + bx^s)^{p/q}$. Hence, it becomes clear that

$$D_{f_0} = D_y \cap D_f.$$

Thus, D_{f_0} is of one of the forms D_f, $D_f \cap [0, \infty)$, $D_f \cap (0, \infty)$ or $D_f \backslash \{0\}$, and as D_f is regular, then D_{f_0} is regular too. In particular, a glance at Tables 10.1.1 and 10.1.2 reveals that D_{f_0} consists of one or two disjoint intervals.

10.2 Changes of variable for binomial differentials

The changes of variable used to find the primitive of a binomial differential are restrictions of binomial differentials.

Through this section, we will consider a binomial differential

$$g(x) = (a + bx^s)^{1/q}, \ x \in D_g$$

and we will investigate which changes of variable may be obtained from $g(x)$.

Following Table 10.1.1, the regular decomposition of D_g is either one single interval $D_g = I_0$, in which case D_g° denotes the open interval I_0°, or it is the union of two disjoint intervals, $D_g = I_1 \cup I_2$, in which case $I_1 = D_g \cap (-\infty, 0)$, $I_2 = D_g \cap (0, \infty)$ and D_g° denotes the set $I_1^\circ \cup I_2^\circ$. This classification of cases enables the following result.

Proposition 10.2.1. *Let $g(x) = (a + bx^s)^{1/q}$ be a binomial differential, so that one of the following cases is true:*

(i) $D_g = I_0$ *is an interval and $0 \notin I_0^\circ$;*
(ii) $D_g = I_0$ *is an interval, $0 \in I_0^\circ$ and $s = 1$;*
(iii) $D_g = I_0$ *is an interval, $0 \in I_0^\circ$ and $s \neq 1$; in this case, let $I_1 := D_g \cap (-\infty, 0)$ and $I_2 := D_g \cap (0, \infty)$;*
(iv) $D_g = I_1 \cup I_2$, *where I_1 and I_2 are a pair of intervals such that $I_1 = D_g \cap (-\infty, 0)$, $I_2 = D_g \cap (0, \infty)$.*

Then, if (i) or (ii) holds, the best change of variable which can be obtained from $g(x)$ is

$$t = g_0(x) := g|_{I_0^\circ}(x), \ x \in I_0^\circ$$

and if (iii) or (iv) holds, the best changes of variable are

$$t = g_1(x) := g|_{I_1^\circ}(x), \ x \in I_1^\circ$$
$$t = g_2(x) := g|_{I_2^\circ}(x), \ x \in I_2^\circ.$$

Proof. Let $\tau := |a/b|^{1/|s|} > 0$. In investigating the differentiability of $g(x)$, the usual rules of differentiation show

$$g'(x) = \frac{bs}{q}(a + bx^s)^{\frac{1}{q}-1} x^{s-1} \neq 0, \ x \in D_g^\circ \setminus \{0\}. \tag{10.6}$$

The elements of the domain of g not considered in (10.6) belong to $N := D_g \cap \{0, \tau, -\tau, \tau^{-1}, -\tau^{-1}\}$.

If $0 \in D_g$, l'Hôpital's rule and (10.6) yield

$$\lim_{x \to 0} \frac{g(x) - g(0)}{x} = \lim_{x \to 0} g'(x) = \lim_{x \to 0} \frac{bs}{q}(a + bx^s)^{\frac{1}{q}-1} x^{s-1}.$$

Thus,

$$g'(0) = \begin{cases} 0 & \text{if } s > 1 \\ \dfrac{b}{q}a^{\frac{1}{q}-1} \neq 0 & \text{if } s = 1 \\ \sharp & \text{if } s < 1. \end{cases}$$

Any other point $x_0 \in (N\backslash\{0\})\backslash D_g^\circ$ is an endpoint of some of the intervals that make the regular decomposition of D_g (see Table 10.1.1). In all these cases, the fact that $a + bx^s \to 0$ when $x \to x_0$ and (10.6) show that g is not differentiable at x_0. Therefore, if $g|_J$ is a change of variable, then $J \subset I_0^\circ$ if hypotheses (i) or (ii) hold, and if (iii) or (iv) hold, then either $J \subset I_1^\circ$ or $J \subset I_2^\circ$. This means that if (i) or (ii) holds, then I_0° is the largest interval on which g is a change of variable, so roughly speaking, $g|_{I_0^\circ}$ is the best possible change of variable. For the same reason, $g|_{I_1^\circ}$ and $g|_{I_2^\circ}$ are the best possible changes of variable whenever (iii) or (iv) holds. □

The inverse g_j^{-1} corresponding to each of the changes listed in Proposition 10.2.1 maps every $t \in g_j(I_j^\circ)$ to

$$x = \delta_j \left(\varepsilon_j \frac{t^q - a}{b} \right)^{1/s}, \quad t \in g(I_j^\circ) \tag{10.7}$$

where

$$s \in \{1\} \cup (\mathbb{Q}\backslash\mathbb{Z}) \implies \delta_0 = 1, \ \varepsilon_0 = 1$$

and if $s \in \mathbb{Z}\backslash\{1\}$,

$$\begin{cases} \sup I_j \leqslant 0 \Rightarrow \delta_j = -1 \\ 0 \leqslant \inf I_j \Rightarrow \delta_j = 1 \end{cases}$$

$$\begin{cases} s \text{ even} \Rightarrow \varepsilon_j = 1 \\ s \text{ odd} \Rightarrow \varepsilon_j = \delta_j. \end{cases}$$

10.3   Classification of binomial differentials

Henceforth, the value $r = 0$ is assigned to any binomial differential of type (10.1). The binomial differentials given in (10.2) and (10.1) are classified into three types in terms of the parameters r, s, p and q as follows:

(I) $\dfrac{r+1}{s} \in \mathbb{Z}$;

(II) $\dfrac{r+1}{s} \notin \mathbb{Z}$ but $\dfrac{r+1}{s} + \dfrac{p}{q} \in \mathbb{Z}$;

(III) neither (I) nor (II) holds.

Every binomial differential is continuous on its natural domain, which is regular. Hence, it possesses a primitive. The primitive of any binomial differential of type I or II can be calculated by means of elementary methods, as explained in the following two sections. However, this is not possible for the binomial differentials of type III. A proof of this last fact is beyond the scope of this book.

10.4   The primitive of a binomial differential of type I

Let $f(x) = x^r(a + bx^s)^{p/q}$ (alternatively, $f(x) = (a + bx^s)^{p/q}$) be a binomial differential of type I.

In order to find $\int f(x)\, dx$, take these steps:

1: Determine the regular decomposition of D_f. In accordance with Section 10.1, this decomposition is either a single interval $D_f = J_0$ or the union $D_f = J_1 \cup J_2$ of two intervals such that $J_1 \subset (-\infty, 0)$ and $J_2 \subset (0, \infty)$.

2: Obtain the changes of variable needed from the function

$$g(x) = (a + bx^s)^{1/q}.$$

Note that $D_f \subset D_g$. Clearly, one and only one of the following cases occurs:

(G1) $D_f = J_0$ and $0 \notin J_0^\circ$.
(G2) $D_f = J_0$, $0 \in J_0$ and $s = 1$.
(G3) $D_f = J_0$, $0 \in J_0$ and $s \neq 1$. In this case, let $J_1 := J_0 \cap (-\infty, 0]$ and $J_2 := J_0 \cap [0, \infty)$.
(G4) $D_f = J_1 \cup J_2$ where $J_1 := J_0 \cap (-\infty, 0)$ and $J_2 := J_0 \cap (0, \infty)$.

Proposition 10.2.1 allows us to carry out the following actions for each of the cases (G1)–(G4).

(G1) and (G2): the implementation of the change $t = g_0(x) := g|_{J_0^\circ}(x)$ transforms $\int f(x)\, dx$ into the primitive of a fraction of polynomials, which can be found as described in Chapter 6. Once an antiderivative $F_0(x)$ of f on J_0° is obtained, then, by virtue of Proposition 4.3.2, its continuous extension onto $I_0 \backslash I_0^\circ$ is an antiderivative of $f(x)$ on the entire interval I_0.

(G3): implement the changes $t = g_j(x) := g|_{J_j^\circ}(x)$ on the respective intervals I_j°, $j = 1, 2$. As in case (G1), this action provides antiderivatives $F_1(x)$ and $F_2(x)$ on J_1° and J_2°. Proposition 4.3.2 ensures the existence of $\lim_{x \to 0+} F_1(x) = \lambda_1 \in \mathbb{R}$ and $\lim_{x \to 0-} F_2(x) = \lambda_2 \in \mathbb{R}$, and Proposition 4.3.3 proves that

$$F(x) := \begin{cases} F_1(x), & x \in I_1 \\ \lambda_1, & x = 0 \\ F_2(x) + \lambda_1 - \lambda_2, & x \in I_2 \end{cases}$$

is an antiderivative of $f(x)$ on I_0°. The continuous extension of $F(x)$ onto $I_0 \backslash I_0^\circ$ provides an antiderivative of $F(x)$ on the complete interval I_0.

(G4): as in case (G3), implement the changes $t = g_j(x) := g|_{J_j^\circ}(x)$ on the respective intervals I_j°, $j = 1, 2$. This action provides antiderivatives $F_1(x)$ and $F_2(x)$ on J_1° and J_2°. Thus, $F(x) := F_i(x)$ for $x \in J_i$, $i = 1, 2$, is an antiderivative of f whose continuous extension to the set of endpoints $(J_1 \cup J_2) \backslash (J_1^\circ \cup J_2^\circ)$ is an antiderivative of f on D_f.

In order to be more specific, formula (10.7) yields the following substitution formulas for all changes g_j $(j = 0, 1, 2)$ considered in cases (G1)–(G4):

$$x = \delta_j \left(\varepsilon_j \frac{t^q - a}{b} \right)^{1/s}, \quad t \in g(J_j^\circ)$$

$$dx = \delta_j \varepsilon_j \frac{q}{bs} \left(\varepsilon_j \frac{t^q - a}{b} \right)^{\frac{1}{s} - 1} t^{q-1} \, dt \tag{10.8}$$

where

$$s \in \{1\} \cup (\mathbb{Q} \backslash \mathbb{Z}) \ \Rightarrow \ \delta_0 = 1, \ \varepsilon_0 = 1$$

and if $s \in \mathbb{Z} \backslash \{1\}$,

$$\begin{cases} \sup I_j \leqslant 0 \Rightarrow \delta_j = -1 \\ 0 \leqslant \inf I_j \Rightarrow \delta_j = 1 \end{cases}$$

$$\begin{cases} s \text{ even} \Rightarrow \varepsilon_j = 1 \\ s \text{ odd} \Rightarrow \varepsilon_j = \delta_j. \end{cases}$$

Thus, the implementation of any of the changes $t = g_j(x)$ turns the primitive of $x^r (a + bx^s)^{p/q}$ on J_j into the following primitive of a fraction of polynomials:

$$\int \varepsilon_j \delta_j^{r+1} \frac{q}{bs} \left(\varepsilon_j \frac{t^q - a}{b} \right)^{\frac{r+1}{s} - 1} t^{p+q-1} \, dt, \ t \in g(I_j).$$

The following examples will make the above procedure clear.

Example 10.4.1. Find $\int f(x) \, dx$ for

(a) $f(x) = \frac{1}{\sqrt{x}} \sqrt[3]{1 + \sqrt[4]{x}}$,
(b) $f(x) = \sqrt[3]{1 + \sqrt[4]{x}}$.

Solution. Both functions in (a) and (b) are of type I. Hence, the change of variable needed to find the antiderivative will be obtained from the increasing function $g: [0, \infty) \longrightarrow \mathbb{R}$ given by

$$t = (1 + \sqrt[4]{x})^{\frac{1}{3}}, \ x \in [0, \infty). \tag{10.9}$$

As explained before, the largest interval on which g is a change of variable is $I_1 := (0, \infty)$. Let $g_1 := g|_{I_1}$. By rearranging (10.9), and bearing in mind that g maps $(0, \infty)$ onto $(1, \infty)$, the inverse of g_1 is given by

$$x = (t^3 - 1)^4, \ t \in (1, \infty) \tag{10.10}$$

and by differentiation on (10.10),

$$dx = 12(t^3 - 1)^3 t^2. \tag{10.11}$$

Let us now integrate (a) and (b).

(a) The natural domain of $f(x) := x^{-1/2}(1 + x^{1/4})^{1/3}$ is $D_f = (0, \infty)$. Since D_f coincides with the domain of g_1, the primitive of $f(x)$ is found directly by implementation of the change g_1. Thus, by means of the substitutions (10.9) and (10.11),

$$\int x^{-\frac{1}{2}}(1 + x^{\frac{1}{4}})^{\frac{1}{3}}, \; x \in (0, \infty) \xrightarrow{t=g_1(x)}, \int (t^3 - 1)^{1}(-\frac{1}{2}), t, 12(t^n - 1)^n t^2 \, dt$$

$$= 12 \int t^3(t^3 - 1) \, dt = \frac{12}{7} t^7 - 3t^4 + C, \; t \in (1, \infty).$$

(10.12)

By plugging (10.9) into (10.12), we obtain

$$\int x^{-\frac{1}{2}}(1 + x^{\frac{1}{4}})^{\frac{1}{3}} = \frac{12}{7}(1 + \sqrt[4]{x})^{\frac{7}{3}} - 3(1 + \sqrt[4]{x})^{\frac{4}{3}} + C, \; x \in (0, \infty)$$

where C is any constant.

(b) The natural domain of $f(x) = (1 + x^{\frac{1}{4}})^{\frac{1}{3}}$ is $D_f = [0, \infty)$. In accordance with the established procedure, the change of variable to be used is g_1, as in part (a), but this time the domain of g_1 misses covering D_f by just the point $x = 0$. Thus, it is necessary to proceed as explained in the domain adjustment method in Section 4.4. Diagram 10.13 represents the steps to be taken in accordance with the guidelines given in Section 4.4.

(10.13)

Removing the point $x = 0$ from D_f we obtain I_1. In order to calculate the primitive of $f|_{I_1}$, implement the change of variable $t = g_1(x)$ by means of the substitutions (10.9) and (10.11), so that

$$\int (1 + x^{\frac{1}{4}})^{\frac{1}{3}}, \; x \in (0, \infty) \xrightarrow{t=g_1(x)} \int t \cdot 12(t^3 - 1)^3 t^2 \, dt$$

$$= 12 \int t^{12} - 3t^9 + 3t^6 - t^3 \, dt$$

$$= \frac{12}{13}t^{13} - \frac{18}{5}t^{10} + \frac{36}{7}t^7 - 3t^4 + C, \; t \in (1, \infty).$$

(10.14)

Thus, by plugging (10.9) into (10.14), we obtain

$$\int (1 + x^{\frac{1}{4}})^{\frac{1}{3}} \, dx =$$

$$\frac{12}{13}(1 + \sqrt[4]{x})^{\frac{13}{3}} - \frac{18}{5}(1 + \sqrt[4]{x})^{\frac{10}{3}} + \frac{36}{7}(1 + \sqrt[4]{x})^{\frac{7}{3}} - 3(1 + \sqrt[4]{x})^{\frac{4}{3}} + C,$$

(10.15)

for $x \in (0, \infty)$, where C is any real constant. But the expression on the right side of (10.15) is well defined and continuous at $x = 0$. Hence, by virtue of Proposition 4.3.1, formula (10.15) for $x \in [0, \infty) = D_f$ is a complete solution to the primitive of $f(x)$ on its entire natural domain.   □

Example 10.4.2. Find $\int f(x) \, dx$ for

(a) $f(x) = \dfrac{1}{x\sqrt[3]{1 - x^2}}$,

(b) $f(x) = \dfrac{x^3}{\sqrt[3]{1 - x^2}}$.

Solution. Both functions in (a) and (b) are of type I. Hence, the change of variable needed for both of them will be obtained from the function $g \colon [-1, 1] \longrightarrow \mathbb{R}$ given by

$$t = (1 - x^2)^{\frac{1}{3}}, \ x \in [-1, 1]. \tag{10.16}$$

As explained at the beginning of this section, the largest intervals on which g is a change of variable are $I_1 := (0, 1)$ and $I_2 := (-1, 0)$. Let $g_1 := g|_{I_1}$ and $g_2 := g|_{I_2}$ and note that they are respectively decreasing and increasing.

Clearly,

$$g(I_1) = \left(\lim_{x \to 1^-} g(x), \ \lim_{x \to 0^+} g(x) \right) = (0, 1)$$

$$g(I_2) = \left(\lim_{x \to -1^+} g(x), \ \lim_{x \to 0^-} g(x) \right) = (0, 1).$$

By rearranging (10.16), the inverses of g_1 and g_2 are respectively obtained as

$$x = g_1^{-1}(t) = (1 - t^3)^{\frac{1}{2}}, \ t \in (0, 1) \tag{10.17}$$

$$x = g_2^{-1}(t) = -(1 - t^3)^{\frac{1}{2}}, \ t \in (0, 1) \tag{10.18}$$

and by differentiation on (10.17) and (10.18),

$$dx = \frac{3}{2}(1 - t^3)^{-\frac{1}{2}} t^2 \, dt, \ x \in I_1 \tag{10.19}$$

$$dx = -\frac{3}{2}(1 - t^3)^{-\frac{1}{2}} t^2 \, dt, \ x \in I_2. \tag{10.20}$$

Let us now integrate (a) and (b).

(a) The natural domain of $f(x) = x^{-1}(1 - x^2)^{-\frac{1}{3}}$ is $D_f = J_2 \cup J_1$ where $J_2 = (-1, 0)$ and $J_1 = (0, 1)$.

In accordance with the established procedure, the changes of variable to be used are g_1 and g_2, and as $J_1 = I_1$ and $J_2 = I_2$, the primitives of f on J_1 and J_2 can be found by direct application of the changes g_1 and g_2 respectively.

In order to find $\int f|_{I_1} \, dx$, the change g_1 is implemented by means of the substitutions (10.17) and (10.19), so that

$$\int x^{-1}(1 - x^2)^{-\frac{1}{3}} \, dx, \ x \in I_1 \xleftarrow{\ t = g_1(x)\ } \int (1 - t^3)^{-\frac{1}{2}} t^{-1} \frac{3}{2}(1 - t^3)^{-\frac{1}{2}} t^2 \, dt$$

$$= -\frac{3}{2} \int \frac{t}{t^3 - 1} \, dt, \ t \in (0, 1). \tag{10.21}$$

To find this last primitive, since the irreducible factorization of $t^3 - 1$ is $(t-1)(t^2 + t + 1)$, the usual techniques of decomposition of fractions given in Chapter 6 yield

$$\frac{t}{t^3 - 1} = \frac{\frac{1}{3}}{t - 1} + \frac{-\frac{1}{3}t + \frac{1}{3}}{t^2 + t + 1}, \tag{10.22}$$

and by adjusting coefficients and completing squares in the last fraction,

$$\frac{t - 1}{t^2 + t + 1} = \frac{1}{2}\frac{2t + 1}{t^2 + t + 1} - 2\frac{1}{\left(\frac{2}{\sqrt{3}}t + \frac{1}{\sqrt{3}}\right)^2 + 1}. \tag{10.23}$$

The combination of (10.22) and (10.23) gives

$$\int \frac{t}{t^3 - 1}\, dt = F(t) + C, \ t \in (0, 1) \tag{10.24}$$

where

$$F(t) := \frac{1}{3}\log(1 - t) - \frac{1}{6}\log(t^2 + t + 1) + \frac{1}{\sqrt{3}}\arctan\left(\frac{2}{\sqrt{3}}t + \frac{1}{\sqrt{3}}\right), \ t \in (0, 1).$$

In order to reverse the change g_1, plug (10.16) into (10.24), then multiply the result by $-3/2$, so that (10.21) yields

$$\int x^{-1}(1 - x^2)^{-\frac{1}{3}}\, dx = -\frac{3}{2}F(\sqrt[3]{1 - x^2}) + C, \ x \in (0, 1) \tag{10.25}$$

where C is any constant.

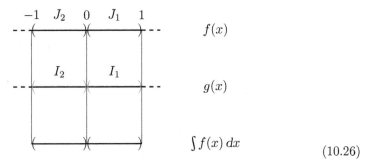

In order to find the primitive of $f(x)$ on I_2, implement the change g_2 by means of the substitutions (10.18) and (10.20), so that

$$\int x^{-1}(1 - x^2)^{-\frac{1}{3}}\, dx, \ x \in I_2 \xleftarrow{t = g_2(x)}$$

$$\int -(1 - t^3)^{-\frac{1}{2}}t^{-1}\frac{(-3)}{2}(1 - t^3)^{-\frac{1}{2}}t^2\, dt = -\frac{3}{2}\int \frac{t}{t^3 - 1}\, dt, \ t \in (0, 1). \tag{10.27}$$

This last primitive is found by means of (10.24), and substituting $t = g_2(x)$, we obtain

$$\int x^{-1}(1 - x^2)^{-\frac{1}{3}}\, dx = -\frac{3}{2}F(\sqrt[3]{1 - x^2}) + C, \ x \in (-1, 0) \tag{10.28}$$

where C is any constant.

Thus, the full solution to $\int f(x)\,dx$ on D_f is obtained from (10.25) and (10.28) as

$$-\frac{3}{2}F(\sqrt[3]{1-x^2}) + C =$$

$$-\frac{1}{2}\log(1 - \sqrt[3]{1-x^2}) + \frac{1}{4}\log(\sqrt[3]{1-x^2}^2 + \sqrt[3]{1-x^2} + 1)$$

$$-\frac{\sqrt{3}}{2}\arctan\left(\frac{2}{\sqrt{3}}\sqrt[3]{1-x^2} + \frac{1}{\sqrt{3}}\right) + C(x), \quad x \in (-1,0) \cup (0,1)$$

where $C(x)$ is any locally constant function on $(-1,0) \cup (0,1)$.

(b) The natural domain of $f(x) = x^3(1-x^2)^{-\frac{1}{3}}$ is $D_f = (-1,1)$.

According to the domain adjustment method, the first action to be taken is the calculation of the primitives of $f(x)$ on $L_1 := (-1,1) \cap I_1$ and $L_2 := (-1,1) \cap I_2$. It is clear that $L_i = I_i$ for $i = 1, 2$. Note that the point $x = 0$ is not covered by $I_1 \cup I_2$; more technically, $D_f \backslash (I_1 \cup I_2) = \{0\}$.

$$\tag{10.29}$$

First, find $\int f|_{I_1}(x)\,dx$. The change g_1 is implemented by means of the substitutions (10.17) and (10.19), so that

$$\int x^3(1-x^2)^{-\frac{1}{3}}\,dx \xrightarrow{t=g_1(x)} \int (1-t^3)^{\frac{3}{2}}t^{-1}\frac{3}{2}(1-t^3)^{-\frac{1}{2}}t^2\,dt$$

$$= \frac{3}{2}\int(1-t^3)t\,dt = \frac{3}{4}t^2 - \frac{3}{10}t^5 + C_1, \quad t \in (0,1).$$

The change g_1 is reversed by means of the substitution (10.16), which gives

$$\int x^3(1-x^2)^{-\frac{1}{2}}\,dx = \frac{3}{4}(1-x^2)^{\frac{2}{3}} - \frac{3}{10}(1-x^2)^{\frac{5}{3}} + C_1, \quad x \in I_1. \tag{10.30}$$

In order to find $\int f|_{I_2}(x)\,dx$, the change g_2 is implemented by means of (10.18) and (10.20):

$$\int x^3(1-x^2)^{-\frac{1}{2}}\,dx \xrightarrow{t=g_2(x)} \int(-1)^3(1-t^3)^{\frac{3}{2}}t^{-1}(-1)\frac{3}{2}(1-t^3)^{-\frac{1}{2}}t^2\,dt$$

$$= \frac{3}{2}\int(1-t^3)t\,dt = \frac{3}{4}t^2 - \frac{3}{10}t^5 + C_2, \quad t \in (0,1).$$

Substituting (10.16) reverses the change g_2, yielding

$$\int x^3 (1-x^2)^{-\frac{1}{2}}\, dx = \frac{3}{4}(1-x^2)^{\frac{2}{3}} - \frac{3}{10}(1-x^2)^{\frac{5}{3}} + C_2, \ x \in I_2. \qquad (10.31)$$

Next, the solutions of (10.30) and (10.31) must be assembled to obtain a global solution on $(-1,1)$. To do that, let

$$F(x) := \frac{3}{4}(1-x^2)^{\frac{2}{3}} - \frac{3}{10}(1-x^2)^{\frac{5}{3}}, \ x \in (-1,1).$$

By virtue of (10.30) and (10.31), $F'(x) = f(x)$ for all $x \in (-1,1)\setminus\{0\}$. But $F(x)$ is continuous at 0. Thus, by Proposition 4.3.3, $F(x)$ is a primitive of $f(x)$ on $(-1,1)$ and

$$\int f(x)\, dx = F(x) + C, \ x \in (-1,1)$$

for any real constant C. $\qquad\qquad\square$

Example 10.4.3. Find $\displaystyle\int x^5 (1+x^3)^{1/2}\, dx$.

Solution. The function $f(x) := x^5(1+x^3)^{1/2}$ is a binomial differential of type I and its domain is $D_f = [-1, \infty)$. Hence, the changes of variable to be used must be obtained from

$$t = g(x) := (1+x^3)^{1/2}, \ x \in (-1, \infty).$$

As explained at the beginning of this section, the largest changes of variable which can be managed from g are $g_1 := g|_{I_1}$ and $g_2 := g|_{I_2}$, where $I_1 := (-1, 0)$ and $I_2 := (0, \infty)$, and their respective inverses are

$$x = (t^2 - 1)^{1/3}, \ t \in g(I_2) = (1, \infty)$$
$$x = -(1 - t^2)^{1/3}, \ t \in g(I_1) = (0, 1)$$

so that

$$dx = \frac{2}{3}(t^2 - 1)^{-2/3} t\, dt, \ t \in (1, \infty)$$
$$dx = \frac{2}{3}(1 - t^2)^{-2/3} t\, dt, \ t \in (0, 1).$$

Thus, the implementation of the change $t = g_2(x)$ yields

$$\int x^5 (1+x^3)^{1/2}\, dx, \ x \in I_2 \ \xrightarrow{t=g_2(x)}\ \int \frac{2}{3}(t^2-1)^{5/3} t^2 (t^2-1)^{-2/3}\, dt$$

$$= \frac{2}{3}\int (t^2 - 1) t^2\, dt = \frac{2}{15}t^5 - \frac{2}{9}t^3 + C_2, \ t \in g(I_2).$$

Reversing the change $t = g_2(x)$, we obtain the local antiderivative of $f(x)$

$$F_2(x) = \frac{2}{15}(1+x^3)^{5/2} - \frac{2}{9}(1+x^3)^{3/2}, \ x \in (0, \infty). \qquad (10.32)$$

Implementation of $t = g_1(x)$ yields

$$\int x^5 (1+x^3)^{1/2} \, dx, \; x \in I_1 \xrightarrow{t=g_1(x)} \int (-1)^5 \frac{2}{3} (1-t^2)^{5/3} t^2 (1-t^2)^{-2/3} \, dt$$

$$= -\frac{2}{3} \int (1-t^2) t^2 \, dt = \frac{2}{15} t^5 - \frac{2}{9} t^3 + C_1, \; t \in g(I_1)$$

and, reversing the change $t = g_1(x)$, we get the local antiderivative of $f(x)$

$$F_1(x) = -\frac{2}{9}(1+x^3)^{3/2} + \frac{2}{15}(1+x^3)^{5/2}, \; x \in (-1,0). \qquad (10.33)$$

From (10.32) and (10.33),

$$F(x) := -\frac{2}{9}(1+x^3)^{3/2} + \frac{2}{15}(1+x^3)^{5/2}, \; x \in [-1,\infty)$$

is an antiderivative of $f(x)$ on $D_f \backslash \{-1, 0\}$, and, as $F(x)$ is continuous on $[-1, \infty)$, Proposition 4.3.3 shows that $F(x)$ is an antiderivative of $f(x)$ on the entire interval $[-1, \infty)$. Thus, the complete primitive of $f(x)$ is $F(x) + C$ for all constants C. $\square$

An alternative way to approach Example 10.4.3 is the following: first note that the function $u(t) := (f \circ g)(t) \cdot g'(t)$ adopts the same expression on both intervals $g(I_1)$ and $g(I_2)$: $2(t^2 - 1)t^2/3$. Moreover, $u(t)$ is well defined on $g([-1, \infty)) = [0, \infty)$ and its antiderivative on this interval is easily calculated as $P(t) := (2/15)t^5 - (2/9)t^3$. An application of Corollary 4.3.5 yields that $P \circ g^{-1}(x)$ is an antiderivative of $f(x)$ on its entire domain $[-1, \infty)$.

As can be deduced, Corollary 4.3.5 has allowed us to skip assembling local antiderivatives at $x = 0$. In general, for a binomial differential as in (10.2) or in (10.1), this trick works only if $0 \in D_f^\circ$ and s is an odd natural number.

10.5   The primitive of a binomial differential of type II

Let $f(x) = x^r (a + bx^s)^{p/q}$ (alternatively, $f(x) = (a + bx^s)^{p/q}$) be a binomial differential of type II. The changes of variable to find the primitive of $f(x)$ will be obtained from the function

$$h(x) = (b + ax^{-s})^{1/q}.$$

In the case when s is an odd integer, then it is necessary to resort to the changes of variable provided by $h(x)$ and by the function

$$l(x) = (-b - ax^{-s})^{1/q}$$

in order to complete the solution of $\int f(x) \, dx$. Be aware that

$$h(x) = (a + bx^s)^{1/q} x^{-s/q} \text{ for } x > 0$$
$$l(x) = (a + bx^s)^{1/q} (-x)^{-s/q} \text{ for } s \in \mathbb{Z} \text{ odd and } x < 0. \qquad (10.34)$$

The difficulty in using $h(x)$ and $l(x)$ is that their domains do not adjust well to D_f, as can be observed from the following discussion.

As $f(x)$ is binomial, either D_f consists of one single interval J_0 or it is the union of two disjoint intervals J_1 and J_2 such that $J_1 = D_f \cap (-\infty, 0)$ and $J_2 = D_f \cap (0, \infty)$.

Analogously, either D_h consists of one single interval I_0 or it is the union of two disjoint intervals I_1 and I_2 such that $I_1 = D_h \cap (-\infty, 0)$ and $I_2 = D_h \cap (0, \infty)$. The regular decomposition of D_f is explicitly given in Tables 10.1.1 and 10.1.2, and the regular decomposition of D_h is given in Table 10.5.1, where $\tau = |a/b|^{1/|s|}$.

Table 10.5.1. Regular decomposition of D_h:

s	$a<0, b>0$	$a>0, b<0$	$a>0, b>0$	$a<0, b<0$
$s \in \mathbb{N}$ even	$(-\infty, -\tau] \cup [\tau, \infty)$	$[-\tau, 0) \cup (0, \tau]$	$(-\infty, 0) \cup (0, \infty)$	$-$
$s \in \mathbb{N}$ odd	$(-\infty, 0) \cup [\tau, \infty)$	$(0, \tau]$	$(-\infty, -\tau] \cup (0, \infty)$	$[-\tau, 0)$
$s \in \mathbb{N}^-$ even	$[-\tau^{-1}, \tau^{-1}]$	$(-\infty, -\tau^{-1}] \cup [\tau^{-1}, \infty)$	$(-\infty, \infty)$	$-$
$s \in \mathbb{N}^-$ odd	$(-\infty, \tau^{-1}]$	$[\tau^{-1}, \infty)$	$[-\tau^{-1}, \infty)$	$(-\infty, -\tau^{-1}]$
$s \in \mathbb{Q}^+ \backslash \mathbb{N}$	$[\tau, \infty)$	$(0, \tau]$	$(0, \infty)$	$-$
$s \in \mathbb{Q}^- \backslash \mathbb{N}^-$	$[0, \tau^{-1}]$	$[\tau^{-1}, \infty)$	$[0, \infty)$	$-$

It can be observed from Table 10.5.1 that $0 \in D_h^\circ$ only if $s \in \mathbb{N}^-$, but in that case, $0 \notin D_f$. Hence, the only worthy changes of variable to find $\int f(x)\, dx$ that can be managed from $h(x)$ are:

(H1) $t = h_0(x) := h|_{I_0^\circ}(x)$ if $D_h = I_0$ and $0 \notin I_0^\circ$,

(H2) $t = h_1(x) := h|_{I_1^\circ}(x)$ and $t = h_2(x) := h|_{I_2^\circ}(x)$ if $D_h = I_0$ and $0 \in I_0^\circ$, where $I_1 := I_0 \cap (-\infty, 0)$ and $I_2 := I_0 \cap (0, \infty)$,

(H3) $t = h_1(x) := h|_{I_1^\circ}(x)$ and $t = h_2(x) := h|_{I_2^\circ}(x)$ if $D_h = I_1 \cup I_2$.

Let us label the case corresponding to the i-th row and j-th column in each of the three mentioned tables as $[i,j]$.

In cases $[2,2]$, $[2,3]$, $[2,4]$, $[4,2]$, $[4,3]$ and $[4,4]$ (where $s \in \mathbb{Z}$ odd) $D_f^\circ \backslash \{0\} \not\subset D_h$, but $D_f^\circ \backslash \{0\} \subset D_h \cup D_l$. Thus, the actions to be done in these cases are as follows.

$[2,2]$: In this case,

$$D_f^\circ = (-\infty, \tau)$$
$$D_h = (0, \tau]$$
$$D_l = (-\infty, 0) \cup [\tau, \infty).$$

Let $L_1 := (-\infty, 0)$ and $L_2 := (0, \tau)$. The implementation of the changes $t = l|_{L_1}(x)$ and $t = h|_{L_2}(x)$ transform $\int f|_{L_i}(x)\, dx$ into the primitive of a rational function (use (10.34) if necessary) which can be found as described in Chapter 6. This provides local antiderivatives $F_1(x)$ and $F_2(x)$ of $f(x)$ on L_1 and L_2. As in case (H3), $F_1(x)$ and $F_2(x)$ are assembled at $x = 0$ by means of Propositions 4.3.2 and 4.3.3 to build an antiderivative $F(x)$ of $f(x)$ on $L_1 \cup \{0\} \cup L_2$. In case $\tau \in D_f$, the continuous extension of $F(x)$ at $x = \tau$ is a global antiderivative of $f(x)$ by virtue of Proposition 4.3.2.

[2,3]: in this case,

$$D_f^\circ = (-\tau, \infty)$$
$$D_h = (-\infty, -\tau] \cup (0, \infty)$$
$$D_l = [-\tau, 0).$$

In a similar way as in case [2,2], let $L_1 := (0, \infty)$, $L_2 := (-\tau, 0)$ and implement the changes $t = h|_{L_1}(x)$, $t = l|_{L_2}(x)$ on L_1 and L_2 so that $\int f|_{L_1}(x)\, dx$ and $\int f|_{L_2}(x)\, dx$ are respectively transformed into primitives of rational functions. This provides a pair of local antiderivatives of $f(x)$ on L_1 and L_2. Next, assemble this pair of antiderivatives at $x = 0$ by means of Propositions 4.3.2 and 4.3.3 in order to build a global antiderivative of $f(x)$ on D_f.

[2,4]: in this case,

$$D_f^\circ = (-\infty, -\tau)$$
$$D_h = [-\tau, 0)$$
$$D_l = (-\infty, -\tau] \cup (0, \infty).$$

Thus, as explained in case [2,2], the implementation of the change $t = l|_{(-\infty,-\tau)}(x)$ provides an antiderivative $F(x)$ of $f(x)$ on $(-\infty, -\tau)$. In case $-\tau \in D_f$, an application of Proposition 4.3.2 ensures that the continuous extension of $F(x)$ at $x = -\tau$ is a global antiderivative of $f(x)$.

[4,2]: in this case,

$$D_f^\circ = (-\infty, 0) \cup (\tau^{-1}, \infty)$$
$$D_h = [\tau^{-1}, \infty)$$
$$D_l = (-\infty, \tau^{-1}].$$

Thus, the implementation of the changes $l|_{(-\infty,0)}$ and $h|_{(\tau^{-1},\infty)}$ provide local antiderivatives $F_1(x)$ and $F_2(x)$ on $(-\infty, 0)$ and (τ^{-1}, ∞). If $\tau^{-1} \in D_f$, $F_2(x)$ has a continuous extension at $x = \tau^{-1}$ that is a local antiderivative of $f(x)$ on $[\tau^{-1}, \infty)$.

[4,3]: in this case

$$D_f^\circ = (-\infty, -\tau^{-1}) \cup (0, \infty)$$
$$D_h = [-\tau^{-1}, \infty)$$
$$D_l = (-\infty, -\tau^{-1}].$$

Thus, the changes of variable $l|_{(-\infty,-\tau^{-1})}$ and $h|_{(0,\infty)}$ provide local antiderivatives of $f(x)$. If $-\tau^{-1} \in D_f$, its local antiderivative on $(-\infty, -\tau^{-1})$ can be extended with continuity at $x = -\tau^{-1}$ in order to manage an antiderivative on $(-\infty, -\tau^{-1}]$.

[4,4]: since $D_f^\circ \subset [-\tau^{-1}, 0) \cup (0, \infty) = D_l$, the implementation of the change $l|_{(-\tau^{-1},0)}$ suffices to obtain an antiderivative of $f(x)$ on D_f° to be extended with continuity at $-\tau^{-1}$ if necessary.

In the remaining cases, $D_f^\circ\setminus\{0\} \subset D_h$, so they can be solved with the changes h_0, h_1 and h_2 using a similar method to that used for the binomial differentials of type I in Section 10.4.

To summarize, the two tips to be considered in finding the primitive of a binomial differential of type II are:

- changes of variable from $l(x)$ are necessary in cases [2,2], [2,3], [2,4], [4,2], [4,3] and [4,4].
- assembling local antiderivatives has to be done only at the point $x = 0$, and only if $0 \in D_f^\circ$.

Example 10.5.1. Find $\displaystyle\int \frac{1}{\sqrt{(1-x^2)^3}}\, dx$.

Solution. Let $f(x) := (1-x^2)^{-3/2}$. Clearly, $D_f = (-1,1)$ and f is a binomial differential of type II [1,2]. Hence, the changes of variable to be used must be obtained from the function $g\colon (-1,1)\setminus\{0\} \longrightarrow \mathbb{R}$ given by

$$t = (-1 + x^{-2})^{\frac{1}{2}}. \tag{10.35}$$

Denote $I_1 := (0,1)$, $I_2 := (-1,0)$ and note that $I_i \cap D_f = I_i$ for $i = 1$ and $i = 2$. Consider the changes of variable $g_i := g|_{I_i}$ for $i = 1$ and $i = 2$. Their inverses are respectively

$$x = g_1^{-1}(t) = (t^2 + 1)^{-1/2},\ t \in (0,\infty) \tag{10.36}$$
$$x = g_2^{-1}(t) = -(t^2 + 1)^{-1/2},\ t \in (0,\infty) \tag{10.37}$$

and by differentiation on (10.36) and (10.37),

$$dx = -(t^2 + 1)^{-3/2}t\, dt, x \in I_1, \tag{10.38}$$
$$dx = (t^2 + 1)^{-3/2}t\, dt, x \in I_2. \tag{10.39}$$

(Diagram 10.29 also works for this example.)

It is time to find the primitive of $f(x)$ on I_1. To do this, implement the change g_1 by means of the substitutions (10.36) and (10.38), so that

$$\int (1-x^2)^{-3/2}\, dx,\ x \in I_1 \xleftarrow{t=g_1(x)} -\int\left(1 - \frac{1}{t^2 + 1}\right)^{-3/2} t(t^2 + 1)^{-3/2}\, dt$$

$$= \int \frac{(t^2)^{-3/2}}{(t^2 + 1)^{-3/2}} t(t^2 + 1)^{-3/2}\, dt$$

$$= -\int |t|^{-3}t\, dt = -\int t^{-2}\, dt = t^{-1} + C_1,\ t \in (0,\infty). \tag{10.40}$$

And by plugging (10.35) into (10.40), we obtain

$$\int (1-x^2)^{-3/2}\, dx = \frac{1}{\sqrt{-1+x^{-2}}} + C_1,\ x \in I_1.$$

Let us select a local antiderivative of $f(x)$ on I_1 from (10.5), for instance, choose

$$F_1(x) := \frac{1}{\sqrt{-1 + x^{-2}}}, \quad x \in I_1.$$

Next, let us find the primitive of $f(x)$ on I_2. To do this, implement the change g_2 by means of the substitutions (10.37) and (10.39). This yields

$$\int (1 - x^2)^{-3/2} \, dx, \; x \in I_2 \xleftrightarrow{\;t = g_2(x)\;} \int \left(1 - \frac{1}{t^2 + 1}\right)^{-3/2} t(t^2 + 1)^{-3/2} \, dt$$

$$= \int |t|^{-3} t \, dt = \int t^{-2} \, dt = -t^{-1} + C_1, \; t \in (0, \infty), \qquad (10.41)$$

and substituting (10.35) into (10.41) gives

$$\int (1 - x^2)^{-3/2} \, dx = -\frac{1}{\sqrt{-1 + x^{-2}}} + C_2, \; x \in I_2.$$

Choose any local antiderivative of $f(x)$ on I_2 from (10.5), for instance, take

$$F_2(x) := -\frac{1}{\sqrt{-1 + x^{-2}}}, \quad x \in I_2.$$

Now we have to assemble the chosen local antiderivatives $F_i(x)$ in order to get a primitive of $f(x)$ on the entire set D_f. In this case, since

$$\lim_{x \to 0^-} F_2(x) = 0 = \lim_{x \to 0^+} F_1(x)$$

Proposition 4.3.3 ensures that the function

$$F(x) := \begin{cases} F_1(x) = \frac{x^2}{\sqrt{1-x^2}}, & x \in I_1 \\ 0, & x = 0 \\ F_2(x) = -\frac{x^2}{\sqrt{1-x^2}}, & x \in I_2 \end{cases}$$

is an antiderivative of $f(x)$ on $(-1, 1)$ and

$$\int f(x) \, dx = F(x) + C, \; x \in (-1, 1)$$

where C is any constant.   □

Example 10.5.2. Find $\displaystyle\int x^{-2}(x^2 - 1)^{1/2} \, dx$.

Solution. The regular decomposition of the domain of $f(x) := x^{-2}(x^2 - 1)^{1/2}$ is $D_f = J_1 \cup J_2$ where $J_1 := [1, \infty)$ and $J_2 := (-\infty, -1]$. Clearly, $f(x)$ is a binomial differential of type II [1,1]. Hence, the changes of variable to be used must be obtained from the function $g(x)$ defined by the expression

$$t = (-x^{-2} + 1)^{1/2}, \; x \in J_1 \cup J_2. \qquad (10.42)$$

Stated precisely, g provides the changes of variable $g_i := g|_{J_i^\circ}$ for $i = 1, 2$. Note that g_1 is increasing and g_2 is decreasing. According to the established guidelines, we have to calculate a local antiderivative for each $f|_{J_i^\circ}$, $i = 1, 2$, and then, to find the primitive of f on the entire D_f.

As $g_i(I_i^\circ) = (0, 1)$, the inverses of g_i are

$$t = g_1^{-1}(t) = (1 - t^2)^{-1/2}, \quad t \in (0, 1) \tag{10.43}$$

$$x = g_2^{-1}(t) = -(1 - t^2)^{-1/2}, \quad t \in (0, 1). \tag{10.44}$$

By differentiation on (10.43) and (10.44),

$$dx = (1 - t^2)^{-3/2} t \, dt, \quad x \in J_1^\circ \tag{10.45}$$

$$dx = -(1 - t^2)^{-3/2} t \, dt, \quad x \in J_2^\circ. \tag{10.46}$$

Let us find $\int f(x) \, dx$ on J_1°. To do this, implement the change g_1 by means of the substitutions (10.43) and (10.45), so that

$$\int x^{-2}(x^2 - 1)^{1/2} \, dx, \quad x \in J_1^\circ \xleftarrow{t = g_1(x)} \int (1 - t^2) \left(\frac{1}{1 - t^2} - 1 \right)^{\frac{1}{2}} (1 - t^2)^{-\frac{3}{2}} t \, dt$$

$$= \int (1 - t^2) \frac{(t^2)^{\frac{1}{2}}}{(1 - t^2)^{\frac{1}{2}}} (1 - t^2)^{-\frac{3}{2}} t \, dt = \int (1 - t^2)^{-1} |t| t \, dt$$

$$= -\int \frac{t^2}{t^2 - 1} \, dt, \quad t \in (0, 1). \tag{10.47}$$

The primitive of $t^2/(t^2 - 1)$ is found as indicated in Chapter 6:

$$\int \frac{t^2}{t^2 - 1} \, dt = \int 1 + \frac{1}{t^2 - 1} \, dt = \int 1 + \frac{\frac{1}{2}}{t - 1} - \frac{\frac{1}{2}}{t + 1} \, dt$$

$$= t + \frac{1}{2} \log \frac{|t - 1|}{|t + 1|}, \quad t \in \mathbb{R} \setminus \{-1, 1\}. \tag{10.48}$$

Thus, the last primitive in (10.47) is

$$-\int \frac{t^2}{t^2 - 1} \, dt = -t - \frac{1}{2} \log \frac{1 - t}{t + 1}, \quad t \in (0, 1) \tag{10.49}$$

and reversing the change g_1 in (10.49) by means of the substitution (10.42),

$$\int f(x) \, dx = -\sqrt{1 - x^{-2}} - \frac{1}{2} \log \frac{1 - \sqrt{1 - x^{-2}}}{1 + \sqrt{1 - x^{-2}}} + C_1 \tag{10.50}$$

for $x \in J_1^\circ$. But the function on the right side of (10.50) is continuous at $x = 1$. Hence, Proposition 4.3.1 ensures that it is also the solution to $\int f(x) \, dx$ on J_1.

In order to find the primitive of $\int f(x) \, dx$ on J_2°, implement the change g_2 by means of the substitutions (10.44) and (10.46), so that, repeating a similar process as that done for the primitive on J_1°,

$$\int x^{-2}(x^2 - 1)^{1/2} \, dx, \quad x \in I_2 \xleftarrow{t = g_2(x)} -\int (1 - t^2) \left(\frac{1}{1 - t^2} - 1 \right)^{\frac{1}{2}} (1 - t^2)^{-\frac{3}{2}} t \, dt$$

$$= -\int (1 - t^2)^{-1} |t| t \, dt = \int \frac{t^2}{t^2 - 1} \, dt, \quad t \in (0, 1) \tag{10.51}$$

and by (10.48), the last primitive in (10.51) is

$$\int \frac{t^2}{t^2 - 1}\,dt = t + \frac{1}{2}\log\frac{1 - t}{t + 1}, \quad t \in (0,1). \tag{10.52}$$

Next, reverse the change g_2 in (10.52) by means of the substitution (10.42), so that

$$\int f(x)\,dx = \sqrt{1 - x^{-2}} + \frac{1}{2}\log\frac{1 - \sqrt{1 - x^{-2}}}{1 + \sqrt{1 - x^{-2}}} + C_2 \tag{10.53}$$

for $x \in J_2^\circ$. Since the function on the right side of (10.53) is continuous at $x = -1$, it is also the solution to $\int f(x)\,dx$ on J_2.

Hence, from (10.50) and (10.53),

$$\int f(x)\,dx = \begin{cases} -\sqrt{1 - x^{-2}} - \dfrac{1}{2}\log\dfrac{1 - \sqrt{1 - x^{-2}}}{1 + \sqrt{1 - x^{-2}}} + C_1, \ x \in [1,\infty) \\[4mm] \sqrt{1 - x^{-2}} + \dfrac{1}{2}\log\dfrac{1 - \sqrt{1 - x^{-2}}}{1 + \sqrt{1 - x^{-2}}} + C_2, \ x \in (-\infty, -1] \end{cases}$$

where C_1 and C_2 are any pair of constants.

Example 10.5.3. Find $\displaystyle\int \frac{1}{\sqrt[4]{1 - x^4}}\,dx$.

Solution. The natural domain of $f(x) := (1 - x^4)^{-1/4}$ is $D_f = (-1,1)$ and $f(x)$ is a binomial differential of type II [1,2]. Hence, the changes of variable to be used must be obtained from the function $g \colon [-1,1]\setminus\{0\} \longrightarrow \mathbb{R}$ given by

$$t = (-1 + x^{-4})^{1/4}. \tag{10.54}$$

The largest changes of variable provided by g are the decreasing change $g_1 := g|_{I_1}$ and the increasing change $g_2 := g|_{I_2}$, where $I_1 := (0,1)$ and $I_2 := (-1,0)$.

Clearly, $g_i(I_i) = (0,\infty)$ for $i = 1,2$. Thus, the inverses of g_i are

$$x = g_1^{-1}(t) = (t^4 + 1)^{-1/4}, \ t \in (0,\infty) \tag{10.55}$$

$$x = g_2^{-1}(t) = -(t^4 + 1)^{-1/4}, \ t \in (0,\infty) \tag{10.56}$$

and by differentiation on (10.55) and (10.56),

$$dx = -(t^4 + 1)^{-5/4}t^3 \, dt, \ x \in I_1 \tag{10.57}$$

$$dx = (t^4 + 1)^{-5/4}t^3 \, dt, \ x \in I_2. \tag{10.58}$$

Following the standard procedure, we first find the primitives of $f(x)$ on $I_1 = D_f \cap I_1$ and $I_2 = D_f \cap I_2$. We will then assemble a pair of local antiderivatives on I_1 and I_2 to obtain the primitive of $f(x)$ on the entire set D_f.

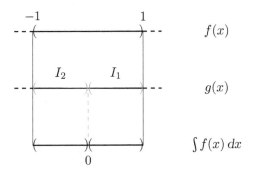

Let us find $\int f|_{I_1}(x) \, dx$. To do that, implement the change g_1 by means of the substitutions (10.55) and (10.57), so that

$$\int (1 - x^4)^{-1/4} \, dx, \ x \in I_1 \xrightarrow{t = g_1(x)} - \int \left(1 - \frac{1}{t^4 + 1}\right)^{-1/4} (t^4 + 1)^{-5/4}t^3 \, dt$$

$$= -\int \frac{t^{-1}}{(t^4 + 1)^{-1/4}}(t^4 + 1)^{-\frac{5}{4}}t^3 \, dt = -\int \frac{t^2}{t^4 + 1} \, dt, \ t \in (0, \infty). \tag{10.59}$$

By virtue of Example 6.5.4,

$$\int \frac{t^2}{t^4 + 1} \, dt = H_1(t) + C \tag{10.60}$$

where C is any constant and

$$H_1(t) := \frac{1}{4\sqrt{2}} \log \frac{t^2 - \sqrt{2}t + 1}{t^2 + \sqrt{2}t + 1} + \frac{1}{2\sqrt{2}} \arctan(\sqrt{2}t - 1)$$

$$+ \frac{1}{2\sqrt{2}} \arctan(\sqrt{2}t + 1), \ t \in \mathbb{R}.$$

Hence, reversing the change g_1 in (10.60) by means of (10.54), and plugging the result into (10.59) yields

$$\int (1 - x^4)^{-1/4} \, dx = -H_1(\sqrt[4]{-1 + x^{-4}}) + C_1 = F_1(x) + C_1, \ x \in I_1 \tag{10.61}$$

where

$$F_1(x) := -H_1(\sqrt[4]{-1 + x^{-4}}), \ x \in I_1$$

is a local antiderivative of $f(x)$ on I_1.

In order to find $\int f|_{I_2}(x)\,dx$, implement the change g_2 by means of the substitutions (10.56) and (10.58), so that

$$\int (1-x^4)^{-1/4}\,dx,\ x \in I_2 \xrightarrow{t=g_2(x)} \int \left(1 - \frac{1}{t^4+1}\right)^{-1/4} (t^4+1)^{-5/4}t^3\,dt$$

$$= \int \frac{t^2}{t^4+1}\,dt,\ t \in (0,\infty). \tag{10.62}$$

Thus, from (10.62) and (10.60), and reversing the change g_2 by means of the substitution (10.54), we have

$$\int (1-x^4)^{-1/4}\,dx = H_1(\sqrt[4]{-1+x^{-4}}) + C_2 = F_2(x) + C_2,\ x \in I_2 \tag{10.63}$$

where

$$F_2(x) := H_1(\sqrt[4]{-1+x^{-4}})\ x \in I_2 \tag{10.64}$$

is a local antiderivative of $f(x)$ on I_2.

The final step is the assembly of $F_1(x)$ and $F_2(x)$ to obtain the primitive of $f(x)$ on $(-1,1)$. To do this, with the help of Proposition 4.2.5 we calculate the limits of the local antiderivatives $F_1(x)$ and $F_2(x)$ at the separating point $x = 0$ as

$$\lim_{x\to 0^+} F_1(x) = \lim_{t\to\infty} -H_1(t)$$

$$= -\lim_{t\to\infty} \frac{1}{4\sqrt{2}} \log \frac{t^2 - \sqrt{2}t + 1}{t^2 + \sqrt{2}t + 1} - \lim_{t\to\infty} \frac{1}{2\sqrt{2}} \arctan(\sqrt{2}t - 1)$$

$$- \lim_{t\to\infty} \frac{1}{2\sqrt{2}} \arctan(\sqrt{2}t + 1)$$

$$= -\frac{\pi}{2\sqrt{2}}$$

and

$$\lim_{x\to 0^-} F_1(x) = \lim_{t\to\infty} H_1(t) = \frac{\pi}{2\sqrt{2}}.$$

Hence, by virtue of Proposition 4.3.3, an antiderivative of $f(x)$ is

$$F(x) := \begin{cases} F_1(x), & x \in (0,1) \\[2mm] -\dfrac{\pi}{2\sqrt{2}}, & x = 0 \\[4mm] F_2(x) - \dfrac{\pi}{\sqrt{2}}, & x \in (-1,0) \end{cases}$$

and the complete solution is

$$\int (1-x^4)^{-1/4}\,dx = F(x) + C$$

for any real constant C. $\qquad\qquad\qquad\qquad\qquad\qquad\qquad\qquad\square$

A graph of the antiderivative $F(x)$ and of the local antiderivatives $F_1(x)$ and $F_2(x)$ corresponding to Example 10.5.3 is depicted below:

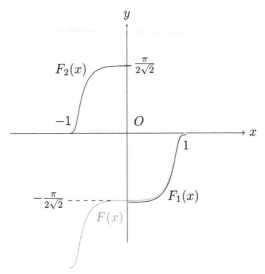

Example 10.5.4. Find $\displaystyle\int \frac{1}{\sqrt[3]{(1-x^3)}}\, dx.$

Solution. The function $f(x) = (1-x^3)^{-1/3}$ is a binomial differential of type II [2,2] and its domain is $D_f = (-\infty, 1)$. Thus, the changes of variable to be used must be obtained from

$$h_1(x) = (-1 + x^{-3})^{1/3}, \ x \in D_{h_1} = (0,1)$$
$$l(x) = (1 - x^{-3})^{1/3}, \ x \in D_l = (-\infty, 0) \cup (1, \infty)$$

The solution of $\int f(x)\, dx$ will take three steps: the calculation of an antiderivative $G_1(x)$ of $f(x)$ on $(0,1)$ by means of the change $t = h_1(x)$, the calculation of an antiderivative $G_2(x)$ on $(-\infty, 0)$ by means of the change $h_2 := l|_{(-\infty, 0)}$, and the assembling of $G_1(x)$ and $G_2(x)$ to build a global antiderivative of $f(x)$.

First step: the substitution formulas for the decreasing change of variable $t = h_1(x)$, $x \in (0,1)$, are

$$x = (1 + t^3)^{-1/3}, \ t \in (0, \infty)$$
$$dx = -(1 + t^3)^{-4/3}t^2\, dt. \tag{10.65}$$

Also note that, as $x > 0$,

$$t = (-x^3 + 1)^{1/3} x^{-1}. \tag{10.66}$$

Implement the change h_1 by means of the substitution formulas (10.65) and (10.66), so that

$$\int (1 - x^3)^{-1/3}\, dx, \ x \in (0,1) \xleftarrow{t = h_1(x)} -\int x^{-1} t^{-1} (1 + t^3)^{-4/3} t^2\, dt$$

$$= -\int t(1 + t^3)^{-1}\, dt, \ t \in (0, \infty). \tag{10.67}$$

As the partial decomposition of $t/(t^3+1)$ is

$$\frac{t}{t^3+1} = \frac{-\frac{1}{3}}{t+1} + \frac{\frac{1}{3}t+\frac{1}{3}}{t^2-t+1}$$

it follows that

$$-\int \frac{t\,dt}{t^3+1} = \frac{1}{3}\int \frac{dt}{t+1} - \frac{1}{6}\int \frac{2t-1}{t^2-t+1}\,dt - \frac{1}{2}\int \frac{dt}{t^2-t+1}$$
$$= F_1(t) + C,$$

where

$$F_1(t) := \frac{1}{3}\log \frac{t+1}{(t^2-t+1)^{1/2}} - \frac{1}{\sqrt{3}}\arctan\left(\frac{2}{\sqrt{3}}t - \frac{1}{\sqrt{3}}\right), \quad t \in (0,\infty).$$

A reversion of the change of variable $t = h_1(x)$ yields the local antiderivative

$$G_1(x) := F_1\big((-1+x^{-3})^{1/3}\big), \quad x \in (0,1).$$

Second step: in order to find an antiderivative $G_2(x)$ of $f(x)$ on $(-\infty,0)$, we will resort to the change of variable $t = h_2(x) = (1-x^{-3})^{1/3}$, $x \in (-\infty,0)$. The substitution formulas for this increasing change of variable are

$$x = -(t^3-1)^{-1/3}, \quad t \in (1,\infty)$$
$$dx = (t^3-1)^{-4/3}t^2\,dt. \tag{10.68}$$

Also note that, as $x < 0$,

$$t = (-x^3+1)^{1/3}(-x)^{-1}. \tag{10.69}$$

Implement the change h_2 by means of the substitution formulas (10.68) and (10.69), so that

$$\int (1-x^3)^{-1/3}\,dx, \ x \in (-\infty,0) \xrightarrow{t=h_2(x)} \int (-x)^{-1}t^{-1}(t^3-1)^{-4/3}t^2\,dt$$
$$= \int t(t^3-1)^{-1}\,dt, \ t \in (1,\infty). \tag{10.70}$$

As the partial decomposition of $t/(t^3-1)$ is

$$\frac{t}{t^3-1} = \frac{\frac{1}{3}}{t-1} + \frac{-\frac{1}{3}t+\frac{1}{3}}{t^2+t+1}$$

it follows that

$$\int \frac{t\,dt}{t^3-1} = \frac{1}{3}\int \frac{dt}{t-1} - \frac{1}{6}\int \frac{2t+1}{t^2+t+1}\,dt + \frac{1}{2}\int \frac{dt}{t^2+t+1}$$
$$= F_2(t) + C,$$

where

$$F_2(t) := \frac{1}{3}\log \frac{t-1}{(t^2+t+1)^{1/2}} + \frac{1}{\sqrt{3}}\arctan\left(\frac{2}{\sqrt{3}}t + \frac{1}{\sqrt{3}}\right), \quad t \in (1,\infty).$$

A reversion of the change of variable $t = h_2(x)$ yields the local antiderivative

$$G_2(x) := F_2\big((1-x^{-3})^{1/3}\big), \quad x \in (-\infty,0).$$

Third step: in order to make a global antiderivative of $f(x)$, we calculate the limits of $G_1(x)$ and $G_2(x)$ at the separating point $x = 0$ with the help of Proposition 4.2.5:

$$\lim_{x \to 0^+} G_1(x) - \lim_{t \to \infty} F_1(t) = -\frac{1}{\sqrt{3}}\frac{\pi}{2}$$

$$\lim_{x \to 0^-} G_2(x) = \lim_{t \to \infty} F_2(t) = \frac{1}{\sqrt{3}}\frac{\pi}{2}.$$

Hence, by virtue of Proposition 4.3.3, a global antiderivative of $f(x)$ is

$$G(x) := \begin{cases} G_1(x) + \frac{\pi}{\sqrt{3}}, & x \in (0,1) \\ \frac{\pi}{2\sqrt{3}}, & x = 0 \\ G_2(x), & x \in (-\infty, 0) \end{cases}$$

and $\int f(x)\,dx = G(x) + C$, $x \in (-\infty, 1)$ (see the graphs of $G_1(x)$, $G_2(x)$ and $G(x)$ below). $\quad\square$

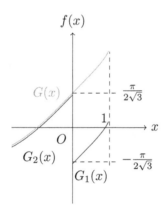

10.6 Exercises

Find the domain and an antiderivative for each of the following functions.

Question 10.6.1

$$f(x) = x^{-1}(8 + x^3)^{1/2}$$

Question 10.6.3

$$f(x) = x^{-1}(2 + 3x^{1/4})^{1/2}$$

Question 10.6.2

$$f(x) = (8 - x^{-3})^{2/3}$$

Question 10.6.4

$$f(x) = (1 - x^{-1})^{1/3}$$

Question 10.6.5

$$f(x) = (1 + x^2)^{1/2} x^3$$

Question 10.6.6

$$f(x) = x^3 \sqrt{1 - x}$$

Question 10.6.7

$$f(x) = x^2 \sqrt{1 - x^2}$$

Question 10.6.8

$$f(x) = (1 - x^3)^{-1/3}$$

Question 10.6.9

$$f(x) = x(x^3 - 1)^{1/3}$$

Question 10.6.10

$$f(x) = (1 + x^3)^{-1/3}$$

Question 10.6.11

$$f(x) = (1 - x^{-2})^{1/2}$$

Answers to the Questions

Answer 10.6.1

$$D_f = [-2, 0) \cup (0, \infty),$$
$$\frac{2}{3}\sqrt{8 + x^3} + \frac{2\sqrt{2}}{3} \log \left| \frac{\sqrt{8 + x^3} - 2\sqrt{2}}{\sqrt{8 + x^3} + 2\sqrt{2}} \right|$$

Answer 10.6.2

$$D_f = (-\infty, 2],$$
$$\begin{cases} F(\sqrt[3]{8x^{-3} - 1}) + \frac{32}{3\sqrt{3}}, & x \in (0, 2) \\ \frac{16}{3\sqrt{3}}, & x = 0 \\ -F(\sqrt[3]{1 - 8x^{-3}}), & x \in (-\infty, 0) \end{cases}$$
where $F(t) = \frac{16}{9} \log |t + 1| - \frac{8}{9} \log (t^2 - t + 1) - \frac{16}{3\sqrt{3}} \arctan \left(\frac{2t-1}{\sqrt{3}} \right) + \frac{8t^2}{3(t^3 + 1)}$

Answer 10.6.3

$$D_f = (0, \infty),$$
$$8(2 + 3x^{1/4})^{1/2} + 4\sqrt{2} \log \frac{\sqrt{2 + 3x^{1/4}} - \sqrt{2}}{\sqrt{2 + 3x^{1/4}} + \sqrt{2}}$$

Answer 10.6.4

$$D_f = (-\infty, 0) \cup [1, \infty),$$
$$F(\sqrt[3]{1 - x^{-1}}) \text{ where}$$

$$F(t) = -\frac{t}{t^3 - 1} + \frac{1}{3}\log|t - 1| - \frac{1}{6}\log(t^2 + t + 1) - \frac{1}{\sqrt{3}}\arctan\left(\frac{2t}{\sqrt{3}} + \frac{1}{\sqrt{3}}\right)$$

Answer 10.6.5

$D_f = (-\infty, \infty)$,

$$\frac{1}{5}(1 + x^2)^{5/2} - \frac{1}{3}(1 + x^2)^{3/2}$$

Answer 10.6.6

$D_f = (-\infty, 1]$,

$$\frac{2}{9}(1 - x)^{9/2} - \frac{6}{7}(1 - x)^{7/2} + \frac{6}{5}(1 - x)^{5/2} - \frac{2}{3}(1 - x)^{3/2}$$

Answer 10.6.7

$D_f = [-1, 1]$

$$\begin{cases} F(\sqrt{-1 + x^{-2}}), & x \in (0, 1] \\ -\frac{\pi}{16}, & x = 0 \\ -F(\sqrt{-1 + x^{-2}}) - \frac{\pi}{8}, & x \in [-1, 0) \end{cases}$$

where

$$F(t) = -\frac{t^3 - t}{8(t^2 + 1)^2} - \frac{1}{8}\arctan t$$

Answer 10.6.8

$D_f = (-\infty, 1)$,

$$\begin{cases} F_1(\sqrt[3]{-1 + x^{-3}}), & x \in (0, 1) \\ -\frac{\pi}{2\sqrt{3}}, & x = 0 \\ F_2(\sqrt[3]{1 - x^{-3}}) - \frac{\pi}{\sqrt{3}}, & x \in (-\infty, 0) \end{cases}$$

where

$$F_1(t) = \frac{1}{3}\log|t + 1| - \frac{1}{6}\log(t^2 - t + 1) - \frac{1}{\sqrt{3}}\arctan\left(\frac{2}{\sqrt{3}}t - \frac{1}{\sqrt{3}}\right)$$

$$F_2(t) = \frac{1}{3}\log|t - 1| - \frac{1}{6}\log(t^2 + t + 1) + \frac{1}{\sqrt{3}}\arctan\left(\frac{2}{\sqrt{3}}t + \frac{1}{\sqrt{3}}\right)$$

Answer 10.6.9

$D_f = [1, \infty)$,

$F(\sqrt[3]{1 - x^{-3}})$ where

$$F(t) = \frac{-t/3}{t^3 - 1} + \frac{1}{9} \log(1 - t) - \frac{1}{18} \log(t^2 + t + 1) - \frac{\sqrt{3}}{9} \arctan\left(\frac{2t}{\sqrt{3}} + \frac{1}{\sqrt{3}}\right)$$

Answer 10.6.10

$D_f = (-1, \infty)$,

$$\begin{cases} F_1(\sqrt[3]{1 + x^{-3}}), & x \in (0, \infty) \\ -\frac{1}{2\sqrt{3}}\pi, & x = 0 \\ F_2(\sqrt[3]{-1 - x^{-3}}) - \frac{1}{\sqrt{3}}\pi, & x \in (-1, 0) \end{cases}$$

where

$$F_1(t) = -\frac{1}{3} \log|t - 1| + \frac{1}{6} \log(t^2 + t + 1) - \frac{1}{\sqrt{3}} \arctan\left(\frac{2t}{\sqrt{3}} + \frac{1}{\sqrt{3}}\right)$$

$$F_2(t) = -\frac{1}{3} \log|t + 1| + \frac{1}{3} \log(\sqrt{t^2 + t + 1}) + \frac{1}{\sqrt{3}} \arctan\left(\frac{2}{\sqrt{3}}t - \frac{1}{\sqrt{3}}\right)$$

Answer 10.6.11

$D_f = (-\infty, -1] \cup [1, \infty)$,

$$\sqrt{-1 + x^2} - \arctan\left(\sqrt{-1 + x^2}\right) \cdot \begin{cases} 1, & x \in [1, \infty) \\ -1, & x \in (-\infty, -1] \end{cases}$$

Chapter 11

Rational functions of trigonometric arguments

This chapter focuses on methods to find the primitive of

$$\varrho(x) := R(\sin x, \cos x) \qquad (11.1)$$

where R is a rational function of its arguments, that is, a function given by a finite combination of sums, products and quotients of scalars and the functions $\sin x$ and $\cos x$.

11.1 The domain of $R(\sin x, \cos x)$

It is straightforward to show that a function $\varrho(x)$ as in (11.1) is continuous on its natural domain since $\sin x$ and $\cos x$ are continuous on $\mathbb{R}$. Let us prove that its domain is regular.

From

$$1 + \tan^2 x = \frac{1}{\cos^2 x}, \qquad x \neq \frac{\pi}{2} + k\pi, \ k \in \mathbb{Z} \qquad (11.2)$$

and from statements (i) and (ii) in Proposition refprop furthertrig, it readily follows that

$$\sin x = 2 \sin \frac{x}{2} \cos \frac{x}{2} = 2 \frac{\tan(x/2)}{1 + \tan^2(x/2)} \qquad (11.3)$$

$$\cos x = \cos^2 \frac{x}{2} - \sin^2 \frac{x}{2} = \frac{1 - \tan^2(x/2)}{1 + \tan^2(x/2)}, \qquad x \neq \pi + 2k\pi, \ k \in \mathbb{Z}. \qquad (11.4)$$

The main result of this section, about the function given in (11.1), can now be established.

Proposition 11.1.1. *The natural domain of $\varrho(x)$ is a regular domain.*

Proof. Since $\sin x$ and $\cos x$ are continuous on $\mathbb{R}$, it follows from (11.3), (11.4) and (11.1) that $\varrho(x)$ is the continuous extension to D_ϱ of

$$\rho(x) := R\left(\frac{2\tan(x/2)}{1 + \tan^2(x/2)}, \frac{1 - \tan^2(x/2)}{1 + \tan^2(x/2)}\right), x \in D_\varrho \backslash \{\pi + 2k\pi : k \in \mathbb{Z}\}. \qquad (11.5)$$

In turn, $\rho(x)$ can be expressed as

$$\rho(x) = Q \circ g|_{D_\rho}(x) \tag{11.6}$$

where $Q(t) = P_1(t)/P_2(t)$ for a pair of polynomials $P_1(t)$ and $P_2(t)$ over $\mathbb{R}$ and where

$$g(x) := \tan\frac{x}{2}, \quad x \in \mathbb{R}\backslash\{(2k-1)\pi \colon k \in \mathbb{Z}\}. \tag{11.7}$$

The natural domain of $Q(t)$ is $\mathbb{R}\backslash Z$ where Z is the set of all real roots of $P_2(t)$. Let $I_k := ((2k-1)\pi, (2k+1)\pi)$, $g_k := g|_{I_k}$ and $A_k := I_k\backslash g_k^{-1}(Z)$ for every $k \in \mathbb{Z}$. Thus,

$$D_\rho = g^{-1}(D_Q) = D_g\backslash g^{-1}(Z)$$
$$= \bigcup_{k\in\mathbb{Z}} A_k.$$

Since Z is finite and g_k is injective, each $g_k^{-1}(Z)$ is finite too. Hence, every A_k is regular and its regular decomposition is finite (more precisely, it is formed by at most $m+1$ intervals, where m is the cardinal of Z).

Of course, D_ρ does not contain any number of the form $(2k-1)\pi$ for $k \in \mathbb{Z}$, but $\varrho(x)$ may do. With more precision, since $\varrho(x)$ is periodic of period 2π, it follows that

$$D_\varrho = D_\rho \cup S \tag{11.8}$$

where $S = \varnothing$ or $S = \{(2k-1)\pi \colon k \in \mathbb{Z}\}$, and for the same reason, $A_k = 2k\pi + A_0$. Then, since the regular decomposition of each A_k is finite and $A_k \subset I_k$, it is clear from (11.8) that D_f is a regular domain. $\qquad\square$

An important sequel to the proof of Proposition 11.1.1 is that $\mathbb{R}\backslash D_\varrho$ is a set of isolated points. Thus, each interval of the regular decomposition of D_ϱ is open.

Once it is proved that D_ϱ is a regular domain, the existence of $\int \varrho(x)\,dx$ on D_ϱ is a consequence of Theorem 4.1.9. The calculation of $\int \varrho(x)\,dx$ often needs the changes of variable described in the following section.

11.2 Trigonometric changes of variable

The standard changes of variable used to find the primitive of a function such as that of (11.1) are certain restrictions of the functions $t = \tan(x/2)$, $t = \tan x$, $t = \sin x$ and $t = \cos x$. Each of the aforementioned functions has been studied in Chapter 3 and in Appendix A.4.

The changes obtained from $t = \tan(x/2)$ work for any rational function $\varrho(x)$ of trigonometric arguments. The changes derived from the other functions give some technical advantage with respect to those obtained from $t = \tan(x/2)$, but they only work if $\varrho(x)$ satisfies some additional hypotheses.

The nature of the changes of variable that will now be introduced require a strict observation of the techniques developed in Sections 4.3 and 4.4.

• $t = \tan{(x/2)}$:

Let $I_k := \left((2k-1)\pi, (2k+1)\pi\right)$ for all $k \in \mathbb{Z}$. The function $g \colon \bigcup_{k=-\infty}^{\infty} I_k \longrightarrow \mathbb{R}$ given by

$$ t = \tan\frac{x}{2} \tag{11.9} $$

is not injective. Furthermore, it is periodic of period 2π. However, the derivative of each restriction

$$ g_k := g|_{I_k} \tag{11.10} $$

is

$$ g_k'(x) = \frac{1}{2}\left(1 + \tan^2\frac{x}{2}\right) > 0, \quad x \in I_k. $$

Hence g_k is an increasing change of variable. Its image is

$$ g_k(I_k) = \left(\lim_{x \to (k-1)\pi^+} \tan(x/2), \lim_{x \to (k+1)\pi^-} \tan(x/2)\right) = (-\infty, \infty). $$

The inverse g_k^{-1} is given by

$$ x = 2\arctan t + 2k\pi, \quad t \in (-\infty, \infty) \tag{11.11} $$

and the substitution formula (4.7) for the change $t = g_k(x)$ is obtained by differentiation on (11.11), which yields

$$ dx = \frac{2}{1+t^2}\, dt, \quad x \in I_k. \tag{11.12} $$

• $t = \tan x$:

Denote $I_k := \left((2k-1)\pi/2, (2k+1)\pi/2\right)$ for $k \in \mathbb{Z}$. The function $g \colon \bigcup_{k=-\infty}^{\infty} I_k \longrightarrow \mathbb{R}$ given by

$$ t = \tan x \tag{11.13} $$

is not injective. Furthermore, it is periodic of period π. However, the derivative of each restriction

$$ g_k := g|_{I_k} \tag{11.14} $$

is

$$ g_k'(x) = 1 + \tan^2 x > 0, \quad x \in I_k. $$

Hence g_k is an increasing change of variable. Its image is

$$ g_k(I_k) = \left(\lim_{x \to -\pi/2^+} \tan x, \lim_{x \to \pi/2^-} \tan x\right) = (-\infty, \infty). $$

The inverse g_k^{-1} is given by

$$ x = \arctan t + k\pi, \quad t \in (-\infty, \infty) \tag{11.15} $$

and the substitution formula (4.7) for the change $t = g_k(x)$ is obtained by differentiation on (11.15), which yields

$$ dx = \frac{1}{1+t^2}\, dt, \quad x \in I_k. \tag{11.16} $$

- $t = \sin x$:

The function $g: \mathbb{R} \longrightarrow \mathbb{R}$ given by

$$t = \sin x \qquad\qquad (11.17)$$

is periodic of period 2π and its derivative, $g'(x) = \cos x$, is null only at $x = \frac{\pi}{2} + k\pi$ for all $k \in \mathbb{Z}$. Denote

$$I_k := \left((2k - 1)\frac{\pi}{2}, (2k + 1)\frac{\pi}{2}\right), \quad k \in \mathbb{Z} \qquad\qquad (11.18)$$

and let

$$g_k := g\big|_{I_k}. \qquad\qquad (11.19)$$

Each function g_k is an increasing change of variable if k is even and is a decreasing change of variable if k is odd. In all cases, $g_k(I_k) = (-1, 1)$.

The inverse g_k^{-1} is given by

$$x = (-1)^k \arcsin t + k\pi, \quad t \in (-1, 1) \qquad\qquad (11.20)$$

and the substitution formula (4.7) for the change $t = g_k(x)$ is obtained by differentiation on (11.20), which gives

$$dx = \frac{(-1)^k}{\sqrt{1 - t^2}} \, dt, \quad x \in I_k. \qquad\qquad (11.21)$$

- $t = \cos x$:

The function $g: \mathbb{R} \longrightarrow \mathbb{R}$ given by

$$t = \cos x \qquad\qquad (11.22)$$

is periodic of period 2π and its derivative, $g'(x) = -\sin x$, is null only at $x = k\pi$ for all $k \in \mathbb{Z}$. Denote

$$I_k := \big(k\pi, (k + 1)\pi\big), \quad k \in \mathbb{Z} \qquad\qquad (11.23)$$

and let

$$g_k := g\big|_{I_k}. \qquad\qquad (11.24)$$

Each function g_k is a decreasing change of variable if k is even and is an increasing change of variable if k is odd. In all cases, $g_k(I_k) = (-1, 1)$.

The inverse g_k^{-1} is given by

$$x = (-1)^k \arccos t + k\pi, \quad t \in (-1, 1) \qquad\qquad (11.25)$$

and the substitution formula (4.7) for the change $t = g_k(x)$ is obtained by differentiation on (11.25), which gives

$$dx = \frac{(-1)^{k+1}}{\sqrt{1 - t^2}} \, dt, \quad x \in I_k. \qquad\qquad (11.26)$$

11.3 The primitive of $R(\sin x, \cos x)$

The primitive of

$$\varrho(x) = R(\sin x, \cos x), \quad x \in D_\varrho \tag{11.27}$$

where $R(t_1, t_2)$ is a rational function of its arguments, can be found by means of the changes of variable derived from

$$t = \tan \frac{x}{2}$$

in accordance with the following procedure.

1: Let $\bigcup_{i \in J} J_i$ be the regular decomposition of D_ϱ. Following the notation in (11.10), for each integer k, I_k denotes the interval $((2k-1)\pi, (2k+1)\pi)$ and $g_k(x)$ is the change of variable given by $t = \tan(x/2)$, $x \in I_k$. Consider only the intervals $L_i^k := J_i \cap I_k$ which are non-empty.

2: Find the primitive of $\varrho(x)$ on each L_i^k by implementing the change of variable $t = g_k(x)$. Thanks to the identities (11.3), (11.4) and (11.12), this implementation can be carried out directly by means of the substitutions

$$\sin x = \frac{2t}{1+t^2} \tag{11.28}$$

$$\cos x = \frac{1-t^2}{1+t^2} \tag{11.29}$$

$$dx = \frac{2}{1+t^2} \, dt \tag{11.30}$$

so that

$$\int \varrho(x) \, dx = \int R(\sin x, \cos x) \, dx, \quad x \in L_i^k$$

$$\xleftarrow{t=g_k(x)} \int R\left(\frac{2t}{1+t^2}, \frac{1-t^2}{1+t^2}\right) \frac{2}{1+t^2} \, dt, \quad t \in g_k(L_i^k). \tag{11.31}$$

Clearly, the function

$$r(t) := R\left(\frac{2t}{1+t^2}, \frac{1-t^2}{1+t^2}\right) \frac{2}{1+t^2} \tag{11.32}$$

can be reduced to a fraction of polynomials, so its primitive can be found using techniques from Chapter 6. Reverse the change $t = g_k(x)$ in order to obtain the primitive of $\varrho(x)$ on L_i^k. Choose a local antiderivative $F_i^k(x)$ of $\varrho(x)$ on L_i^k.

3: Following the procedure explained in Section 4.4, carefully apply Propositions 4.3.1 and 4.3.3 on the points where the local antiderivatives $F_i^k(x)$ have to be assembled to make a local antiderivative $F_i(x)$ of $\varrho(x)$ on each interval J_i. This requires the choice of adequate constants C_i^k so that the function defined on $J_i \backslash S$ and given by

$$H_i(x) := F_i^k(x) + C_i^k, \quad \text{if } x \in L_i^k \tag{11.33}$$

is extended to a continuous function F_i on J_i. Then, the global primitive of $\varrho(x)$ is given by

$$\int \varrho(x) \, dx = F_i(x) + C_i, \quad x \in J_i, \ i \in J$$

for any choice of real constants C_i.

Let us elaborate on the process of assembling the antiderivatives F_i^k. First of all, realize that this process must be done at the endpoints of the intervals I_k that belong to D_ϱ, that is, at the elements of the subset S introduced in (11.8). Assume, then, that $S \neq \varnothing$; equivalently, $S = \{(2k-1)\pi \colon k \in \mathbb{Z}\}$.

As shown in Sections 6.6 and 6.8 (see (6.48)) the function $r(t)$ given in (11.32) can be decomposed as

$$r(t) = P_1(t) + \left(\frac{P_0(t)}{Q_0(t)}\right)' + \sum_{i=1}^{m} \frac{A_i}{t - a_i} + \sum_{i=1}^{n} \frac{M_i t + N_i}{t^2 + \mu_i t + \nu_i} \tag{11.34}$$

where $P_1(t)$ is a polynomial over $\mathbb{R}$ (identically null if the fraction in (11.32) has negative degree), $P_0(t)/Q_0(t)$ is a fraction of polynomials over $\mathbb{R}$ of negative degree, and A_i, M_i, N_i, a_i, μ_i and ν_i are real numbers. Choose a polynomial $P(t)$ whose constant term is null and such that $P'(t) = P_1(t)$. Note that under these conditions, if $P_1(t)$ is identically null, then $P(t)$ must be chosen identically null too. It follows from formulas (6.1) and (6.3) that $r(t)$ has a unique antiderivative of the form

$$G(t) = P(t) + \frac{P_0(t)}{Q_0(t)} + \mathcal{L}(t) + \mathcal{A}(t), \ t \in D_G \tag{11.35}$$

where

$$\mathcal{L}(t) = \sum_{i=1}^{m} A_i \log|t - a_i| + \sum_{i=1}^{n_1} B_i \log(t^2 + \mu_i t + \nu_i) \tag{11.36}$$

$$\mathcal{A}(t) = \sum_{i=1}^{n_2} C_i \arctan(\alpha_i t + \beta_i) \tag{11.37}$$

for certain real numbers B_i, C_i, α_i and β_i, and certain natural numbers $n_1 \leqslant n$ and $n_2 \leqslant n$. Let $P(t)$, $P_0(t)/Q_0(t)$, $\mathcal{L}(t)$ and $\mathcal{A}(t)$ be respectively called the *polynomial part*, *fractional part*, the *logarithmic part* and the *arctan part* of $G(t)$. The local antiderivatives of $\varrho(x)$ on each L_i^k can be chosen as

$$F_i^k(x) = G\left(\tan\frac{x}{2}\right), \quad x \in L_i^k.$$

Pick any element $(2k+1)\pi \in S$. This number is the separating point between I_k and I_{k+1} and in turn, between L_j^k and L_j^{k+1} for certain j. Since $\varrho(x)$ is continuous at $(2k+1)\pi$, Proposition 4.3.2 shows that both F_j^k and F_j^{k+1} can be extended with continuity at $(2k+1)\pi$. Hence, there exist

$$l_j^k := \lim_{x \to (2k+1)\pi^-} F_j^k(x) \in \mathbb{R}$$

$$\lambda_j^{k+1} := \lim_{x \to (2k+1)\pi^+} F_j^{k+1}(x) \in \mathbb{R}.$$

In order to find the constants C_j^i in (11.33) to compound the antiderivative F_j on J_j in accordance with Proposition 4.3.3, it is necessary to know the values of l_j^k and λ_j^{k+1}. This can be done by means of Proposition 4.2.5, which in turn reveals that the values of l_j^k and λ_j^{k+1} are independent from j and k:

$$l := l_j^k = \lim_{t \to \infty} G(t) \tag{11.38}$$

$$\lambda := \lambda_j^{k+1} = \lim_{t \to -\infty} G(t). \tag{11.39}$$

Since the fraction $P_0(t)/Q_0(t)$ is of negative degree, its limit at ∞ is null. Hence,

$$l = \lim_{t \to \infty} P(t) + \mathcal{L}(t) + \mathcal{A}(t). \tag{11.40}$$

Moreover, as $\lim_{t \to \infty} \mathcal{A}(t) = \frac{\pi}{2} \sum_{i=1}^{n_2} C_i \in \mathbb{R}$, it follows from (11.40) that there exists $\lim_{t \to \infty} P(t) + \mathcal{L}(t) \in \mathbb{R}$. But $\mathcal{L}(t)/t \to 0$ as $t \to \infty$. Hence, $P(t)$ must be identically constant, and since its constant term is null, then $P(t) \equiv 0$. This means that $\mathcal{L}(t)$ must have finite limit at ∞ too, and by (11.36), its value is the logarithm of

$$\lim_{t \to \infty} \prod_{i=1}^{m} |t - a_i|^{A_i} \cdot \prod_{i=1}^{n_1} (t^2 + \mu_i t + \nu_i)^{B_i} = \lim_{t \to \infty} t^c$$

where $c = \sum_{i=1}^{m} A_i + 2 \sum_{i=1}^{n_1} B_i$. But $\lim_{t \to \infty} t^c$ is finite if and only if $c = 0$, in which case $\lim_{t \to \infty} t^c = 1$ and subsequently, $\lim_{t \to \infty} \mathcal{L}(t) = 0$. Thus, from (11.40),

$$l = \lim_{t \to \infty} \mathcal{A}(t) = \frac{\pi}{2} \sum_{i=1}^{n_2} C_i, \tag{11.41}$$

that is, the limit at ∞ of the arctan part of $G(t)$.

With a similar argument, λ can be calculated as the limit at $-\infty$ of the arctan part of $G(t)$:

$$\lambda = \lim_{t \to -\infty} \mathcal{A}(t) = -\frac{\pi}{2} \sum_{i=1}^{n_2} C_i = -l. \tag{11.42}$$

Note that from (11.41) and (11.42), it follows that the length of the jump of the local antiderivatives $F_i^k(x)$ at each point of S is the same: $|l_j^k - \lambda_j^{k+1}| = \pi |\sum_{i=1}^{n_2} C_i| = 2l$.

Observation: Formulas (11.41) and (11.42) are a practical shortcut to calculate the values of the sided limits l_j^k and λ_j^{k+1}.

Example 11.3.1. Calculate $\displaystyle\int \frac{dx}{1 + \cos x}$.

Solution. Denote $f(x) := 1/(1 + \cos x)$. The regular decomposition of D_f is $\bigcup_{n=-\infty}^{\infty} J_n$ where $J_n := \big((2n-1)\pi, (2n+1)\pi\big)$.

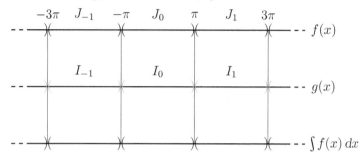

Since each interval J_n coincides with the domain of the change $g_n(x)$ defined in (11.10), this change can be directly implemented on J_n. To do that, by means of the substitutions (11.28), (11.29) and (11.30)

$$\int \frac{dx}{1 + \cos x}, \ x \in J_n \xleftarrow{t=\tan(x/2)} 2 \int \frac{1}{1 + \frac{1-t^2}{1+t^2}} \frac{dt}{1+t^2}$$

$$= \int dt = t + C, \ t \in (-\infty, \infty).$$

(11.43)

Reversing the change $t = g_k(x)$, we obtain

$$\int \frac{dx}{1 + \cos x} = \tan \frac{x}{2} + C, \ x \in J_n.$$

Hence, the complete solution is

$$\int \frac{dx}{1 + \cos x} = \tan \frac{x}{2} + C(x), \quad x \in \bigcup_{n=-\infty}^{\infty} \big((2n-1)\pi, (2n+1)\pi\big)$$

for all functions $C(x)$ locally constant on D_f. □

Example 11.3.2. Find $\displaystyle\int \frac{dx}{3 + \sin x + \cos x}$.

Solution. It is observable at a glance that the natural domain of $f(x) := 1/(3 + \sin x + \cos x)$ is $D_f = (-\infty, \infty)$. Since the calculation of $\int f(x)\, dx$ requires the changes of variable $g_n \colon I_n \longrightarrow \mathbb{R}$ given in (11.10), we first find the primitive of $f(x)$ on each interval $I_n := \big((2n-1)\pi, (2n+1)\pi\big)$, $n \in \mathbb{Z}$. This will provide an antiderivative $F_n(x)$ of $f(x)$ on I_n which will be assembled later to get a primitive of $f(x)$ on the entire interval $(-\infty, \infty)$.

Following the outlined plan, the change $t = g_n(x)$ is implemented by means of the substitutions (11.28), (11.29) and (11.30), so that

$$\int \frac{dx}{3 + \sin x + \cos x} \, dx, \; x \in I_n \xleftarrow{t = \tan \frac{x}{2}} \int \frac{1}{3 + \frac{2t+1-t^2}{1+t^2}} \frac{2\,dt}{1+t^2}$$

$$= \int \frac{dt}{t^2 + t + 2}$$

$$= \frac{4}{7} \int \frac{dt}{\left(\frac{2}{\sqrt{7}}t + \frac{1}{\sqrt{7}}\right)^2 + 1}$$

$$= G(t) + C_n, \; t \in (-\infty, \infty) \qquad (11.44)$$

where

$$G(t) := \frac{2}{\sqrt{7}} \arctan\left(\frac{2}{\sqrt{7}}t + \frac{1}{\sqrt{7}}\right), \; t \in (-\infty, \infty).$$

Thus, by reversing the change g_n in (11.44), we obtain a local antiderivative of $f(x)$ on I_n given by $F_n(x) := G\big(\tan(x/2)\big)$, $x \in I_n$. Next, we need to choose adequate constants C_n so that the function $H(x)$ defined on $\bigcup_{n \in \mathbb{Z}} I_n$ by

$$H(x) := F_n(x) = G\left(\tan\frac{x}{2}\right) + C_n, \; \text{if } x \in I_n \qquad (11.45)$$

can be extended to a continuous function $F(x)$ on $\mathbb{R}$. This means that we have to calculate the one-sided limits of $G(x)$ at the endpoints of the intervals I_n:

$$l_n := \lim_{x \to (2n+1)\pi^-} G\left(\tan\frac{x}{2}\right) = \lim_{t \to \infty} G(t) = \frac{\pi}{\sqrt{7}}$$

$$\lambda_n := \lim_{x \to (2n-1)\pi^+} G\left(\tan\frac{x}{2}\right) = -\frac{\pi}{\sqrt{7}}.$$

Thus, letting $C_n := 2\pi n/\sqrt{7}$ in (11.45), the local antiderivatives F_n so chosen satisfy

$$\lim_{x \to (2n+1)\pi^-} F_n(x) + C_n = \frac{\pi}{\sqrt{7}} + \frac{2\pi}{\sqrt{7}}n = \lim_{x \to (2n+1)\pi^+} F_{n+1}(x) + C_{n+1}.$$

Hence, the function

$$F(x) := \begin{cases} G\left(\tan\frac{x}{2}\right) + \frac{2\pi}{\sqrt{7}}n, & x \in I_n, \; n \in \mathbb{Z} \\ \frac{\pi}{\sqrt{7}} + \frac{2\pi}{\sqrt{7}}n, & x = (2n+1)\pi, \; n \in \mathbb{Z} \end{cases}$$

is continuous on $\mathbb{R}$, $F|_{I_n} = F_n$ for all $n \in \mathbb{Z}$, and by virtue of Proposition 4.3.3, it is an antiderivative of $f(x)$ on $\mathbb{R}$. Therefore, $\int f(x)\,dx = F(x) + C$ for all $C \in \mathbb{R}$. $\quad\square$

For the convenience of the reader, the graphs of $F(x)$ and $G(\tan x/2)$ of Example 11.3.2 are drawn next.

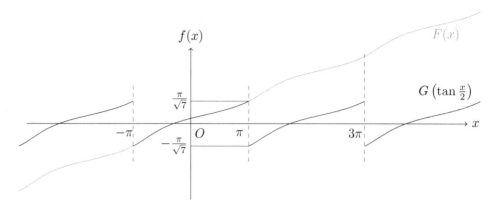

Example 11.3.3. Find $\displaystyle\int \frac{dx}{\sin x + \cos x}$.

Solution. Denote $f(x) := 1/(\sin x + \cos x)$. Since $\sin x + \cos x = 0$ if and only if $x = -\frac{\pi}{4} + n\pi$ for all $n \in \mathbb{Z}$, the regular decomposition of the natural domain of f is $D_f = \bigcup_{n=-\infty}^{\infty} J_n$, where

$$J_n := \left(-\frac{1}{4}\pi + n\pi, \frac{3}{4}\pi + n\pi \right), \quad n \in \mathbb{Z}.$$

Let $G_n := J_{2n-1} \cup J_{2n}$ for every $n \in \mathbb{Z}$. Since $f(x)$ is periodic of period 2π, the task of finding the primitive of $f(x)$ on D_f can be reduced to looking for a local antiderivative $H_0(x)$ on G_0 so that, by replication of $H_0(x)$ on each set G_n, the function so obtained,

$$F(x) := H_0(x - 2n\pi), \quad x \in G_n, \ n \in \mathbb{Z} \tag{11.46}$$

is a complete antiderivative of $f(x)$ on D_f and consequently,

$$\int \frac{dx}{\sin x + \cos x} = F(x) + C(x), \quad x \in \bigcup_{n=-\infty}^{\infty} J_n \tag{11.47}$$

for all functions $C(x)$ locally constant on $\bigcup_{n=-\infty}^{\infty} J_n$.

Bearing in mind the observation just made, we first calculate $\int f(x)\,dx$ on G_0.

Since the regular decomposition of G_0 is $G_0 = J_{-1} \cup J_0$ and since the intervals J_i do not adjust well to the intervals $I_k = ((2k-1)\pi, (2k+1)\pi)$ where the changes of variable g_k given in (11.10) are defined, the first step is the delimitation of the intervals where the changes g_k can be safely applied. These intervals are

$$L_{-1}^{-1} := J_{-1} \cap I_{-1} = \left(-\frac{\pi}{4} - \pi, -\pi \right),$$

$$L_{-1}^{0} := J_{-1} \cap I_0 = \left(-\pi, -\frac{\pi}{4} \right),$$

$$L_0^0 := J_0 \cap I_0 = \left(-\frac{\pi}{4}, \frac{3\pi}{4} \right).$$

Observe that $G_0\backslash(L_{-1}^{-1} \cup L_{-1}^0 \cup L_0^0) = \{-\pi\}$.

The implementation of g_k on each L_i^k is carried out by means of the corresponding substitutions (11.28), (11.29) and (11.30), so that

$$\int \frac{dx}{\sin x + \cos x}, \ x \in L_i^k \xleftarrow{t=\tan\frac{x}{2}} \int \frac{1}{\frac{2t}{1+t^2} + \frac{1-t^2}{1+t^2}} \cdot \frac{2\,dt}{1+t^2}$$

$$= 2 \int \frac{dt}{-t^2 + 2t + 1}, \ t \in g_k(L_i^k). \tag{11.48}$$

The last primitive in (11.48) is solved as

$$\int \frac{2\,dt}{-t^2 + 2t + 1} = -\int \frac{1/\sqrt{2}}{t - 1 - \sqrt{2}}\,dt + \int \frac{1/\sqrt{2}}{t - 1 + \sqrt{2}}\,dt$$

$$= G(t) + C(t), \ t \in \mathbb{R}\backslash\{1 \pm \sqrt{2}\} \tag{11.49}$$

where

$$G(t) := \frac{1}{\sqrt{2}} \log \frac{|t - 1 + \sqrt{2}|}{|t - 1 - \sqrt{2}|}$$

and $C(t)$ is any locally constant function on $\mathbb{R}\backslash\{1 \pm \sqrt{2}\}$.

Thus, from (11.48) and (11.49), a reversion of the change g_k yields the set of all local antiderivatives on L_i^k as

$$F_i^k(x) := G\left(\tan \frac{x}{2}\right) + C_i^k, \ x \in L_i^k \tag{11.50}$$

where C_i^k is any real number.

Next, we have to find the local antiderivatives of $f(x)$ on J_{-1} and J_0.

Since $J_0 = J_0 \cap L_0^0$, the expression (11.50) gives the primitive of $f(x)$ on J_0 directly.

In order to obtain the primitive on J_{-1}, since $-\pi$ belongs to J_{-1} and separates L_{-1}^{-1} from L_{-1}^0, it is necessary to calculate the one-sided limits of $F_{-1}^{-1}(x)$ and F_{-1}^0 at $-\pi$. To do this, as the arctan part of $G(t)$ is null, it readily follows from (11.41) and (11.42) that

$$\lim_{x \to -\pi^+} F_{-1}^{-1}(x) = 0 = \lim_{x \to -\pi^+} F_{-1}^0(x).$$

Hence, letting $C_{-1}^{-1} = C_{-1}^{0} = C_0^0 = 0$ in (11.50), the resulting antiderivatives F_{-1}^{-1}, F_{-1}^0 and F_0^0 can be assembled to make a continuous function $H_0(x)$ on $J_{-1} \cup J_0$ defined as

$$H_0(x) := \begin{cases} G\left(\tan \frac{x}{2}\right), & x \in J_{-1} \cup J_0 \backslash \{-\pi\} \\ 0, & x = -\pi. \end{cases}$$

Thus, $H_0(x)$ is an antiderivative of $f(x)$ on G_0 by virtue of Proposition 4.3.3, and the primitive of $f(x)$ on the entire set D_f can be built by replication as indicated in (11.46) and (11.47). □

The solution on $(-5\pi/4, -\pi/4) \cup (-\pi/4, 3\pi/4)$ of Example 11.3.3 is represented in the Diagram below.

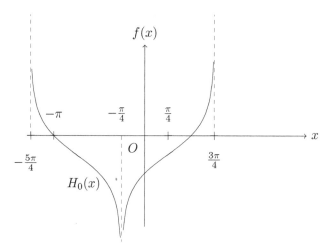

Example 11.3.4. Find $\displaystyle\int \frac{2\,dx}{(1 - \cos x)(2 + \sin x)}$.

Solution. Denote $f(x) := 2/((1 - \cos x)(2 + \sin x))$. Clearly, the regular decomposition of D_f is $\bigcup_{n=-\infty}^{\infty} J_n$ where $J_n := (2n\pi, 2(n+1)\pi)$.

Since $f(x)$ is periodic of period 2π, it will suffice to calculate a local antiderivative $F_0(x)$ of $f(x)$ on J_0, so that, by replication of $F_0(x)$ on each J_n, the function thus obtained,

$$F(x) := F_0(x - 2n\pi), \quad x \in J_n, \ n \in \mathbb{Z} \tag{11.51}$$

is a complete antiderivative of $f(x)$ on D_f and therefore,

$$\int \frac{2\,dx}{(1 - \cos x)(2 + \sin x)} = F(x) + C(x), \quad x \in \bigcup_{n=-\infty}^{\infty} J_n \tag{11.52}$$

for any locally constant function $C(x)$ on $\bigcup_{n=-\infty}^{\infty} J_n$.

Thus, let us calculate $\int f(x)\,dx$ on J_0. The appropriate changes of variable to solve the primitive of $f(x)$ are the functions $g_n : I_n \longrightarrow \mathbb{R}$ given in (11.10).

Since the intervals $I_n = \big((2n-1)\pi, (2n+1)\pi\big)$ do not adjust well with J_0, the changes g_0 and g_1 will be respectively implemented on $L_0^0 := J_0 \cap I_0 = (0, \pi)$ and on $L_0^1 := J_0 \cap I_1 = (\pi, 2\pi)$ by means of the corresponding substitutions (11.28), (11.29) and (11.30), so that

$$\int \frac{2\,dx}{(1-\cos x)(2+\sin x)}, \ x \in L_0^j \ (j=1,2) \xleftarrow{\ t=\tan\frac{x}{2}\ }$$

$$\int \frac{1+t^2}{t^2(1+t+t^2)}\,dt, \ t \in g_j(L_0^j) \tag{11.53}$$

where $g_0(L_0^0) = (0, \infty)$ and $g_1(L_0^1) = (-\infty, 0)$.

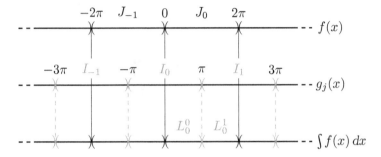

By means of the usual methods given in Chapter 6, the fraction on the right side of (11.53) is decomposed as

$$\frac{1+t^2}{t^2(1+t+t^2)} = -\frac{1}{t} + \frac{1}{t^2} + \frac{t+1}{1+t+t^2}$$

$$= -\frac{1}{t} + \frac{1}{t^2} + \frac{1}{2}\frac{2t+1}{t^2+t+1} + \frac{2/3}{\left(\frac{2}{\sqrt{3}}t + \frac{1}{\sqrt{3}}\right)^2 + 1}$$

so that one of its antiderivatives on its natural domain is

$$G(t) := -\log|t| - \frac{1}{t} + \frac{1}{2}\log(t^2+t+1) + \frac{1}{\sqrt{3}}\arctan\left(\frac{2}{\sqrt{3}}t + \frac{1}{\sqrt{3}}\right), \ t \in \mathbb{R}\backslash\{0\}. \tag{11.54}$$

Thus, by combination of (11.53) and (11.54) and substituting $t = g_j(x)$, we obtain the function

$$F_0^j(x) := G\left(\tan\frac{x}{2}\right), \ x \in L_0^j$$

which is an antiderivative of $f(x)$ on L_0^j for $j=0$ and $j=1$.

In order to assemble a local antiderivative of $f(x)$ on J_0, note that π belongs to J_0 and separates L_0^0 from L_0^1. Hence, we need to calculate the corresponding one-sided limits of F_0^0 and F_0^1 at that point. Since the arctan part of $G(t)$ is

$$\mathcal{A}(t) := \frac{1}{\sqrt{3}}\arctan\left(\frac{2}{\sqrt{3}}t + \frac{1}{\sqrt{3}}\right)$$

it follows from (11.41) and (11.42) that

$$l := \lim_{x \to \pi^-} F_0^0(x) = \lim_{t \to \infty} \mathcal{A}(t) = \frac{1}{2\sqrt{3}}\pi$$

$$\lambda := \lim_{x \to \pi^+} F_0^1(x) = \lim_{t \to -\infty} \mathcal{A}(t) = -\frac{1}{2\sqrt{3}}\pi.$$

Hence, a local antiderivative of $f(x)$ on J_0 is

$$F_0(x) := \begin{cases} G\left(\tan \frac{x}{2}\right), & x \in (0, \pi) \\ \frac{\pi}{2\sqrt{3}}, & x = \pi \\ G\left(\tan \frac{x}{2}\right) + \frac{\pi}{\sqrt{3}}, & x \in (\pi, 2\pi). \end{cases}$$

This is continuous on $(0, 2\pi)$, and as it extends both local antiderivatives $F_0^0(x)$ and $F_0^1(x)$, then, by virtue of Proposition 4.3.3, $F_0(x)$ is a local antiderivative of $f(x)$ on $(0, 2\pi)$ and the complete primitive of $f(x)$ can be obtained following (11.51). □

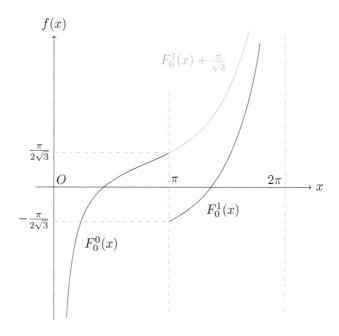

11.4   The case $R(\sin x, \cos x) = R(-\sin x, -\cos x)$

The particular case where the functions given in (11.27) are formed by summands of the form $\sin^{2n} x \cos^{2m} x$ and $\sin^{2n-1} x \cos^{2m-1} x$ for all integers n and m can be recognized because they satisfy the identity

$$R(\sin x, \cos x) = R(-\sin x, -\cos x).$$

In that case, the implementation of the changes obtained from

$$t = \tan x$$

transforms the primitive of $\varrho(x)$ into the primitive of a fraction of polynomials where the degree of its numerator and denominator is reduced by half with respect to the changes derived from $t = \tan(x/2)$. This happens because

$$\sin x = \text{sgn}(\cos x)\frac{\tan x}{\sqrt{1 + \tan^2 x}}, \qquad x \neq \frac{\pi}{2} + k\pi, \ k \in \mathbb{Z} \tag{11.55}$$

$$\cos x = \text{sgn}(\cos x)\frac{1}{\sqrt{1 + \tan^2 x}}, \qquad x \neq \frac{\pi}{2} + k\pi, \ k \in \mathbb{Z}. \tag{11.56}$$

Thus, the functions $\sin x$ and $\cos x$ raised to an even power are

$$\sin^{2n}(x) = \frac{\tan^{2n} x}{(1 + \tan^2 x)^n} \tag{11.57}$$

$$\cos^{2n}(x) = \frac{1}{(1 + \tan^2 x)^n}, \qquad x \neq \frac{\pi}{2} + k\pi, \ k \in \mathbb{Z} \tag{11.58}$$

and

$$\sin^{2n-1}(x)\cos^{2n-1}(x) = \frac{\tan^{2n-1} x}{(1 + \tan^2 x)^n}, \qquad x \neq \frac{\pi}{2} + k\pi, \ k \in \mathbb{Z}. \tag{11.59}$$

The procedure to implement the changes derived from $t = \tan x$ in these cases is similar to that given in Section 11.3:

0: First, if every occurrence of $\sin x$ and of $\cos x$ is raised to an even power in the expression $R(\sin x, \cos x)$, plug the substitutions

$$\cos^2 x = \frac{1 + \cos(2x)}{2} \tag{11.60}$$

$$\sin^2 x = \frac{1 - \cos(2x)}{2} \tag{11.61}$$

into $R(\sin x, \cos x)$. This action reduces the power of each occurrence of $\sin x$ and of $\cos x$ in $R(\sin x, \cos x)$ by half and, consequently, the complexity of the function is also reduced. For convenience, the linear change of variable $z = 2x$ may be implemented. Repeat this action as many times as possible. If after this action the resulting expression does not satisfy the identity that titles this section, then find its primitive by applying the procedure in Section 11.3. Otherwise, proceed with the following steps.

1: Determine the regular decomposition $\bigcup_{i \in J} J_i$ of D_ϱ. Following the notation in (11.14), let $I_k := ((2k-1)\pi/2, (2k+1)\pi/2)$ for all $k \in \mathbb{Z}$ and let $g_k(x)$ be the change of variable given by $t = \tan x$, $x \in I_k$. Consider the intervals $L_i^k := J_i \cap I_k$ that are non-empty.

2: The implementation of the change $t = g_k(x)$ to solve the primitive of $\varrho(x)$ on L_i^k should be carried out by means of the substitutions (11.55), (11.56) and (11.16). However, in practice, the substitutions

$$\sin x = \frac{t}{\sqrt{1 + t^2}} \tag{11.62}$$

$$\cos x = \frac{1}{\sqrt{1 + t^2}} \tag{11.63}$$

$$dx = \frac{1}{1 + t^2} \, dt, \ x \in L_i^k \tag{11.64}$$

leave exactly the same expression as that produced by the substitutions (11.55), (11.56) and (11.16), thanks to the validity of (11.57), (11.58) and (11.59).[1] Thus, it is safe to write

$$\int \varrho(x)\,dx = \int R(\sin x, \cos x)\,dx, \quad x \in L_i^k$$

$$\xleftarrow{t=g_k(x)} \int R\left(\frac{t}{\sqrt{1+t^2}}, \frac{1}{\sqrt{1+t^2}}\right)\frac{1}{1+t^2}\,dt, \quad t \in g_k(L_j^k) \tag{11.65}$$

and the function

$$r(t) := R\left(\frac{t}{\sqrt{1+t^2}}, \frac{1}{\sqrt{1+t^2}}\right)\frac{1}{1+t^2} \tag{11.66}$$

can be reduced to a fraction of polynomials whose primitive can be found as described in Chapter 6. Once its primitive is found, the change $t = g_k(x)$ is reversed in order to obtain the primitive of $\varrho(x)$ on L_i^k. Choose a local antiderivative $F_i^k(x)$ of $\varrho(x)$ on L_i^k.

3: Following the procedure explained in Section 4.4, a careful application of Propositions 4.3.1 and 4.3.3 allows the local antiderivatives $F_i^k(x)$ to be assembled to make a local antiderivative $F_i(x)$ of $\varrho(x)$ on each interval J_i. This means that we have to find suitable constants C_i^k so that the function

$$H_i(x) := F_i^k(x) + C_i^k, \quad x \in L_i^k \tag{11.67}$$

defined on $J_i \backslash T$ is extended to a continuous function F_i on J_i. Then, the global primitive of $\varrho(x)$ is given by

$$\int \varrho(x)\,dx = F_i(x) + C_i, \quad x \in J_i, \ i \in J$$

for any choice of real constants C_i.

The process of assembling the functions F_i^k admits similar indications as those given for the general case treated in Section 11.3: first realize that the assembly must be done at the points of the set T of endpoints of I_k that belong to D_ϱ, and since ϱ is periodic of period π, either $T = \varnothing$ or $T = \{\pi/2 + n\pi : n \in \mathbb{Z}\}$. Assume $T \neq \varnothing$.

Reasoning as in Section 11.3, the fraction of polynomials $r(t)$ given in (11.66) has a unique antiderivative of the form

$$G(t) = P(t) + \frac{P_0(t)}{Q_0(t)} + \mathcal{L}(t) + \mathcal{A}(t), \quad t \in D_G \tag{11.68}$$

where $P(t)$ is a polynomial whose constant term is null, $P_0(t)/Q_0(t)$ is a fraction of polynomials of negative degree,

$$\mathcal{L}(t) = \sum_{i=1}^{m} A_i \log|t - a_i| + \sum_{i=1}^{n_1} B_i \log(t^2 + \mu_i t + \nu_i) \tag{11.69}$$

$$\mathcal{A}(t) = \sum_{i=1}^{n_2} C_i \arctan(\alpha_i t + \beta_i) \tag{11.70}$$

---

[1] Indeed, (11.62) and (11.63) do not hold in general because of a misplacement of signs, but are easier to recall than (11.55) and (11.56).

for certain real numbers B_i, C_i, α_i and β_i, and certain natural numbers m, n_1 and n_2. Clearly,

$$F_i^k(x) = G(\tan x), \quad x \in L_i^k$$

is a local antiderivative of $\varrho(x)$ on L_i^k.

Pick any element $\pi/2 + k\pi \in T$. This number is the separating point between I_k and I_{k+1}, and in turn between L_j^k and L_j^{k+1} for some j.

Since $\varrho(x)$ is continuous at $\pi/2 + k\pi$, Proposition 4.3.2 shows that both F_j^k and F_j^{k+1} can be extended with continuity at this point. Hence, there exist

$$l_j^k := \lim_{x \to \frac{\pi}{2} + k\pi^-} F_j^k(x) = \lim_{t \to \infty} G(t) =: l \in \mathbb{R}$$

$$\lambda_j^{k+1} := \lim_{x \to \frac{\pi}{2} + k\pi^+} F_j^{k+1}(x) = \lim_{t \to \infty} G(t) =: \lambda \in \mathbb{R}.$$

The construction of the antiderivative F_j on J_j in accordance with Proposition 4.3.3 requires finding the suitable constants C_j^i mentioned in (11.67), and in turn, that means the calculation of the values of l and λ. With similar arguments as those given in the section above,

$$l = \lim_{t \to \infty} G(t) = \lim_{t \to \infty} \mathcal{A}(t) \tag{11.71}$$

$$\lambda = \lim_{t \to -\infty} G(t) = \lim_{t \to -\infty} \mathcal{A}(t) = -l. \tag{11.72}$$

From (11.71) and (11.72), it follows that the length of the jump of the local antiderivatives $F_i^k(x)$ at each point of T is the same.

Observation: Formulas (11.71) and (11.72) are a practical shortcut to calculate the values of the sided limits l_j^k and λ_j^{k+1}.

The following examples show in detail the application of the procedure just described.

Example 11.4.1. Find $\displaystyle\int \frac{dx}{\sin x \cos x}$.

Solution. The natural domain of $f(x) := 1/(\sin x \cos x)$ is $D_f = \bigcup_{n=-\infty}^{\infty} J_n$ where

$$J_n := \left(n\frac{\pi}{2}, (n+1)\frac{\pi}{2} \right).$$

The appropriate changes of variable for $f(x)$ are the functions $g_k(x)$ given in (11.14). Each g_k is defined on

$$I_k := \big((2k-1)\pi/2, (2k+1)\pi/2 \big).$$

Since $J_{2k-1} \cup J_{2k} \subset I_k$ for all $k \in \mathbb{Z}$, each change g_k must be implemented on J_{2k-1} and on J_{2k} separately. Thus, by means of the substitutions (11.62), (11.63) and (11.64), we obtain

$$\int \frac{dx}{\sin x \cos x}, \quad x \in J_{2k} \xleftarrow{\ t=\tan x\ } \int \frac{1}{\frac{t}{1+t^2}} \cdot \frac{dt}{1+t^2}$$

$$= \int \frac{dt}{t} = \log t + C_{2k}, \quad t \in g_k(J_{2k}) \subset (0, \infty). \tag{11.73}$$

In the same manner

$$\int \frac{dx}{\sin x \cos x}, \quad x \in J_{2k-1} \xleftarrow{\ t=\tan x\ } \log(-t) + C_{2k-1}, \quad t \in g_k(J_{2k-1}) \subset (-\infty, 0). \tag{11.74}$$

From (11.73) and (11.74), and reversing the change g_k, the complete solution is

$$\int \frac{dx}{\sin x \cos x} = \log|\tan x| + C(x), \quad x \in \bigcup_{k\in\mathbb{Z}} J_k$$

for all functions $C(x)$ locally constant on $\bigcup_{k\in\mathbb{Z}} J_k$. ☐

Example 11.4.2. Find $\int \dfrac{\cos^2 x - 2\sin x \cos x}{1 + \cos^2 x - 2\sin x \cos x} \, dx$.

Solution. Clearly, $1 + \cos^2 x > \sin(2x) = 2\sin x \cos x$ for all x, which proves that the natural domain of $f(x) := [\cos^2 x - 2\sin x \cos x]/[1 + \cos^2 x - 2\sin x \cos x]$ is $D_f = \mathbb{R}$.

The appropriate changes of variable are the functions g_k given in (11.14). Since the domain of g_k is $I_k := \big((2k-1)\pi/2, (2k+1)\pi/2\big)$, we first calculate a local antiderivative $F_k(x)$ of $f(x)$ on each I_k. Next, these antiderivatives will be assembled at the points $\pi/2 + k\pi$ to make an antiderivative of $f(x)$ on its entire domain.

$$(11.75)$$

First, implement the change g_k on I_k by means of the substitutions (11.62), (11.63) and (11.64), so that

$$\int \frac{\cos^2 x - 2\sin x \cos x}{1 + \cos^2 x - 2\sin x \sin x} \, dx, \quad x \in I_k \xleftarrow{\ t=\tan x\ }$$

$$\int \frac{\frac{1}{1+t^2} - 2\frac{t}{1+t^2}}{1 + \frac{1}{1+t^2} - 2\frac{t}{1+t^2}} \cdot \frac{dt}{1+t^2} = \int \frac{1 - 2t}{(t^2 - 2t + 2)(t^2 + 1)} \, dt, \quad t \in (-\infty, \infty). \tag{11.76}$$

Following the techniques shown in Chapter 6, we obtain the partial decomposition

$$\frac{1 - 2t}{(t^2 - 2t + 2)(t^2 + 1)} = -\frac{1}{(t-1)^2 + 1} + \frac{1}{t^2 + 1}$$

which readily yields

$$\int \frac{1 - 2t}{(t^2 - 2t + 2)(t^2 + 1)} \, dt = G(t) + C, \ t \in \mathbb{R}, \tag{11.77}$$

where $G(t) := -\arctan(t - 1) + \arctan t$.

By reversing the change of variable $t = g_k(x)$ in (11.77), formula (11.76) yields the local antiderivative

$$F_k(x) := -\arctan(\tan x - 1) + x, \ x \in I_k$$

of $f(x)$.

The next step consists in choosing constants C_k so that the function

$$H(x) := F_k(x) + C_k, \ x \in I_k, \ k \in \mathbb{Z} \tag{11.78}$$

can be extended to a continuous function on $\mathbb{R}$. To do this, we need to calculate the sided limits of the functions F_k at the separating points between their domains, that is, at the points $\pi/2 + k\pi$ $(k \in \mathbb{Z})$. By means of (11.71) and (11.72),

$$\lim_{x \to \frac{\pi}{2} + k\pi^-} F_k(x) = \lim_{t \to \infty} G(t) = 0$$

$$\lim_{x \to \frac{\pi}{2} + k\pi^+} F_k(x) = \lim_{t \to -\infty} G(t) = 0.$$

Hence, the choices $C_k := 0$ for all k in (11.78) allow $H(x)$ to be extended to the continuous function

$$F(x) := \begin{cases} -\arctan(\tan x - 1) + x, & x \in \bigcup_{k=-\infty}^{\infty} I_k \\ 0, & x = \frac{\pi}{2} + k\pi, \ k \in \mathbb{Z}. \end{cases}$$

By virtue of Proposition 4.3.3, $F(x)$ is an antiderivative of $f(x)$ on $\mathbb{R}$, and therefore, $\int f(x) \, dx := F(x) + C$ for all constants C. $\square$

Example 11.4.3. Find $\dfrac{dx}{(1 + \sin x \cos x)(1 + \cos^2 x)}$.

Solution. Since $\sin x \cos x = \sin(2x)/2 \geqslant -1/2$ for all x, it is obvious that the natural domain of $f(x) := 1/((1 + \sin x \cos x)(1 + \cos^2 x))$ is $\mathbb{R}$. Moreover, the intervals $I_k := ((2k - 1)\pi/2, (2k + 1)\pi/2)$ are the respective domains of the changes of variable g_k given in (11.14), which are the appropriate changes for $f(x)$. Hence, we first calculate a local antiderivative $F_k(x)$ of $f(x)$ on each I_k, and then we assemble these antiderivatives at the points $\pi/2 + k\pi$ to make an antiderivative of $f(x)$ on its entire domain (Diagram 11.75 also applies to this example).

In order to calculate a local antiderivative on I_k, implement the change g_k on I_k by means of the substitutions (11.62), (11.63) and (11.64), so that

$$\int \frac{dx}{(1 + \sin x \cos x)(1 + \cos^2 x)}, \quad x \in I_k \xleftarrow{\;t=\tan x\;}$$

$$\int \frac{1}{\left(1 + \frac{t}{1+t^2}\right)\left(1 + \frac{1}{1+t^2}\right)} \frac{dt}{1 + t^2} = \int \frac{1 + t^2}{(t^2 + t + 1)(t^2 + 2)} \, dt, \quad t \in \mathbb{R}. \tag{11.79}$$

By means of the usual techniques of Chapter 6 for partial decompositions,

$$\frac{1 + t^2}{(t^2 + t + 1)(t^2 + 2)} = -\frac{1}{6}\frac{2t - 2}{t^2 + t + 1} + \frac{1}{6}\frac{2t + 2}{t^2 + 2} \tag{11.80}$$

and

$$\frac{2t - 2}{t^2 + t + 1} = \frac{2t + 1}{t^2 + t + 1} - \frac{4}{\left(\frac{2}{\sqrt{3}}t + \frac{1}{\sqrt{3}}\right)^2 + 1} \tag{11.81}$$

$$\frac{2t + 2}{t^2 + 2} = \frac{2t}{t^2 + 2} + \frac{1}{\left(t/\sqrt{2}\right)^2 + 1}. \tag{11.82}$$

Hence, from (11.80), (11.81) and (11.82),

$$\int \frac{1 + t^2}{(t^2 + t + 1)(t^2 + 2)} \, dt = G(t) + C, \quad t \in \mathbb{R} \tag{11.83}$$

where

$$G(t) = -\frac{1}{6}\log(t^2 + t + 1) + \frac{1}{\sqrt{3}} \arctan\left(\frac{2}{\sqrt{3}}t + \frac{1}{\sqrt{3}}\right)$$

$$+ \frac{1}{6}\log(t^2 + 2) + \frac{\sqrt{2}}{6} \arctan(t/\sqrt{2}) + C, \quad t \in \mathbb{R}.$$

Thus, by combination of (11.79) and (11.83), a reversion of the change $t = \tan x$ yields the local antiderivative

$$F_n(x) := G(\tan x), \quad x \in I_n$$

of $f(x)$ on I_n. Next, we need to find a constant C_n for each $n \in \mathbb{Z}$ so that the function defined on $\bigcup_{n \in \mathbb{Z}} I_n$ by

$$H(x) := F_n(x) + C_n, \quad \text{if } x \in I_n \tag{11.84}$$

can be enlarged to a continuous function $F(x)$ on $\mathbb{R}$. This task requires the calculation of the sided limits of the functions F_n at the separating points $\pi/2 + n\pi$ ($n \in \mathbb{Z}$). According to formulas (11.71) and (11.72), since the arctan part of $G(t)$ is

$$\mathcal{A}(t) = \frac{1}{\sqrt{3}} \arctan\left(\frac{2}{\sqrt{3}}t + \frac{1}{\sqrt{3}}\right) + \frac{\sqrt{2}}{6} \arctan(t/\sqrt{2})$$

it follows that

$$\lim_{x \to \frac{\pi}{2} + n\pi^-} F_n(x) = \lim_{t \to \infty} \mathcal{A}(t) = \frac{2\sqrt{3} + \sqrt{2}}{12}\pi$$

$$\lim_{x \to \frac{\pi}{2} + (n+1)\pi^+} F_{n+1}(x) = -\frac{2\sqrt{3} + \sqrt{2}}{12}\pi.$$

Thus, by choosing $C_n := (2\sqrt{3} + \sqrt{2})\pi n/6$ in (11.84), we get the continuous extension

$$F(x) := \begin{cases} G(\tan x) + \frac{2\sqrt{3}+\sqrt{2}}{6}\pi n, & x \in \left(-\frac{\pi}{2}, \frac{\pi}{2}\right) + n\pi \\ \frac{2\sqrt{3}+\sqrt{2}}{12}\pi + \frac{2\sqrt{3}+\sqrt{2}}{6}\pi n, & x = n\pi \end{cases}$$

of $H(x)$ on $\mathbb{R}$, and by Proposition 4.3.3,

$$\int \frac{dx}{(1 + \sin x \cos x)(1 + \cos^2 x)} = F(x) + C, \ x \in \mathbb{R}$$

for all constants C. $\qquad\square$

Some of the functions occurring in Example 11.4.3 are depicted next.

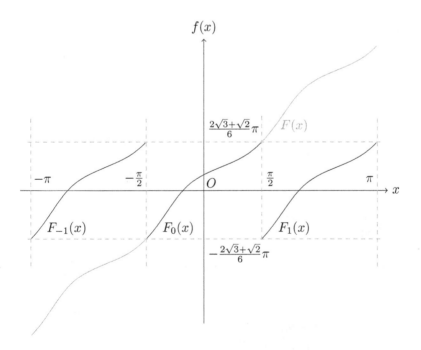

11.5 The case $R(-\sin x, \cos x) = -R(\sin x, \cos x)$

The function $\varrho(x) = R(\sin x, \cos x)$ given in (11.1) satisfies the identity

$$R(-\sin x, \cos x) = -R(\sin x, \cos x) \qquad (11.85)$$

if $\varrho(x)$ is a linear combination of functions of the form

$$\sin^{2k-1} x \cdot S(\sin x, \cos x)$$

where $k \in \mathbb{Z}$ and $S(t_1, t_2)$ is a rational function of its arguments for which each occurrence of t_1 is raised to an even power.

Since $\sin x$ is odd and $\cos x$ is even, a function $\varrho(x)$ which satisfies the identity (11.85) is odd and its natural domain is symmetric. Indeed,

$$\varrho(-x) = R(-\sin x, \cos x) = -R(\sin x, \cos x) = -\varrho(x), \quad x \in D_\varrho.$$

The most appropriate changes of variable to find the primitive of this type of function are the changes derived from $t = \cos x$. Indeed, this change allows us to skip the process of assembling local antiderivatives, as shown below.

The procedure to perform the changes derived from $t = \cos x$ is as follows.

1: Let $\bigcup_{i \in J} J_i$ be the regular decomposition of D_ϱ. Following the notation in (11.23), let $I_k = (k\pi, (k+1)\pi)$ and let $g_k(x)$ be the change of variable determined by $t := \cos x$ for $x \in I_k$. Consider only the intervals $L_i^k := J_i \cap I_k$ that are non-empty.

2: The change g_k is implemented on every L_i^k by means of the substitutions

$$t = \cos x \tag{11.86}$$

$$\sin\left(g_k^{-1}(t)\right) = \sin x \tag{11.87}$$

$$-\frac{dt}{\sin\left(g_k^{-1}(t)\right)} = dx \tag{11.88}$$

so that

$$\int \varrho(x)\, dx = \int R(\sin x, \cos x)\, dx, \, x \in L_i^k \xrightarrow{t = g_k(x)} \int r(t)\, dt$$

where

$$r(t) = -R\left(\sin\left(g_k^{-1}(t)\right), t\right) \cdot \frac{1}{\sin\left(g_k^{-1}(t)\right)}$$

can be reduced to a fraction of polynomials because of the particular hypotheses on $\varrho(x)$. Next, by means of the methods described in Chapter 6, calculate an antiderivative $G(t)$ of $r(t)$. Then, a reversion of the change $t = g_k(x)$ in $G(t)$ yields the local antiderivative $G(\cos x)$ of $\varrho(x)$ on L_i^k. Hence, the function

$$H(x) := G(\cos x), \quad x \in \bigcup_{i \in J} \bigcup_{k \in \mathbb{Z}} L_i^k$$

is a local antiderivative of $\varrho(x)$ on $\bigcup_{i \in J} \bigcup_{k \in \mathbb{Z}} L_i^k$.

3: The function $H(x)$ admits a continuous extension $F(x)$ on D_ϱ and consequently, by virtue of Proposition 4.3.3, the primitive of $\varrho(x)$ is

$$\int \varrho(x)\, dx = F(x) + C(x), \quad x \in D_\varrho$$

for all functions $C(x)$ locally constant on $\bigcup_{i \in J} J_i$.

In order to prove the existence of the continuous extension $F(x)$ of $H(x)$, it suffices to prove that there exists $\lim_{x \to z} G(\cos x) \in \mathbb{R}$ at each $z \in S$ where S is the set of endpoints of the intervals I_k that belong to D_ϱ, that is,

$$S := D_u \setminus \bigcup_{k \in \mathbb{Z}} I_k$$

Note that D_ϱ is symmetric and ϱ is periodic of period 2π, so S must be one of the following sets:

(a) $\varnothing$
(b) $\{2k\pi : k \in \mathbb{Z}\}$
(c) $\{(2k + 1)\pi : k \in \mathbb{Z}\}$
(d) $\{k\pi : k \in \mathbb{Z}\}$.

In case (a), there is nothing to be proved.

(b) Since $0 \in S$, by Proposition 4.3.2, there exist both limits

$$l = \lim_{x \to 0^-} G(\cos x) \in \mathbb{R}$$

$$\lambda = \lim_{x \to 0^+} G(\cos x) \in \mathbb{R}.$$

But $G(\cos x)$ is even. Hence, $l = \lambda$, and this proves that there exists $\lim_{x \to 0} H(x) = l \in \mathbb{R}$ as desired. Since $H(x)$ is periodic of period 2π, it is straightforward that there exists $\lim_{x \to 2k\pi} G(\cos x) = l$ for all $k \in \mathbb{Z}$.

(c) By hypothesis, $\pi \in S$, and by virtue of Proposition 4.3.2, there exists $\lim_{x \to \pi^-} G(\cos x) = l \in \mathbb{R}$. Thus, as $G(\cos x)$ is even, then

$$\lim_{x \to \pi^-} G(\cos x) = l = \lim_{x \to -\pi^+} G(\cos x).$$

But $G(\cos x)$ is periodic of period 2π, so that

$$\lim_{x \to -\pi^+} G(\cos x) = \lim_{x \to \pi^+} G(\cos x)$$

and this leads to

$$\lim_{x \to \pi^-} G(\cos x) = l = \lim_{x \to \pi^+} G(\cos x)$$

which proves that there exists $\lim_{x \to \pi} H(x) \in \mathbb{R}$ and therefore, $H(x)$ is extendable at π with continuity. Using again that $G(\cos x)$ is periodic of period 2π, it follows that there exists $\lim_{x \to (2k+1)\pi^\pm} G(\cos x) = l$ for each $k \in \mathbb{Z}$ and this proves the case (c).

(d) This case is fulfilled by cases (b) and (c).

Let us offer some examples to illustrate this procedure.

Example 11.5.1. Find $\int \sin^3 x \cos^4 x \, dx$.

Solution. The natural domain of $f(x) := \sin^3 x \cos^4 x$ is $\mathbb{R}$. Let us find a local antiderivative of $f(x)$ on each interval $I_n = (n\pi, (n+1)\pi)$ by means of the change of variable $g_n \colon I_n \longrightarrow \mathbb{R}$ given in (11.24).

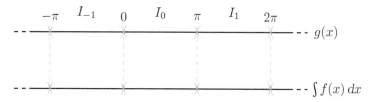

This is done by means of the substitutions (11.86), (11.87) and (11.88), so that

$$\int \sin^3 x \cos^4 x \, dx, \ x \in I_n \xleftarrow{t=g_n(x)} -\int \sin^3\left(g_n^{-1}(t)\right) \cdot t^4 \frac{dt}{\sin\left(g_n^{-1}(t)\right)}$$

$$= -\int (1-t^2) t^4 \, dt = -\frac{1}{5}t^5 + \frac{1}{7}t^7 + C_n, \ t \in (-1,1).$$

(11.89)

Substitute $t = \cos x$ in (11.90) in order to reverse the change g_n, so

$$\int f(x) \, dx = -\frac{1}{5}\cos^5 x + \frac{1}{7}\cos^7 x + C_n, \ x \in I_n.$$

(11.90)

Clearly, the function

$$F(x) := -\frac{1}{5}\cos^5 x + \frac{1}{7}\cos^7 x, \ x \in \mathbb{R}$$

is a local primitive of $f(x)$ on $\bigcup_{n \in \mathbb{Z}} I_n$ and is continuous on $\mathbb{R}$. Hence, by virtue of Proposition 4.3.3, $G(t)$ is also an antiderivative of $f(x)$ on $\mathbb{R}$ and we can conclude that

$$\int \sin^3 x \cos^4 x \, dx = F(x) + C, \quad x \in \mathbb{R}$$

for all real constants C. □

Note that no explicit form for $g_n^{-1}(t)$ has been used to solve the primitive of Example 11.5.1.

Example 11.5.2. Find $\displaystyle\int \frac{\sin x}{\sqrt{2}^{-1} - \cos x} \, dx$.

Solution. The natural domain of $f(x) := \sin x / \left(\sqrt{2}^{-1} - \cos x\right)$ is

$$(-\infty, \infty) \setminus \left\{\frac{\pi}{4} + 2n\pi\right\}_{n=-\infty}^{\infty} \cup \left\{-\frac{\pi}{4} + 2n\pi\right\}_{n=-\infty}^{\infty}$$

and its regular decomposition is $D_f = \bigcup_{n \in \mathbb{Z}} J_n$ where

$$J_{2n} := \left(-\frac{\pi}{4}, \frac{\pi}{4}\right) + 2\pi n,$$

$$J_{2n-1} := \left(\frac{\pi}{4}, \frac{7\pi}{4}\right) + 2\pi(n-1).$$

In order to implement the changes of variable $g_n : I_n \longrightarrow \mathbb{R}$ given in (11.24), where $I_n = (n\pi, (n+1)\pi)$, let $I_k := (k\pi, (k+1)\pi)$ for every $k \in \mathbb{Z}$ and consider the intervals $L_i^k := J_i \cap I_k$ that are non-empty (see Diagram below).

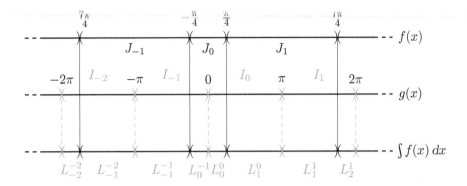

Implement the change of variable $t = \cos x$ on each L_i^k by means of the substitutions (11.86), (11.87) and (11.88), to get

$$\int \frac{\sin x}{\sqrt{2}^{-1} - \cos x} \, dx, \ x \in L_i^k \xleftarrow{t = \cos x} -\int \frac{\sin\left(g_n^{-1}(t)\right)}{\sqrt{2}^{-1} - t} \frac{dt}{\sin\left(g_n^{-1}(t)\right)}$$
$$= \int \frac{dt}{t - \sqrt{2}^{-1}} = \log|t - \sqrt{2}^{-1}| + C, \ t \in \cos(L_i^k).$$

Reverse the change of variable $t = \cos x$ in the last function of (11.5) so that

$$\log|\cos x - \sqrt{2}^{-1}|, \ x \in L_i^k.$$

Thus, the function

$$\log|\cos x - \sqrt{2}^{-1}|, \ x \in \mathcal{D}_f$$

is a local antiderivative of $f(x)$ on each L_i^k. Moreover, it is continuous at the points $k\pi$ ($k \in \mathbb{Z}$), that is, at the endpoints of all intervals I_k, so it is an antiderivative of $f(x)$ on the entire $\mathcal{D}_f$. This means that

$$\int f(x) \, dx = \log|\cos x - \sqrt{2}^{-1}| + C(x), \ x \in \bigcup_{n \in \mathbb{Z}} J_n$$

where $C(x)$ is a locally constant function on $\bigcup_{n \in \mathbb{Z}} J_n$. $\quad\square$

The graph of the solution to Example 11.5.2 is depicted next.

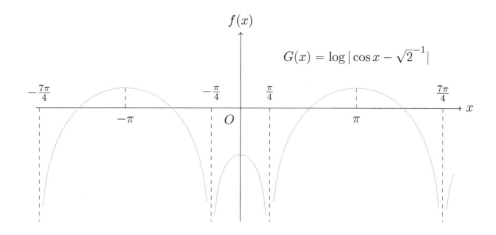

11.6   The case $R(\sin x, -\cos x) = -R(\sin x, \cos x)$

The function $\varrho(x) = R(\sin x, \cos x)$ given in (11.1) satisfies the identity $R(\sin x, -\cos x) = -R(\sin x, \cos x)$ if $\varrho(x)$ is a linear combination of functions of the form

$$\cos^{2k-1} x \cdot S(\sin x, \cos x)$$

where $k \in \mathbb{Z}$ and $S(t_1, t_2)$ is a rational function of its arguments for which each occurrence of t_2 is raised to an even power. These kind of functions are odd with respect to $\pi/2$, that is, $-\pi/2 + D_\varrho$ is symmetric and

$$\varrho(\pi/2 - x) = -\varrho(\pi/2 + x), \quad x \in -\frac{\pi}{2} + D_\varrho. \tag{11.91}$$

For these functions, the changes of variable derived from $t = \sin x$ are very appropriate and their implementation follows similar steps as in Section 11.5:

1: Let $\bigcup_{i \in J} J_i$ be the regular decomposition of D_ϱ. Following the notation in (11.18), let $I_k = \big((k-1)\pi/2, (k+1)\pi/2\big)$ and let $g_k(x)$ be the change of variable determined by $t := \sin x$ for $x \in I_k$. Consider only the intervals $L_i^k := J_i \cap I_k$ which are non-empty.

2: The change g_k is implemented on every L_i^k by means of the substitutions

$$t = \sin x \tag{11.92}$$

$$\cos\left(g_k^{-1}(t)\right) = \cos x \tag{11.93}$$

$$\frac{dt}{\cos\left(g_k^{-1}(t)\right)} = dx \tag{11.94}$$

This action turns $\int f(x)\, dx$ into $\int r(t)\, dt$ where

$$r(t) = R\big(t, \cos(g_k^{-1}(t))\big) \cdot \frac{dt}{\cos\left(g_k^{-1}(t)\right)}$$

can be reduced to a fraction of polynomials because of the particular hypotheses assumed on $\varrho(x)$.

The methods explained in Chapter 6 provide an antiderivative $G(t)$ of $r(t)$, and then, a reversion of the change $t = g_k(x)$ in $G(t)$ shows that $G(\sin x)$ is a local antiderivative of $\varrho(x)$ on L_i^k. Therefore,

$$H(x) := G(\sin x), \quad x \in \bigcup_{i \in J} \bigcup_{k \in \mathbb{Z}} L_i^k$$

is a local antiderivative of $\varrho(x)$ on $\bigcup_{i \in J} \bigcup_{k \in \mathbb{Z}} L_i^k$.

3: The function $H(x)$ admits a continuous extension $F(x)$ defined on D_ϱ and consequently, by virtue of Proposition 4.3.3, the primitive of $\varrho(x)$ is

$$\varrho(x)\, dx = F(x) + C(x), \quad x \in D_\varrho$$

for all functions $C(x)$ locally constant on D_ϱ.

In order to prove the existence of the continuous extension $F(x)$, it is sufficient to prove that there exists $\lim_{x \to z} G(\sin x) \in \mathbb{R}$ at each $z \in S$ where S is the set of endpoints of the intervals I_k that belong to D_ϱ, that is,

$$S := D_\varrho \setminus \bigcup_{k \in \mathbb{Z}} I_k.$$

Since $\pi/2 + D_\varrho$ is symmetric and $\varrho(x)$ is periodic of period 2π, it readily follows that S is of one of the following forms:

(a) $\varnothing$
(b) $\{\pi/2 + 2\pi n : n \in \mathbb{Z}\}$
(c) $\{-\pi/2 + 2\pi n : n \in \mathbb{Z}\}$
(d) $\{\pi/2 + \pi n : n \in \mathbb{Z}\}$.

Bearing in mind that $G(\sin x)$ is symmetric with respect to $\pi/2$ and periodic of period 2π, a similar reasoning to that given in Section 11.5 shows that if $\pi/2 \in S$ then there exist

$$\lim_{x \to \frac{\pi}{2} + 2k\pi^-} G(\sin x) = \lim_{x \to \frac{\pi}{2} + 2k\pi^+} G(\sin x) \in \mathbb{R} \quad \text{for all } k \in \mathbb{Z}.$$

And if $-\pi/2 \in S$ then

$$\lim_{x \to \frac{-\pi}{2} + 2k\pi^-} G(\sin x) = \lim_{x \to \frac{-\pi}{2} + 2k\pi^+} G(\sin x) \in \mathbb{R}, \quad \text{for all } k \in \mathbb{Z}.$$

This proves the existence of the continuous extension $F(x)$ on D_ϱ for all cases (a), (b), (c) and (d).

Example 11.6.1. Solve $\displaystyle\int \frac{\sin^2 x}{\cos x}\, dx.$

Solution. Denote $f(x) := \sin^2 x / \cos x$. The regular decomposition of D_f is $\bigcup_{n=-\infty}^{\infty} J_n$ where $J_n := \left((2n-1)\pi/2, (2n+1)\pi/2\right)$. Since each J_n coincides with the domain of the change of variable $g_n(x)$ given in (11.19), this change can be directly implemented on J_n. To do this, the substitutions (11.92), (11.93) and (11.94) yield

$$\int \frac{\sin^2 x}{\cos x}\, dx, \; x \in I_n \xleftarrow{t=g_n(x)} \int \frac{t^2}{\cos\left(g_n^{-1}(t)\right)} \frac{dt}{\cos\left(g_n^{-1}(t)\right)}$$

$$= \int \frac{t^2}{\cos^2\left(g_k^{-1}(t)\right)}\, dt = \int \frac{t^2}{1-t^2}\, dt, \; t \in (-1,1). \tag{11.95}$$

The last primitive in (11.95) is solved by separating into parts:

$$\int \frac{t^2}{1-t^2}\, dt = -\int dt - \int \frac{1/2}{t-1}\, dt + \int \frac{1/2}{t+1}\, dt$$

$$= -t + \frac{1}{2}\log\frac{1+t}{1-t} + C, \; t \in (-1,1). \tag{11.96}$$

By reversing the change g_n in (11.96), we obtain the following local antiderivative of $f(x)$ on I_n:

$$-\sin x + \frac{1}{2}\log\frac{1+\sin x}{1-\sin x}, \; x \in J_n.$$

It is straightforward that the primitive on the entire set $D_f = \bigcup_{n \in \mathbb{Z}} I_n$ is

$$\int \frac{\sin^2 x}{\cos x}\, dx = -\sin x + \frac{1}{2}\log\frac{1+\sin x}{1-\sin x} + C(x), \; x \in D_f$$

where $C(x)$ is locally constant on $\bigcup_{n \in \mathbb{Z}} I_n$.

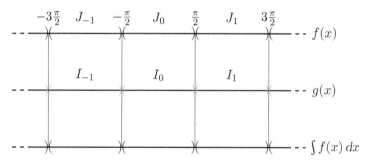

Example 11.6.2. Solve $\displaystyle\int \frac{\cos x}{1+\cos^2 x}\, dx$.

Solution. The natural domain of

$$f(x) := \frac{\cos x}{1+\cos^2 x} = \frac{\cos x}{2-\sin^2 x}$$

is $(-\infty, \infty)$. To solve its primitive by means of the changes of variable $g_n : I_n \longrightarrow \mathbb{R}$ given in (11.19), first find a local antiderivative of $f(x)$ on each I_n (Diagram 11.75 is also valid for this example). Implement the change g_n by means of the substitutions (11.92), (11.93) and (11.94) so that

$$\int \frac{\cos x}{2 - \sin^2 x} \, dx, \; x \in I_n \xrightarrow{t=g_n(x)}$$

$$\int \frac{\cos(g_n^{-1}(t))}{2 - t^2} \frac{dt}{\cos(g_n^{-1}(t))} = \int \frac{dt}{2 - t^2}, \; t \in (-1, 1). \tag{11.07}$$

The last primitive in (11.97) is solved as follows:

$$\int \frac{dt}{2 - t^2} = -\int \frac{1/(2\sqrt{2})}{t - \sqrt{2}} \, dt + \int \frac{1/(2\sqrt{2})}{t + \sqrt{2}} \, dt$$

$$= -\frac{1}{2\sqrt{2}} \log(\sqrt{2} - t) + \frac{1}{2\sqrt{2}} \log(\sqrt{2} + t) + C$$

$$= \frac{1}{2\sqrt{2}} \log \frac{\sqrt{2} + t}{\sqrt{2} - t} + C, \; t \in (-1, 1). \tag{11.98}$$

Let

$$G(t) := \frac{1}{2\sqrt{2}} \log \frac{\sqrt{2} + t}{\sqrt{2} - t}, \; t \in (-1, 1).$$

By reversing the change of variable $t = \sin x$, it follows from (11.98) that $G(\sin x)$ is a local antiderivative of $f(x)$ on I_n, and therefore, it readily follows that

$$G(\sin x) = \frac{1}{2\sqrt{2}} \log \frac{\sqrt{2} + \sin x}{\sqrt{2} - \sin x}, \; x \in \mathbb{R}$$

is a local antiderivative of $f(x)$ on $\bigcup_{n \in \mathbb{Z}} I_n$. But $G(\sin x)$ is continuous on $\mathbb{R}$. Hence, by Proposition 4.3.3, it is also an antiderivative of $f(x)$ on $\mathbb{R}$, and we conclude that

$$\int \frac{\cos x}{1 + \cos^2 x} \, dx = \frac{1}{2\sqrt{2}} \log \frac{\sqrt{2} + \sin x}{\sqrt{2} - \sin x} + C, \; x \in \mathbb{R}$$

for all constants C. $\qquad \square$

11.7 Exercises

Find the domain and an antiderivative for each of the following functions.

Question 11.7.1

$$f(x) = \frac{1}{1 - \cos x}$$

Question 11.7.3

$$f(x) = \frac{\cos x \sin x}{\cos^4 x + \sin^4 x}$$

Question 11.7.2

$$f(x) = \frac{\cos x + \sin x}{1 - \cos x}$$

Question 11.7.4

$$f(x) = \frac{\sin x}{\sin^2 x + \cos x - 2}$$

Question 11.7.5

$f(x) = \sin^4 x \cos^2 x$

Question 11.7.6

$f(x) = \dfrac{\sec x}{2\tan x + \sec x - 1}$

Question 11.7.7

$f(x) = \dfrac{1}{\sin x - \cos x + 2}$

Question 11.7.8

$f(x) = \dfrac{1}{\sin x - \cos x}$

Question 11.7.9

$f(x) = \dfrac{1}{3\sin^2 x + 5\cos^2 x}$

Question 11.7.10

$f(x) = \dfrac{1}{\sin x}$

Question 11.7.11

$f(x) = \sin^3 x \cos^3 x$

Question 11.7.12

$f(x) = \dfrac{\sin x \cos x}{2\sin^2 x + \cos^2 x}$

Question 11.7.13

$f(x) = \dfrac{\cos x}{2\sin^2 x + \cos^2 x}$

Question 11.7.14

$f(x) = \dfrac{\sin x}{2 - \cos^2 x}$

Question 11.7.15

$f(x) = \dfrac{\sin x}{2 + \cos x}$

Question 11.7.16

$f(x) = \dfrac{\sin^3(x)/\cos(x)}{\left(1 + \cos^2(x) - 2\sin(x)\cos(x)\right)^2}$

Question 11.7.17

$f(x) = \dfrac{\sin^2 x \cos x - \cos^3 x}{\sin^3 x}$

Answers to the Questions

Answer 11.7.1

$D_f = \mathbb{R}\backslash\{2n\pi : n \in \mathbb{Z}\}$,

$\begin{cases} -\cot(x/2), & x \in D_f\backslash\{n\pi : n \in \mathbb{Z}\} \\ 0, & x \in \{n\pi : n \in \mathbb{Z}\} \end{cases}$

Answer 11.7.2

$D_f = \mathbb{R}\backslash\{2n\pi : n \in \mathbb{Z}\}$,

$$\begin{cases} F\big(\tan(x/2))\big) - n\pi, & x \in (0,\pi) + n\pi \\ -\pi - n\pi, & x = \pi + n\pi \\ F\big(\tan(x/2))\big) - 2\pi - n\pi, & x \in (\pi, 2\pi) + n\pi \end{cases}$$

for all $n \in \mathbb{Z}$, where

$$F(t) = 2\log|t| - \frac{1}{t} - \log(t^2 + 1) - 2\arctan t$$

Answer 11.7.3

$D_f = (-\infty, \infty)$,

$$\begin{cases} \frac{1}{2}\arctan(\tan^2 x), & x \ne \frac{\pi}{2} + k\pi \\ \frac{\pi}{4}, & x = \frac{\pi}{2} + k\pi, \ k \in \mathbb{Z} \end{cases}$$

Answer 11.7.4

$D_f = (-\infty, \infty)$,

$$\begin{cases} -\frac{2}{3}\arctan\big(\sqrt{3}\tan^2(x/2)\big), & x \ne (2n+1)\pi, \ n \in \mathbb{Z} \\ -\pi/3, & x = (2n+1)\pi, \ n \in \mathbb{Z} \end{cases}$$

Answer 11.7.5

$D_f = (-\infty, \infty)$,

$$\frac{\cos x \sin^5 x}{16} - \frac{\cos^3 x \sin^3 x}{6} - \frac{\cos^5 x \sin x}{16} + \frac{x}{16}$$

Answer 11.7.6

$D_f = \mathbb{R}\backslash\{\frac{\pi}{2} + k\pi, \ 2k\pi, \ -2\arctan(2) + 2k\pi : k \in \mathbb{Z}\}$,

$$\begin{cases} \frac{1}{2}\log\left|\frac{\tan(x/2)}{\tan(x/2)+2}\right|, & x \in D_f\backslash\{(2k+1)\pi : k \in \mathbb{Z}\} \\ 0, & x \in \{(2k+1)\pi : k \in \mathbb{Z}\} \end{cases}$$

Answer 11.7.7

$D_f = (-\infty, \infty)$,

$$\begin{cases} \sqrt{2}\arctan\big(\frac{3}{\sqrt{2}}\tan(x/2) + \frac{1}{\sqrt{2}}\big) + n\sqrt{2}\pi, & x \in (-\pi, \pi) + 2n\pi, \ n \in \mathbb{Z} \\ \frac{\pi}{\sqrt{2}} + n\sqrt{2}\pi, & x = (1 + 2n)\pi, \ n \in \mathbb{Z} \end{cases}$$

Answer 11.7.8

$D_f = \mathbb{R}\backslash\{\frac{\pi}{4} + k\pi : k \in \mathbb{Z}\}$,

$$\begin{cases} \frac{1}{\sqrt{2}} \log \left| \frac{\tan(x/2)+1-\sqrt{2}}{\tan(x/2)+1+\sqrt{2}} \right|, & x \in D_f \backslash \{(2k+1)\pi : k \in \mathbb{Z}\} \\ 0, & x \in \{(2k+1)\pi : k \in \mathbb{Z}\} \end{cases}$$

Answer 11.7.9

$D_f = (-\infty, \infty),$

$$\begin{cases} \frac{1}{\sqrt{15}} \arctan(\sqrt{3/5} \tan x) + \frac{\pi}{\sqrt{15}} n, & x \in \left(-\frac{\pi}{2}, \frac{\pi}{2}\right) + n\pi \\ \frac{\pi}{2\sqrt{15}} + \frac{\pi}{\sqrt{15}} n, & x = \frac{\pi}{2} + n\pi, \ n \in \mathbb{Z} \end{cases}$$

Answer 11.7.10

$D_f = \bigcup_{k \in \mathbb{Z}} (k\pi, (k+1)\pi),$

$$\frac{1}{2} \log \left| \frac{1 - \cos x}{1 + \cos x} \right|$$

Answer 11.7.11

$D_f = (-\infty, \infty),$

$$\frac{1}{4} \sin^4 x - \frac{1}{6} \sin^6 x$$

Answer 11.7.12

$D_f = (-\infty, \infty),$

$$\frac{1}{2} \log \left(1 + \sin^2 x\right)$$

Answer 11.7.13

$D_f = (-\infty, \infty),$

$\arctan(\sin x)$

Answer 11.7.14

$D_f = (-\infty, \infty),$

$$\frac{1}{2\sqrt{2}} \log \frac{\sqrt{2} - \cos x}{\sqrt{2} + \cos x}$$

Answer 11.7.15

$D_f = (-\infty, \infty),$

$-\log (2 + \cos x)$

Answer 11.7.16

$D_f = \bigcup_{n \in \mathbb{Z}} \left(-\frac{\pi}{2} + n\pi, \frac{\pi}{2} + n\pi \right),$

$$\frac{\tan(x)}{\tan^2(x) - 2\tan(x) + 2} + \frac{1}{3} \log(\tan^2(x) - 2\tan(x) + 2) + 2\arctan(\tan(x) - 1)$$

Answer 11.7.17

$D_f = \bigcup_{n \in \mathbb{Z}} (n\pi, (n+1)\pi),$

$$\begin{cases} 2\log|\tan x| + \frac{1}{2\tan^2 x} - \log(1 + \tan^2 x), & x \in D_f \backslash \{n\pi + \pi/2 \colon n \in \mathbb{Z}\} \\ 0, & x \in \{n\pi + \pi/2 \colon n \in \mathbb{Z}\} \end{cases}$$

Chapter 12

Functions $R(x, \sqrt{ax^2 + bx + c})$

This chapter deals with the calculation of primitives of

$$\varrho(x) := R(x, \sqrt{ax^2 + bx + c}), \quad x \in D_\varrho \tag{12.1}$$

where $R(t_1, t_2)$ is a rational function of its arguments and the coefficients a, b and c satisfy the conditions (i), (ii) and (iii) given at the beginning of Chapter 7, so that $ax^2 + bx + c$ has only two simple roots. Therefore the natural domain of

$$r(x) := \sqrt{ax^2 + bx + c}$$

is of one of the forms

(i') $[x_0, x_1]$ with $x_0 < x_1$,
(ii') $(-\infty, \infty)$,
(iii') $(-\infty, x_0] \cup [x_1, \infty)$ with $x_0 < x_1$

where (i'), (ii') or (iii') holds if and only if the aforementioned condition (i), (ii) or (iii), respectively, holds. In cases (i') and (iii'), x_0 and x_1 are the real roots of $ax^2 + bx + c$ and $x_0 < x_1$. In case (ii'), $r(x)$ has no real root.

Clearly, D_r is a regular domain in any of the cases (i'), (ii') or (iii').

12.1 The domain of $R(x, \sqrt{ax^2 + bx + c})$

The function $\varrho(x)$ given in (12.1) is continuous on D_ϱ. Hence, the existence of its primitive is a consequence of Theorem 4.1.9 and the following proposition.

Note that the function $\varrho(x)$ can be rearranged so that

$$\varrho(x) = \frac{P_1(x) + Q_1(x)\sqrt{ax^2 + bx + c}}{P_2(x) + Q_2(x)\sqrt{ax^2 + bx + c}}, \quad x \in D_\varrho \tag{12.2}$$

where $P_1(x)$, $P_2(x)$, $Q_1(x)$ and $Q_2(x)$ are polynomials over $\mathbb{R}$. Throughout this chapter, at least one of the polynomials $Q_i(x)$ will be assumed to be non-identically null in order to avoid the trivial case where $\varrho(x)$ is a fraction of polynomials.

Proposition 12.1.1. *The natural domain of the function $\varrho(x)$ given in (12.1) is a regular domain.*

213

Proof. Assume (12.2) is a rearrangement for $\varrho(x)$, so at least one of the polynomials $Q_i(x)$ is non-identically null.

It readily follows from (12.2) that $D_\varrho = D_r \backslash Z$, where

$$Z := \{x \in \mathbb{R} : P_2(x) + Q_2(x)\sqrt{ax^2 + bx + c} = 0\}.$$

Since the hypotheses on a, b and c yield that D_r is regular, D_ϱ will be proved to be regular as soon as it is proved that Z is finite.

Each element of Z is a solution of the equation

$$P_2^2(x) = Q_2^2(x)(ax^2 + bx + c). \tag{12.3}$$

Since (12.3) is a polynomial identity, it has only a finite number of solutions. Hence Z is finite too, and in turn, D_ϱ is a regular domain. □

By completing squares, the function $r(x) = \sqrt{ax^2 + bx + c}$ can be rewritten as one of the following functions for certain real coefficients α, β and γ:

(i") $\gamma\sqrt{1 - (\alpha x + \beta)^2}$, $x \in [x_0, x_1]$;
(ii") $\gamma\sqrt{1 + (\alpha x + \beta)^2}$, $x \in (-\infty, \infty)$;
(iii") $\gamma\sqrt{-1 + (\alpha x + \beta)^2}$, $x \in (-\infty, x_0] \cup [x_1, \infty)$

where (i") or (iii") holds if $ax^2 + bx + c$ has two different real roots, x_0 and x_1; (ii") holds if $ax^2 + bx + c$ has no real root.

By means of the linear change of variable $z = \alpha x + \beta$, $\int R(x, \sqrt{ax^2 + bx + c})\, dx$ is transformed into one of the following expressions in correspondence with (i"), (ii") and (iii"):

(a) $\displaystyle\int \frac{1}{\alpha} R\left(\frac{z - \beta}{\alpha}, \sqrt{1 - z^2}\right) dz,$

(b) $\displaystyle\int \frac{1}{\alpha} R\left(\frac{z - \beta}{\alpha}, \sqrt{1 + z^2}\right) dz,$

(c) $\displaystyle\int \frac{1}{\alpha} R\left(\frac{z - \beta}{\alpha}, \sqrt{-1 + z^2}\right) dz.$

Consequently, only the cases

- $R(x, \sqrt{1 - x^2})$,
- $R(x, \sqrt{1 + x^2})$,
- $R(x, \sqrt{-1 + x^2})$

deserve attention and will be separately treated in the last sections of this chapter.

12.2   Changes of variable associated with $R(x, \sqrt{ax^2 + bx + c})$

Three changes of variable are used:

- $x = \sin t$:

The function $g\colon (-1, 1) \longrightarrow (-\pi/2, \pi/2)$ given by

$$g(x) := \arcsin x \tag{12.4}$$

is an increasing bijection for which $g'(x) = 1/\sqrt{1 - x^2} \neq 0$ for all $x \in (-1, 1)$. Hence, it is a change of variable and its inverse, $g^{-1}\colon (-\pi/2, \pi/2) \longrightarrow (-1, 1)$, is given by $g^{-1}(t) = \sin t$.

- $x = \tan t$:

The function $g\colon (-\infty, \infty) \longrightarrow (-\pi/2, \pi/2)$ given by

$$g(x) := \arctan x \tag{12.5}$$

is an increasing bijection for which $g'(x) = 1/(1 + x^2) \neq 0$ for all x. Hence, it is a change of variable and its inverse, $g^{-1}\colon (-\pi/2, \pi/2) \longrightarrow (-\infty, \infty)$, is given by $g^{-1}(t) = \tan t$.

- $x = 1/\cos t$:

Consider the bijective function $g\colon (-\infty, -1) \cup (1, \infty) \longrightarrow (0, \pi/2) \cup (\pi/2, \pi)$ given by $g(x) := \arccos(1/x)$. The restrictions

$$g_0 := g|_{(-\infty, -1)} \tag{12.6}$$

$$g_1 := g|_{(1, \infty)} \tag{12.7}$$

are increasing. The function g_0 maps $I_0 := (-\infty, -1)$ onto $(\pi/2, \pi)$, and g_1 maps $I_1 := (1, \infty)$ onto $(0, \pi/2)$.

Their derivatives are $g_i'(x) = \sqrt{1 - x^{-2}}^{-1} x^{-2} \neq 0$ for all $x \in I_i$ and $i = 0, 1$. Clearly, each function g_i is a change of variable and g_i^{-1} is given by the expression $x = g_i^{-1}(t) = 1/\cos t$, with $t \in (\pi/2, \pi)$ for $i = 0$ and $t \in (0, \pi/2)$ for $i = 1$.

12.3 Functions $R(x, \sqrt{1 - x^2})$

For $\varrho(x) = R(x, \sqrt{1 - x^2})$, the change of variable to be implemented is that given in (12.4) (the change $x = \cos t$ also works).

The associated formulas of substitution with this change are

$$x = \sin t, \ t \in (-\pi/2, \pi/2) \tag{12.8}$$

$$dx = \cos t \, dt \tag{12.9}$$

and from (12.8),

$$\cos t = \sqrt{1 - \sin^2 t}. \tag{12.10}$$

The practical procedure to solve $\int \varrho(x)\, dx$ is the following:

1: Let $D_\varrho = \bigcup_{n=1}^{m} J_n \subset [-1,1]$ be the regular decomposition of the natural domain of ϱ where the intervals J_n are labeled so that $\sup J_i \leqslant \inf J_{i+1}$ for all i. Since the domain of g is $(-1,1)$, the intervals on which this change can be applied are $L_n := J_n \cap (-1,1)$.

2: The change g is implemented on each interval L_n by means of the substitution formulas (12.8), (12.9) and (12.10). This yields

$$\int R(x, \sqrt{1-x^2})\, dx, \ \ x \in L_n \ \xleftarrow{\ x=\sin t\ } \ \int R(\sin t, \cos t) \cos t\, dt, \ \ t \in \arcsin(L_n).$$

The function $R(t_1, t_2)t_2$ is a rational function of its arguments which can be solved in accordance with Chapter 11. Next, a reversion of the change of variable provides a local antiderivative $F_n(x)$ of $\varrho(x)$ on L_n.

3: In the case $-1 \in J_1$, the function F_1 is extended to the entire J_1 by letting $F_1(-1) := \lim_{x \to -1+} F_1(x)$. In a similar manner, if $1 \in J_m$, F_m is extended to J_m by letting $F_m(1) := \lim_{x \to 1-} F_m(x)$. Thus, by virtue of Proposition 4.3.2 the function $F(x) := F_n(x)$ for all $x \in J_n$ and all $1 \leqslant n \leqslant m$ is an antiderivative of $\varrho(x)$ on the entire D_ϱ, and therefore, its complete primitive is $F(x) + C(x)$ for any locally constant function on D_ϱ.

Example 12.3.1. Find $\displaystyle\int x^4 \sqrt{1-x^2}\, dx$.

Solution. The natural domain of $x^4\sqrt{1-x^2}$ is $[-1,1]$. Since the change of variable $t = \arcsin x$ is valid only on $x \in (-1,1)$, the primitive will be solved on this interval and then the solution will be extended to $[-1,1]$ by virtue of Proposition 4.3.1.

To do this, the aforementioned change is implemented by means of the substitutions (12.8), (12.9) and (12.10) so that

$$\int x^4 \sqrt{1-x^2}\, dx, \ \ x \in (-1,1) \ \xleftarrow{\ x=\sin t\ } \ \int \sin^4 t \cos^2 t\, dt, \ \ t \in (-\pi/2, \pi/2). \quad (12.11)$$

The primitive on the right side of (12.11) can be found by means of the change of variable $u = \tan(t/2)$, but following the recommendation suggested in Section 11.4, step **0**, the application of the reduction formulas (Section 13.2) lead to a direct solution as follows:

$$\int \sin^4 t \cos^2 t \, dt = \int \left(\frac{1 - \cos(2t)}{2} \right)^2 \left(\frac{1 + \cos(2t)}{2} \right) dt$$

$$-\frac{1}{8} \int 1 - \cos^2(2t) + \cos^3(2t) - \cos(2t) \, dt$$

$$= \frac{1}{8} \int 1 - \cos^2(2t) + \cos(2t) \cdot \left(1 - \sin^2(2t)\right) - \cos(2t) \, dt$$

$$= \frac{1}{8} \int 1 - \frac{1 + \cos(4t)}{2} - \sin^2(2t) \cos(2t) \, dt$$

$$= \frac{1}{8} \int dt - \frac{1}{16} \int dt - \frac{1}{16} \int \cos(4t) \, dt - \frac{1}{8} \int \sin^2(2t) \cos(2t) \, dt$$

$$= \frac{1}{16} t - \frac{1}{64} \sin(4t) - \frac{1}{48} \sin^3(2t) + C, \ t \in \left(-\frac{\pi}{2}, \frac{\pi}{2} \right). \tag{12.12}$$

The change $x = \sin t$ is reversed in (12.12), so that

$$\int x^4 \sqrt{1 - x^2} dx \tag{12.13}$$

$$= \frac{1}{16} \arcsin x - \frac{1}{64} \sin(4 \arcsin x) - \frac{1}{48} \sin^3(2 \arcsin x) + C$$

for $x \in (-1, 1)$. But as the function on the right side of (12.13) is continuous at -1 and at 1, it also yields the primitive of $f(x)$ on $[-1, 1]$.

For a better presentation, applying some of the trigonometric identities given in Section 13.2 allows us to rewrite (12.13) as

$$\int x^4 \sqrt{1 - x^2} dx$$

$$= \frac{1}{16} \arcsin x - \left[\frac{1}{16} x(1 - 2x^2) + \frac{1}{6} x^3 (1 - x^2) \right] \sqrt{1 - x^2} + C, \ x \in [-1, 1]$$

where C is any real constant. $\square$

Example 12.3.2. Find $\displaystyle\int \frac{dx}{1 + \sqrt{1 - x^2}}$.

Solution. The natural domain of $f(x) := 1/(1 + \sqrt{1 - x^2})$ is $[-1, 1]$. By means of the change of variable $x = \sin u$ given in (12.4), we will find the primitive of $f(x)$ on $(-1, 1)$ and any further matter about the endpoints of $[-1, 1]$ will be fixed by resorting to Proposition 4.3.1.

Thus, by means of the substitutions (12.8), (12.9) and (12.10),

$$\int \frac{dx}{1 + \sqrt{1 - x^2}}, \ x \in (-1, 1) \xleftarrow{\ x = \sin u\ } \int \frac{\cos u}{1 + \cos u} \, du, \ u \in \left(-\frac{\pi}{2}, \frac{\pi}{2} \right). \tag{12.14}$$

Next, the change $t = \tan(u/2)$ is performed by means of the substitutions (11.28), (11.29) and (11.30), so that

$$\int \frac{\cos u}{1 + \cos u}\, du,\ u \in \left(-\frac{\pi}{2}, \frac{\pi}{2}\right) \xleftarrow{t=\tan(u/2)} \int \frac{\frac{1-t^2}{1+t^2}}{1 + \frac{1-t^2}{1+t^2}} \frac{2\, dt}{1 + t^2}$$

$$= \int \frac{1 - t^2}{1 + t^2}\, dt = \int -1 + \frac{2}{1 + t^2}\, dt$$

$$= -t + 2\arctan t + C,\ t \in (-1, 1). \tag{12.15}$$

Reversing the change $t = \tan(u/2)$ in (12.15) yields

$$\int \frac{\cos u}{1 + \cos u}\, du = -\tan\left(\frac{u}{2}\right) + u + C,\ u \in \left(-\frac{\pi}{2}, \frac{\pi}{2}\right). \tag{12.16}$$

Now, combining (12.16) with (12.14), and reversing the first change $x = \sin u$,

$$\int \frac{dx}{1 + \sqrt{1 - x^2}} = -\tan\left(\frac{1}{2}\arcsin x\right) + \arcsin x + C \tag{12.17}$$

for $x \in (-1, 1)$. But the function on the right side of (12.17) is continuous at -1 and at 1. Hence, by virtue of Proposition 4.3.1, formula (12.17) also determines the primitive of $1/(1 + \sqrt{1 - x^2})$ on the entire domain $[-1, 1]$.

By choosing the adequate identity from Section 13.2, the desired primitive can be rewritten as

$$\int \frac{dx}{1 + \sqrt{1 - x^2}} = \arcsin x + C + \begin{cases} -\sqrt{\frac{1 - \sqrt{1 - x^2}}{1 + \sqrt{1 - x^2}}}, & x \in [0, 1] \\[2ex] \sqrt{\frac{1 - \sqrt{1 - x^2}}{1 + \sqrt{1 - x^2}}}, & x \in [-1, 0] \end{cases}$$

for any constant C.   □

12.4   Functions $R(x, \sqrt{1 + x^2})$

For $\varrho(x) = R(x, \sqrt{1 + x^2})$, the change of variable to be implemented is that given in (12.5).

The associated formulas of substitution with this change are

$$x = \tan t,\ t \in (-\pi/2, \pi/2) \tag{12.18}$$

$$dx = \frac{dt}{\cos^2 t} \tag{12.19}$$

and from (12.18),

$$\frac{1}{\cos t} = \sqrt{1 + \tan^2 t}. \tag{12.20}$$

The practical procedure to find $\int \varrho(x)\, dx$ is as follows:

1: Find the regular decomposition $D_\varrho = \bigcup_{n=1}^{m} J_n$. It follows from the proof of Proposition 12.1.1 that each J_n is open.

2: The change g is implemented on each interval J_n by means of the substitution formulas (12.18), (12.19) and (12.20). This yields

$$\int R(x, \sqrt{1 + x^2})\, dx, \ x \in J_n \xrightarrow{x=\tan t} \int R(\tan t, \cos^{-1} t) \cos^{-2} t\, dt, \ t \in \arctan(J_n).$$

The function $R(t_1, t_2^{-1}) t_2^{-2}$ is a rational function of its arguments which can be solved in accordance with Chapter 11. Next, a reversion of the change of variable provides a local antiderivative $F_n(x)$ of $\varrho(x)$ on J_n.

3: A global antiderivative of $\varrho(x)$ is $F(x) := F_n(x)$ for $x \in J_n$ and $1 \leqslant n \leqslant m$, and consequently, its primitive is

$$\int \varrho(x)\, dx = F(x) + C(x), \ x \in \bigcup_{n=1}^{m} J_n$$

for any locally constant function on $\bigcup_{n=1}^{m} J_n$.

Example 12.4.1. Find $\displaystyle\int \frac{\sqrt{1 + x^2}}{x^2}\, dx$.

Solution. The natural domain of $f(x) := \sqrt{1 + x^2}/x^2$ is $J_0 \cup J_1$ where $J_0 := (-\infty, 0)$ and $J_1 := (0, \infty)$.

Apply the change of variable $x(u) = \tan u$ on each J_i by means of the substitutions (12.18), (12.19) and (12.20) so that

$$\int \frac{\sqrt{1 + x^2}}{x^2}\, dx, \ x \in J_i \xrightarrow{x=\tan u} \int \frac{1/\cos u}{\tan^2 u} \cdot \frac{du}{\cos^2 u}$$
$$= \int \frac{du}{\sin^2 u \cos u}, \ u \in G_i \tag{12.21}$$

where $G_0 := (-\pi/2, 0)$ and $G_1 := (0, \pi/2)$.

Next, an application of the change $u(t) = \arcsin t$ by means of the substitutions (11.92), (11.93) and (11.94) in the primitive on the right side of (12.21) gives

$$\int \frac{du}{\sin^2 u \cos u}, \ u \in G_i \xrightarrow{\sin u=t} \int \frac{dt}{t^2(1 - t^2)}, \ t \in I_i \tag{12.22}$$

where $I_0 := (-1, 0)$ and $I_1 := (0, 1)$. Solving the primitive on the right side of (12.22) yields

$$\int \frac{-\,dt}{t^2(t^2 - 1)} = \int \frac{0}{t} + \frac{1}{t^2} + \frac{-1/2}{t - 1} + \frac{1/2}{t + 1}\, dt$$
$$= -\frac{1}{t} + \frac{1}{2} \log \frac{1 + t}{1 - t} + C, \ t \in I_i. \tag{12.23}$$

By reversion of the change $t = \sin u$ in (12.23), we obtain

$$\int \frac{du}{\sin^2 u \cos u} = -\frac{1}{\sin u} + \frac{1}{2} \log \frac{1 + \sin u}{1 - \sin u} + C, \ u \in G_i. \tag{12.24}$$

A further reversion of $x = \tan u$ in the two solutions (12.24) on G_1 and on G_2 yields

$$\int \frac{\sqrt{1+x^2}}{x^2}\, dx = -\frac{1}{\sin(\arctan x)} + \frac{1}{2}\log\frac{1+\sin(\arctan x)}{1-\sin(\arctan x)} + C(x)$$

$$= -\frac{\sqrt{1+x^2}}{x} + \frac{1}{2}\log\frac{\sqrt{1+x^2}+x}{\sqrt{1+x^2}-x} + C(x)$$

$$= -\frac{\sqrt{1+x^2}}{x} + \frac{1}{2}\log(1+2x^2+2\sqrt{1+x^2}) + C(x), \quad x \in J_1 \cup J_2$$

where $C(x)$ is any locally constant function on $J_1 \cup J_2$. $\square$

12.5   Functions $R(x, \sqrt{x^2-1})$

For $\varrho(x) = R(x, \sqrt{x^2-1})$, the changes of variable to be implemented are those given in (12.6) and (12.7).

Following the same notation given in (12.6) and (12.7). Let $I_0 := (-\infty, -1)$, $I_1 := (1, \infty)$, $K_0 := g_0(I_0) = (\pi/2, \pi)$ and $K_1 := g_1(I_1) = (0, \pi/2)$.

The associated formulas of substitution with g_0 and g_1 are

$$x = \frac{1}{\cos t}, \quad t \in K_i \tag{12.25}$$

$$dx = \frac{\sin t}{\cos^2 t}\, dt \tag{12.26}$$

and from (12.25),

$$\tan t = (-1)^{i+1}\sqrt{\cos^{-2} x - 1}, \quad t \in K_i, i = 0, 1. \tag{12.27}$$

The practical procedure to solve $\int \varrho(x)\, dx$ is as follows:

1: Find the regular decomposition of the domain of ϱ: $D_\varrho = \bigcup_{n=1}^{m} J_n$. Observe that for every i, either $J_i^\circ \subset I_1$ or $J_i^\circ \subset I_0$. Assume the intervals J_n are labeled so that $\sup J_{i+1} \leqslant \inf J_i$ for all i. Thus $J_1^\circ \subset I_1$ and $J_m^\circ \subset I_0$.

Since the domain of g_i is I_i, the intervals on which this change can be applied are $L_n^i := J_n \cap I_i = J_n^\circ$.

2: Each change g_i is implemented on each interval L_n^i by means of the substitution formulas (12.25), (12.26) and (12.27). This yields

$$\int R(x, \sqrt{x^2-1})\, dx, \quad x \in L_n^i \xleftarrow{\ x=1/\cos t\ } \int R(\cos^{-1} t, (-1)^{i+1}\tan t)\frac{\sin t}{\cos^2 t}\, dt, \quad t \in g_i(L_n^i).$$

The function $R(t_1, (-1)^{i+1}t_2t_1)t_2t_1^2$ is a rational function of its arguments which can be solved in accordance with Chapter 11. Next, a reversion of the change of variable g_i provides a local antiderivative $F_n(x)$ of $\varrho(x)$ on L_n^i.

3: In the case $1 \in J_1$, the function F_1 is extended to the entire J_1 by letting $F_1(0) := \lim_{x\to 1+} F_1(x)$. In a similar manner, if $-1 \in J_m$, F_m is extended to J_m by letting $F_m(1) := \lim_{x\to -1-} F_m(x)$. Thus, the function $F(x) := F_n(x)$ for all $x \in J_n$ and all $1 \leqslant n \leqslant m$ is an antiderivative of $\varrho(x)$ on D_ϱ by virtue of Proposition 4.3.2

and therefore its complete primitive is $F(x) + C(x)$ for any locally constant function on D_ϱ.

Example 12.5.1. Find $\displaystyle\int \frac{\sqrt{x^2 - 1}}{x^2}\, dx$.

Solution. The natural domain of $f(x) = \sqrt{x^2 - 1}/x^2$ is $J_0 \cup J_1$, where $J_0 = (-\infty, -1]$ and $J_1 = [1, \infty)$. The change of variable $x = 1/\cos u$ will be separately implemented on J_1° and on J_0°.

In order to perform the change on J_1°, the substitutions (12.25), (12.26) and (12.27) give

$$\int \frac{\sqrt{x^2 - 1}}{x^2}\, dx, \ x \in J_1^\circ \xleftarrow{x = 1/\cos u} \int \cos^2 u \, |\tan u| \frac{\sin u}{\cos^2 u}\, du$$

$$= \int \frac{\sin^2 u}{\cos u}\, du, \ u \in \left(0, \frac{\pi}{2}\right). \tag{12.28}$$

Next, the change $t = \sin u$ is implemented in the last primitive occurring at (12.28) by means of the substitutions (11.92), (11.93) and (11.94), so that

$$\int \frac{\sin^2 u}{\cos u}\, du, \ u \in \left(0, \frac{\pi}{2}\right) \xleftarrow{t = \sin u} \int \frac{t^2}{1 - t^2}\, dt, \ t \in (0, 1). \tag{12.29}$$

The usual techniques of Chapter 6 allows us to solve the last primitive in (12.29) as

$$-\int \frac{t^2}{t^2 - 1}\, dt = \int -1 - \frac{1/2}{t - 1} + \frac{1/2}{t + 1}\, dt$$

$$= -t + \frac{1}{2} \log \frac{1 + t}{1 - t} + C_1, \ t \in (0, 1). \tag{12.30}$$

Let us reverse the change $t = \sin u$ in (12.30) so that

$$\int \frac{\sin^2 u}{\cos u}\, du = -\sin u + \frac{1}{2} \log \frac{1 + \sin u}{1 - \sin u} + C_1, \ u \in (0, \pi/2). \tag{12.31}$$

Next, substitute $u = \arccos\left(x^{-1}\right)$ in (12.31) to reverse the first change of variable, so

$$\int \frac{\sqrt{x^2 - 1}}{x^2}\, dx = -\sin\left(\arccos\left(x^{-1}\right)\right) + \frac{1}{2} \log \frac{1 + \sin\left(\arccos\left(x^{-1}\right)\right)}{1 - \sin\left(\arccos\left(x^{-1}\right)\right)}$$

$$= -\sqrt{1 - x^{-2}} + \frac{1}{2} \log \frac{1 + \sqrt{1 - x^{-2}}}{1 - \sqrt{1 - x^{-2}}} + C_1$$

$$= -\frac{\sqrt{x^2 - 1}}{x} + \frac{1}{2} \log \frac{x + \sqrt{x^2 - 1}}{x - \sqrt{x^2 - 1}} + C_1, \ x \in (1, \infty). \tag{12.32}$$

Next, solve the primitive of $f(x)$ on J_0°. This time, the implementation of $x = 1/\cos u$ on J_0° means that $\tan u < 0$. Hence, the substitutions (12.25), (12.26) and (12.27) give

$$\int \frac{\sqrt{x^2 - 1}}{x^2}\, dx, \ x \in J_0^\circ \xleftarrow{x = 1/\cos u} \int \cos^2 u \, |\tan u| \frac{\sin u}{\cos^2 u}\, du$$

$$= -\int \frac{\sin^2 u}{\cos u}\, du, \ u \in \left(\frac{\pi}{2}, \pi\right). \tag{12.33}$$

The change $t = \sin u$ on the last primitive of (12.33) yields

$$-\int \frac{\sin^2 u}{\cos u}\, du, \ u \in \left(\frac{\pi}{2}, \pi\right) \xleftarrow{\ t=\sin u\ } \int \frac{t^2}{t^2 - 1}\, dt$$

$$= t + \frac{1}{2} \log \frac{1-t}{1+t} + C_0, \ t \in (0,1). \tag{12.34}$$

The reversion in (12.34) of the change $t = \sin u$ leads to

$$-\int \frac{\sin^2 u}{\cos u}\, du = \sin u + \frac{1}{2} \log \frac{1 - \sin u}{1 + \sin u} + C_0, \ u \in \left(\frac{\pi}{2}, \pi\right) \tag{12.35}$$

and the reversion in (12.35) of the change $x = 1/\cos u$ finally yields

$$\int \frac{\sqrt{x^2 - 1}}{x^2}\, dx = \frac{\sqrt{x^2 - 1}}{x} + \frac{1}{2} \log \frac{x - \sqrt{x^2 - 1}}{x + \sqrt{x^2 - 1}} + C_0, \ x \in (-\infty, -1). \tag{12.36}$$

Since the expressions given on the right sides of (12.32) and of (12.36) are continuous at 1 and at -1 respectively, they also determine the respective primitives of $f(x)$ on J_1 and J_0. The primitive of $f(x)$ on $J_1 \cup J_0$ can now be easily obtained as

$$-\frac{\sqrt{x^2 - 1}}{x} + \frac{1}{2} \log(2x^2 - 1 + 2x\sqrt{x^2 - 1}) + C(x), \ x \in J_0 \cup J_1$$

where $C(x)$ is locally constant on $J_0 \cup J_1$. $\square$

12.6   Exercises

Find the domain and an antiderivative for each of the following functions.

Question 12.6.1

$$f(x) = \frac{\sqrt{2 - x^2}}{x}$$

Question 12.6.2

$$f(x) = \frac{\sqrt{1 - x^2}}{x^2}$$

Question 12.6.3

$$f(x) = \frac{1}{1 + \sqrt{2x - x^2}}$$

Question 12.6.4

$$f(x) = x^3 \sqrt{1 - x^2}$$

Question 12.6.5

$$f(x) = \frac{\sqrt{1 + x^2}}{1 + \sqrt{1 + x^2}}$$

Question 12.6.6

$$f(x) = \frac{\sqrt{1 + x^2}}{x}$$

Question 12.6.7

$$f(x) = \sqrt{x^2 + x + 1}$$

Question 12.6.8

$$f(x) = \sqrt{x^2 - 1}$$

Answers to the Questions

Answer 12.6.1

$D_f = [-\sqrt{2}, 0) \cup (0, \sqrt{2}]$,

$$\sqrt{2 - x^2} + \frac{1}{\sqrt{2}} \log \frac{\sqrt{2} - \sqrt{2 - x^2}}{\sqrt{2} + \sqrt{2 - x^2}}$$

Answer 12.6.2

$D_f = [-1, 0) \cup (0, 1]$,

$$- \arcsin x - \frac{\sqrt{1 - x^2}}{x}$$

Answer 12.6.3

$D_f = [0, 2]$,

$$\arcsin (x - 1) + \begin{cases} -\sqrt{\frac{1 - \sqrt{2x - x^2}}{1 + \sqrt{2x - x^2}}}, & x \in [1, 2] \\[4mm] +\sqrt{\frac{1 - \sqrt{2x - x^2}}{1 + \sqrt{2x - x^2}}}, & x \in [0, 1] \end{cases}$$

Answer 12.6.4

$D_f = [-1, 1]$,

$$-\frac{1}{3}\sqrt{1 - x^2}^3 + \frac{1}{5}\sqrt{1 - x^2}^5$$

Answer 12.6.5

$D_f = (-\infty, \infty)$,

$$\begin{cases} -\sqrt{\frac{-1 + \sqrt{1 + x^2}}{1 + \sqrt{1 + x^2}}} + \log (\sqrt{1 + x^2} + x), & x \in [0, \infty) \\[4mm] \sqrt{\frac{-1 + \sqrt{1 + x^2}}{1 + \sqrt{1 + x^2}}} - \log (\sqrt{1 + x^2} + x), & x \in (-\infty, 0) \end{cases}$$

Answer 12.6.6

$D_f = (-\infty, 0) \cup (0, \infty)$,

$$\frac{1}{\sqrt{1 + x^2}} + \frac{1}{2} \log \frac{\sqrt{1 + x^2} - 1}{\sqrt{1 + x^2} + 1}$$

Answer 12.6.7

$D_f = (-\infty, \infty),$

$\dfrac{3}{16} \log(5 + 8x^2 + 8x + 4\sqrt{3}\sqrt{x^2 + x + 1}) + \dfrac{1}{4}(2x + 1)\sqrt{x^2 + x + 1}$

Answer 12.6.8

$D_f = (-\infty, 1] \cup [1, \infty),$

$\dfrac{1}{4} \log \dfrac{x - \sqrt{x^2 - 1}}{x + \sqrt{x^2 - 1}} + \dfrac{1}{2}x\sqrt{x^2 - 1}$

Chapter 13

Tables

The methods of antidifferentiation explained in this book are condensed in the tables of Section 13.1. Some trigonometric identities which are frequent in antidifferentiation are collected in Section 13.2.

13.1 Elementary methods of antidifferentiation

The following table summarizes the types of functions and their associated methods of antidifferentiation explained through Chapters 5–12.

The first column on the left indicates the chapter where the type of function occurring in the second column has been studied. The third column indicates which function must be chosen to obtain the corresponding change of variable, in which case, the fourth column shows the transformation obtained after performing the indicated change.

An asterisk ∗ in an entry of the second column means that the corresponding function and the recommended change of variable may need an adjustment of domains as explained in Section 4.4.

A star ⋆ means that the values ε and δ may be 1 or −1: its correct value depend on each specific function considered in the second column (see (10.8)).

Letters q, q_i denote natural numbers; letters p, p_i denote integers; letters r, s denote rational numbers; in the first row, u and v denote functions $u = u(x)$ and $v = v(x)$, so that $du(x) = u'(x)\,dx$ and $dv(x) = v'(x)\,dx$; $P(x)$ and $Q(x)$ denote polynomials over $\mathbb{R}$; the notation $P_n(x)$ denotes a polynomial over $\mathbb{R}$ of degree n. A function $R(z_1, \ldots, z_n)$ is a rational function of its arguments.

Ch.	Function	Change	Transformation
5	$\int u\,dv$		$uv - \int v\,du$
6	$\int \dfrac{P(x)}{Q(x)}\,dx$		Tables 13.1.1, 13.1.2
7	$\int \dfrac{1}{\sqrt{ax^2+bx+c}}\,dx$		direct
	$\int \dfrac{Ax+B}{\sqrt{ax^2+bx+c}}\,dx$		$\dfrac{A}{a}\sqrt{ax^2+bx+c}+$ $+\int \dfrac{B-(A/a)b}{\sqrt{ax^2+bx+c}}\,dx$
	$\int \dfrac{P_n(x)}{\sqrt{ax^2+bx+c}}\,dx$		$P_{n-1}(x)\sqrt{ax^2+bx+c}+$ $+\int \dfrac{\lambda}{\sqrt{ax^2+bx+c}}$
8	$\int \dfrac{1}{(\alpha x+\beta)^p \sqrt{ax^2+bx+c}}\,dx$	$t=1/(\alpha x+\beta)$ $dx=-\,dt/(\alpha t^2)$	$\int \dfrac{P(t)}{\sqrt{a't^2+b't+c'}}\,dt$
9	$\int R\!\left(x,\left(\dfrac{ax+b}{cx+d}\right)^{\frac{p_1}{q_1}},...,\left(\dfrac{ax+b}{cx+d}\right)^{\frac{p_n}{q_n}}\right)dx$	$t^q = \dfrac{ax+b}{cx+d}$ $dx=q(ad-cb)\dfrac{t^{q-1}}{(ct^q-a)^2}\,dt$ $q=\mathrm{LCM}\{q_1,...,q_n\}$	$\int \dfrac{P(t)}{Q(t)}\,dt,\ \ \text{Ch. 6}$
10	$\int x^r (a+bx^s)^{p/q}\,dx$		
	$q=1,\ p\in\mathbb{N}$		direct
	$q=1,\ p\in\mathbb{N}^-$		Ch. 9
	$q>1,\ \dfrac{r+1}{s}\in\mathbb{Z}*$	$\star \qquad t^q=a+bx^s$ $dx=\delta\varepsilon\dfrac{q}{bs}\left(\varepsilon\dfrac{t^q-a}{b}\right)^{\frac{1}{s}-1}t^{q-1}\,dt$	$\int \dfrac{P(t)}{Q(t)}\,dt,\ \ \text{Ch. 6}$
	$q>1,\ \dfrac{r+1}{s}+\dfrac{p}{q}\in\mathbb{Z}*$	$\star \qquad t^q=\pm b\pm ax^{-s}$ $dx=-\delta\varepsilon\dfrac{q}{as}\left(\varepsilon\dfrac{a}{t^q-b}\right)^{\frac{1}{s}+1}t^{q-1}\,dt$	
	otherwise		no elementary sol.
11	$\int R(\sin x,\cos x)\,dx$ *	$t=\tan(x/2)$ $dx=\dfrac{2\,dt}{1+t^2}$ $\cos x=\dfrac{1-t^2}{1+t^2}\ \ \sin x=\dfrac{2t}{1+t^2}$	$\int \dfrac{P(t)}{Q(t)}\,dt,\ \ \text{Ch. 6}$
	$R(-\sin x,-\cos x)=$ $=R(\sin x,\cos x)$ *	$t=\tan x \qquad dx=\dfrac{dt}{1+t^2}$ $\cos x=\dfrac{1}{\sqrt{1+t^2}}\ \ \sin x=\dfrac{t}{\sqrt{1+t^2}}$	
	$R(\sin x,-\cos x)=$ $=-R(\sin x,\cos x)$	$t=\sin x$ $dx=dt/\cos x^{-1}(t)$	
	$R(-\sin x,\cos x)=$ $=-R(\sin x,\cos x)$	$t=\cos x$ $dx=-\,dt/\sin x^{-1}(t)$	
12	$\int R(x,\sqrt{1-x^2})\,dx$	$x=\sin t$ $dx=\cos t\,dt$	$\int R(\sin t,\cos t)\,dt$ Ch. 11
	$\int R(x,\sqrt{1+x^2})\,dx$	$x=\tan t$ $dx=dt/\cos^2 t$	
	$\int R(x,\sqrt{-1+x^2})\,dx$	$x=1/\cos t$ $dx=\sin t\,dt/\cos^2 t$	

* adjustments of domains required; $\star$ $\varepsilon=\pm1$, $\delta=\pm1$, see (10.8).

The following tables show the main methods of antidifferentiation for irreducible fractions of polynomials over $\mathbb{R}$ of negative degree included in Chapter 6. In these tables, $P(x)$ and $Q(x)$ are polynomials over $\mathbb{R}$ such that $\deg P(x) < \deg Q(x) \geqslant 1$ and $F(x) := P(x)/Q(x)$. All factorizations appearing in the table below are assumed to be irreducible over $\mathbb{R}$.

Table 13.1.1 gives the primitives of irreducible fractions with irreducible denominator over $\mathbb{R}$. Table 13.1.2 shows the partial decomposition of irreducible fractions with reducible denominator over $\mathbb{R}$; its left column gives the denominator of the fraction $F(x)$ and the right column shows the decomposition of $F(x)$ into fractions of irreducible denominator whose primitive can be calculated in accordance with Table 13.1.1:

Table 13.1.1. Irreducible fractions with irreducible denominator over $\mathbb{R}$:

$F(x)$	Action	Primitive
$\frac{1}{x-a}$	immediate	$\log|x-a| + C(x)$
$\frac{1}{ax^2+bx+c}$, $a>0$	$ax^2+bx+c=\gamma\left[\left(\frac{\sqrt{a}}{\sqrt{\gamma}}x+\frac{b}{2\sqrt{\gamma a}}\right)^2+1\right]$ $\gamma=c-\frac{b^2}{4a}>0$	$\frac{\gamma^{3/2}}{\sqrt{a}}\arctan\left(\frac{\sqrt{a}}{\sqrt{\gamma}}x+\frac{b}{2\sqrt{\gamma}}\right)$
$\frac{Ax+B}{ax^2+bx+c}$, $a>0$	$F(x) = \lambda\frac{2ax+b}{ax^2+bx+c} + \frac{\mu}{ax^2+bx+c}$ $\lambda=\frac{A}{2a}$ $\mu=B-\lambda b$	$\lambda\log(ax^2+bx+c)+$ $+\int\frac{\mu\,dx}{ax^2+bx+c}$

Table 13.1.2. Irreducible fractions with reducible denominator over $\mathbb{R}$:

$Q(x)$	Decomposition of $F(x)$
$\prod_{i=1}^{m}(x-a_i)$	$\sum_{i=1}^{m}\frac{A_i}{x-a_i}$
$\prod_{i=1}^{n}(x^2+\mu_i x+\nu_i)$	$\sum_{i=1}^{n}\frac{M_i x+N_i}{x^2+\mu_i x+\nu_i}$
$\prod_{i=1}^{m}(x-a_i)\prod_{j=1}^{n}(x^2+\mu_j x+\nu_j)$	$\sum_{i=1}^{m}\frac{A_i}{x-a_i}+\sum_{j=1}^{n}\frac{M_j x+N_j}{x^2+\mu_j x+\nu_j}$
$\prod_{i=1}^{m}(x-a_i)^{m_i}\prod_{j=1}^{n}(x^2+\mu_j x+\nu_j)^{n_j}$	$\left(\frac{P_0(x)}{\prod_{i=1}^{m}(x-a_i)^{m_i-1}\prod_{j=1}^{n}(x^2+\mu_j x+\nu_j)^{n_j-1}}\right)'+$ $+\sum_{i=1}^{m}\frac{A_i}{x-a_i}+\sum_{j=1}^{n}\frac{M_j x+N_j}{x^2+\mu_j x+\nu_j}$ $\deg P_0(x)\leqslant\deg Q(x)-m-n-1$

All a_i, μ_i, ν_i are real numbers.
The polynomials $x-a_i$, $x^2+\mu_i x+\nu_i$ are pairwise different.

13.2   Trigonometric identities

This section shows some trigonometric identities useful in antidifferentiation. All of them are proved or can be easily proved from Appendix A.4.

$$\cos(-x) = \cos x$$
$$\sin(-x) = -\sin x$$

$$\sin(y + \pi/2) = \cos y$$
$$\sin(y - \pi/2) = -\cos y$$

$$\frac{1}{\cos^2 x} = 1 + \tan^2 x$$

$$\cos(x + y) = \cos x \cos y - \sin x \sin y$$
$$\sin(x + y) = \sin x \cos y + \cos x \sin y$$
$$\cos(x - y) = \cos x \cos y + \sin x \sin y$$
$$\sin(x - y) = \sin x \cos y - \cos x \sin y$$

$$\sin x + \sin y = 2 \sin \left(\frac{x + y}{2}\right) \cos \left(\frac{x - y}{2}\right)$$
$$\cos x + \cos y = 2 \cos \left(\frac{x + y}{2}\right) \cos \left(\frac{x - y}{2}\right)$$

$$\sin x \sin y = \frac{1}{2}\left(\cos(x - y) - \cos(x + y)\right)$$
$$\cos x \cos y = \frac{1}{2}\left(\cos(x - y) + \cos(x + y)\right)$$
$$\sin x \cos y = \frac{1}{2}\left(\sin(x - y) + \sin(x + y)\right)$$

Formulas involving the double of a number x:

$$\sin(2x) = 2 \sin x \cos x$$
$$\cos(2x) = \cos^2 x - \sin^2 x$$

which leads to

$$\cos^2 x = \frac{1 + \cos(2x)}{2}$$
$$\sin^2 x = \frac{1 - \cos(2x)}{2}$$

Formulas involving the half of a number x:

$$\sin\left(\frac{x}{2}\right) = \sqrt{\frac{1 - \cos x}{2}}, \qquad x \in [0, 2\pi]$$

$$\cos\left(\frac{x}{2}\right) = \begin{cases} \sqrt{\frac{1+\cos x}{2}} & x \in [0, \pi] \\[2ex] -\sqrt{\frac{1+\cos x}{2}} & x \in [\pi, 2\pi] \end{cases}$$

$$\tan\left(\frac{x}{2}\right) = \begin{cases} \sqrt{\frac{1-\cos x}{1+\cos x}} & x \in [0, \pi] \\[2ex] -\sqrt{\frac{1-\cos x}{1+\cos x}} & x \in [\pi, 2\pi] \end{cases}$$

The following formulas connect the trigonometric functions with their inverses. They are useful for the final form of certain primitives and can be readily obtained from the formulas given above. In all cases, $x \in [-1, 1]$:

$$\cos(\arcsin x) = \sqrt{1 - x^2}$$
$$\sin(\arccos x) = \sqrt{1 - x^2}$$

$$\cos\left(\frac{1}{2}\arccos x\right) = \sqrt{\frac{1 + x}{2}}$$

$$\sin\left(\frac{1}{2}\arccos x\right) = \sqrt{\frac{1 - x}{2}}$$

$$\tan\left(\frac{1}{2}\arccos x\right) = \sqrt{\frac{1 - x}{1 + x}}$$

$$\cos\left(\frac{1}{2}\arcsin x\right) = \sqrt{\frac{1 + \sqrt{1 - x^2}}{2}}$$

$$\sin\left(\frac{1}{2}\arcsin x\right) = \begin{cases} \sqrt{\frac{1-\sqrt{1-x^2}}{2}} & x \in [0, 1] \\[2ex] -\sqrt{\frac{1-\sqrt{1-x^2}}{2}} & x \in [-1, 0] \end{cases}$$

$$\tan\left(\frac{1}{2}\arcsin x\right) = \begin{cases} \sqrt{\frac{1-\sqrt{1-x^2}}{1+\sqrt{1-x^2}}} & x \in [0, 1] \\[2ex] -\sqrt{\frac{1-\sqrt{1-x^2}}{1+\sqrt{1-x^2}}} & x \in [-1, 0] \end{cases}$$

and for all the following formulas, $x \in (-\infty, \infty)$:

$$\cos(\arctan x) = \frac{1}{\sqrt{1 + x^2}}$$

$$\sin(\arctan x) = \frac{x}{\sqrt{1 + x^2}}$$

$$\cos\left(\frac{1}{2}\arctan x\right) = \sqrt{\frac{1 + \sqrt{1 + x^2}}{2\sqrt{1 + x^2}}}$$

$$\sin\left(\frac{1}{2}\arctan x\right) = \begin{cases} \sqrt{\frac{-1+\sqrt{1+x^2}}{2\sqrt{1+x^2}}} & x \in [0, \infty) \\[2ex] -\sqrt{\frac{-1+\sqrt{1+x^2}}{2\sqrt{1+x^2}}} & x \in (-\infty, 0] \end{cases}$$

Appendix A

Complements

A.1 Some observations on the real-valued n-th root functions in connection with the complex numbers

Throughout this book, given any natural number n, the natural domain of the real-valued function $\sqrt[n]{\cdot}$ is $[0, \infty)$. Although some authors extend its domain to $\mathbb{R}$ in the case where n is odd, others do not.

Let us discuss the reasons which move us to confine the domain of all these functions to $[0, \infty)$:

In order to preserve the typical properties of the real-valued n-root functions like

$$\sqrt[p]{\sqrt[q]{x}\sqrt[r]{y}} = \sqrt[pq]{x}\ \sqrt[pr]{y} \tag{A.1}$$

it is necessary to restrict the domain of the real-valued n-root functions to $[0, \infty)$. Indeed, the even root functions $\sqrt[2n]{\cdot}$ are not defined over $(-\infty, 0)$ because $x^{2n} \geqslant 0$ for all real numbers x. That means that any coherent extension of $\sqrt[2n]{\cdot}$ to $\mathbb{R}$ should map $(-\infty, 0)$ into $\mathbb{C}\backslash\mathbb{R}$, which shows that such an extension could not be real-valued. In particular, any reasonable extension of $\sqrt{\cdot}$ to $\mathbb{R}$ would map -1 either to i or to $-i$.

Moreover, although the assignation $\sqrt{-1} := i$ is accepted among some scholars, it is not compatible with (A.1) because it leads to the following nonsense identity:

$$-1 = i^2 = \sqrt{-1}\sqrt{-1} = \sqrt{(-1)(-1)} = \sqrt{1} = 1. \tag{A.2}$$

For odd real-valued root functions $\sqrt[2n-1]{\cdot}$ the situation is a little different. Indeed, since $f(x) = x^{2n-1}$ is an increasing function that maps $(-\infty, \infty)$ onto itself, it admits an inverse function defined on the whole interval $(-\infty, \infty)$ which is called the $(2n-1)$-root real function by some authors and is denoted $\sqrt[2n-1]{\cdot}$. However, this inverse function does not satisfy (A.1), otherwise,

$$1 = \sqrt{\sqrt[3]{-1}\sqrt[3]{-1}} = \sqrt[6]{-1}\sqrt[6]{-1}$$

which does not make sense; following a similar argument to that of (A.2), it can be proved that the expression $\sqrt[6]{-1}$ cannot be defined.

But there are more reasons discouraging us from defining the odd root of a negative number: of course, the only reasonable choice for a real value for $\sqrt[2n-1]{-1}$ is -1 because the only real number x for which $x^{2n-1} = -1$ is $x = -1$. Thus, let us accept for a while that $\sqrt[2n-1]{-1} := -1$ for all $n \in \mathbb{N}$: if (q_n) is an increasing sequence of odd natural numbers, then $q_n \xrightarrow[n]{} \infty$. It would be desirable that

$$(-1)^{1/q_n} \xrightarrow[\quad n \quad]{} (-1)^{1/\lim_n q_n} = (-1)^0 = 1$$

but this is not possible, because $(-1)^{1/q_n} = -1$ for all n. This observation is particularly relevant when seeking the continuity of the general exponential functions $f(x) = a^x$ as treated in Section 3.4.

All these inconveniences vanish if the domain of every real-valued function $\sqrt[n]{\cdot}$ is restricted to $[0, \infty)$ no matter if n is an even or an odd natural number. Under this restriction, formula (A.1) holds, and of course, this choice must be accompanied by the elimination of the identity $i = \sqrt{-1}$, which must be replaced by the identity $i^2 = -1$ adopted in (1.2). The only price to be paid is that the typical formula

$$x = \frac{-b \pm \sqrt{b^2 - 4ac}}{2a}$$

for the roots of the second degree polynomial $ax^2 + bx + c$ over the reals must be replaced by the one given in Proposition 3.2.5.

It is remarkable that the complex-valued n-root functions cannot be defined with continuity on the whole field $\mathbb{C}$, but continuity and complex differentiability of the mentioned functions can be preserved if their domain is restricted to $\mathbb{C} \backslash R_z$, with R_z any radius of the form $R_z := \{tz : t \in [0, \infty)\}$ and z a fixed non-null complex number. All complex analysts agree on removing the radius R_{-1}. In this way, the restriction of each complex-valued n-root to $(0, \infty)$ agrees with the real-valued n-root function restricted to $(0, \infty)$. We refer to [Bak (1993)] or [Chakraborty et al. (2016)] for further discussion on Complex Analysis; that subject is not included in this book.

We note that the real-valued n-root functions are continuous but not differentiable at $x = 0$. We also point out that the interval $(0, \infty)$ endowed with the product of real numbers is a group, and since (A.1) works for all x and y in $(0, \infty)$, $\sqrt[n]{\cdot}$ is a group automorphism on $(0, \infty)$ for each $n \in \mathbb{N}$. This fact can be viewed as an algebraic interpretation preventing us from extending the domain of the functions $\sqrt[n]{\cdot}$ to the entire set $\mathbb{R}$.

A.2   The role of the fundamental functions in Real Analysis

If the order structure of $\mathbb{R}$ is stripped away and the sum and product of real numbers are considered separately, we come across two basic algebraic structures: the additive group $G_1 := (\mathbb{R}, +)$ and the multiplicative group $G_2 = ((0, \infty), \cdot)$. It is almost immediate that any automorphism on G_1 is of the form $f(x) = ax$ with a a non-null real number. In turn, e^x and $\log x$ are the most basic group isomorphisms between G_1 and G_2, and x^n, $\sqrt[n]{x}$ (with $n \in \mathbb{N}$) are the most basic group automorphisms on G_2. These properties place the mentioned functions in the core of Real Analysis, which is why they are called *fundamental*.

But behind the cold word *automorphism* there is more that can be said; assume that we are given two real numbers x and y expressed in decimal digits and we need to know the value of x/y with an error smaller than one thousandth. If no calculator is available and y has too many digits, the task of manually dividing x by y can be very painful. However, if we had some way to find out the logarithm of any real number and to reverse that operation, we would only need to find the logarithms of x and y so that, by the simple subtraction operation $r := \log x - \log y$, and reversing the log operation for r, we would obtain $x/y = e^r$. This is exactly the technical problem that mathematicians such as Napier, Bürgi, Briggs and Bissaker solved in the seventeenth century. They did not have any sort of calculator machine for products or divisions, but they constructed tables that let them find the logarithm of a given real number and its reverse with great accuracy, which let them transform products and divisions into sums and subtractions: in technical words, they used the isomorphic property of the log function between the groups G_1 and G_2.

The discovery of the logarithms was a hit of the seventeenth century. Thanks to them, Kepler made his theory about planetary movements, and logarithmic tables were used extensively until the emergence of modern calculators in the second half of the twentieth century.

The circular functions are also considered as fundamental real-valued functions, but unlike logarithms, they originated in the field of Geometry. The circular functions are intimately connected with the complex-valued exponential function e^z, which is in turn a group isomorphism between the groups $(\mathbb{C}, +)$ and $(\mathbb{C} \backslash \{0\}, \cdot)$.

Because of the role of the fundamental functions in antidifferentiation, they deserve some words about their construction, which is not elementary at all. For the construction of x^n and $\sqrt[n]{x}$ ($n \in \mathbb{N}$), the content of Section 3.1 is sufficient. Let us outline in the subsequent sections a construction for the remaining fundamental functions.

A.3 Notes on the construction of the functions $\log x$ and e^x

There are other ways to introduce the functions $\log x$ and e^x than that given in Section 3.3. For instance, if these functions are wanted right after the introduction of sequences of real numbers then the exponential function can be defined as

$$e^x = \lim_n \left(1 + \frac{x}{n}\right)^n, \quad x \in \mathbb{R}. \tag{A.3}$$

And if a good background on series of real numbers is available, it is also possible to define the exponential function as the convergent series

$$e^x = \sum_{n=0}^{\infty} \frac{x^n}{n!}, \quad x \in \mathbb{R}. \tag{A.4}$$

The ways indicated in formulas (A.3) and (A.4) allow the introduction of the exponential function e^x at an early stage of a first course of Real Mathematical Analysis, but that requires a lot of work in order to prove all the usual properties of the function e^x, including the existence of its inverse, the log function.

Of course, all these constructions must lead to the same functions $\log x$ and e^x discussed in Section 3.3. Since any method of producing the function $\log x$ should produce a differentiable function whose derivative is $1/x$ and whose value at 1 is 0, the following result shall convince the reader that all those constructive methods produce exactly the same log function:

Proposition A.3.1. *There exists a unique function $L\colon (0, \infty) \longrightarrow \mathbb{R}$, differentiable on its domain and satisfying the following conditions:*

(i) $L'(x) = 1/x$ for all $x \in (0, \infty)$,
(ii) $L(1) = 0$.

Proof. The function $\log x$ defined in Section 3.3 satisfies the conditions (i) and (ii). This proves the existence.

Let $L\colon (0, \infty) \longrightarrow \mathbb{R}$ be a function which satisfies (i) and (ii). Thus

$$L'(x) = \frac{1}{x} = \log' x, \quad \text{all } x \in (0, \infty)$$

and therefore, by Proposition 2.3.10, there exists a constant K such that $L(x) \equiv \log x + K$. The evaluation of this last identity at $x = 0$ yields $K = 0$, so $L(x) \equiv \log x$, and this proves that $\log x$ is the only function which satisfies (i) and (ii). □

The uniqueness of the exponential function e^x as the inverse of $\log x$ is a consequence of the uniqueness of the function $\log x$ proved in A.3.1. The uniqueness of e^x can be also derived from the following result.

Proposition A.3.2. *There exists a unique function $\mathcal{E}\colon \mathbb{R} \longrightarrow \mathbb{R}$, differentiable on its domain and satisfying the following conditions:*

(i) $\mathcal{E}'(x) = \mathcal{E}(x)$ for all x,
(ii) $\mathcal{E}(0) = 1$.

Proof. The function $E(x) = e^x$ defined in Section 3.3 satisfies the conditions (i) and (ii). This proves the existence.

In order to prove the uniqueness, let $\mathcal{E}$ be a function which satisfies (i) and (ii). Thus, $\mathcal{E}^{(n)}(0) = \mathcal{E}(0) = 1$ for all n. Hence, for every $x \in \mathbb{R}$ and every $n \in \mathbb{N}$, Theorem 2.3.14 provides a number $\zeta_{n,x} \in I_x$ for which

$$\mathcal{E}(x) = \sum_{k=0}^{n} \frac{1}{k!} x^k + \frac{\mathcal{E}(\zeta_{n,x})}{(n+1)!} x^{n+1} \tag{A.5}$$

where I_x denotes the open interval whose endpoints are 0 and x (recall $0! = 0$).

Fix $x \in \mathbb{R}$ and choose a positive integer m so that $|x/m| < 1$. Thus

$$\frac{|x|^{n+1}}{(n+1)!} \leqslant \left(\frac{|x|}{m}\right)^{n+1-m} \frac{|x|^m}{m!} \xrightarrow[n]{} 0. \tag{A.6}$$

Since $\mathcal{E}$ is continuous on the compact interval $[-|x|, |x|]$, Weierstrass' theorem yields that the sequence $(\mathcal{E}(\zeta_{n,x}))_{n=1}^{m}$ is bounded. This fact and (A.6) show

$$\frac{\mathcal{E}(\zeta_{n,x})}{(n+1)!} x^{n+1} \xrightarrow[n]{} 0$$

for any fixed number x, and in turn, (A.5) yields

$$\mathcal{E}(x) = \lim_n \sum_{k=0}^n \frac{1}{k!} x^k, \quad \text{for all } x \in \mathbb{R}$$

which means that $\mathcal{E}(x)$ is univocally determined by the convergent sequence $(\sum_{k=0}^n \frac{1}{k!} x^k)_{n=1}^\infty$. In particular, $\mathcal{E}(x) = e^x$, and the uniqueness is proved. $\qquad\square$

A.4 Notes on the construction of the circular functions

There are many ways to introduce the circular functions analytically. As in the case of the exponential function, sine and cosine over the reals can be defined in terms of convergent series as

$$\sin x := \sum_{n=1}^\infty \frac{(-1)^{n+1}}{(2n-1)!} x^n, \quad x \in \mathbb{R} \tag{A.7}$$

$$\cos x := \sum_{n=0}^\infty \frac{(-1)^n}{(2n)!} x^n, \quad x \in \mathbb{R}. \tag{A.8}$$

All usual properties of the circular functions can be managed from (A.7), (A.8) if the reader possesses certain background on functional series. But as in the case of the construction of $\log x$ and e^x, the integration techniques work too. For instance, one possibility is departing from an integral function of $1/\sqrt{1-x^2}$, but this choice implies the use of improper integrals.[1] Instead, the construction given in Section 3.5 is based upon the integral function of $\sqrt{1-x^2}$, following a procedure borrowed in part from [Fulks (1961)].

Theorem A.4.1 will make clear that all conceivable constructions of the circular functions produce exactly the same functions $\sin x$ and $\cos x$ introduced in Section 3.5.

Theorem A.4.1. *There exists a unique pair of differentiable functions on $\mathbb{R}$, $c(x)$ and $s(x)$, which satisfy the following conditions:*

(i) $s'(x) = c(x)$ and $c'(x) = -s(x)$ for all x,
(ii) $c(x)^2 + s(x)^2 = 1$ for all x,
(iii) $c(0) = 1$ and $s(0) = 0$.

---

[1]The theory of improper integrals generalizes the Riemann integral to functions defined on non-compact intervals.

Proof. The functions $c(x) = \cos x$ and $s(x) = \sin x$ constructed in Section 3.5 satisfy (i), (ii) and (iii), so the existence is proved.

For the uniqueness, note that condition (i) means that the n-th derivatives $c^{(n)}(x)$ and $s^{(n)}(x)$ do exist for all $n \in \mathbb{N}$ and that $s^{(2k-1)}(0) = (-1)^{k+1}$, $s^{(2k)}(0) = 0$ for all $k \in \mathbb{N}$. Thus, Theorem 2.3.14 provides a real number $\xi_{x,n}$ for every $n \in \mathbb{N}$ and every $x \in \mathbb{R}$ such that

$$s(x) = \sum_{k=1}^{n} \frac{(-1)^{k+1}}{(2k-1)!} x^{2k-1} + \frac{s^{(2n)}(\xi_{x,n})}{(2n)!} x^{2n}. \tag{A.9}$$

A similar argument to that given in the proof of Proposition A.3.2 proves that

$$\frac{s^{(2n)}(\xi_{x,n})}{(2n)!} x^{2n} \xrightarrow[n]{} 0, \quad \text{for all } x \in \mathbb{R}. \tag{A.10}$$

Thus, taking limits on both sides of (A.9) when $n \to \infty$, (A.10), we get that the sequence $\left(\sum_{k=1}^{n}(-1)^{k+1}x^{2k-1}/(2k-1)!\right)_{n=1}^{\infty}$ is convergent for all x and its limit is

$$s(x) = \lim_{n} \sum_{k=1}^{n} \frac{(-1)^{k+1}}{(2k-1)!} x^{2k-1}. \tag{A.11}$$

This proves that $s(x)$ is univocally determined by (A.11), which proves the uniqueness of $s(x)$.

A similar argument shows the uniqueness of $c(x)$. □

A.5 Some facts about the relation between differentiability and integrability

As seen in Theorem 2.4.7, for every continuous function defined on a non-degenerate interval, the relation between its integral functions and its antiderivatives is perfect. However, in spite of Barrow's rule given in Theorem 2.4.8, the relation between Riemann-integrable functions and derivatives is not so good. Indeed, there are integrable functions which are not derivatives, and there are derivatives which are not integrable, as the following examples show.

Example A.5.1. The function $f \colon [0,1] \longrightarrow \mathbb{R}$ given by

$$f(x) = \begin{cases} 0 & \text{if } 0 \leqslant x < 1 \\ 1 & \text{if } x = 1 \end{cases}$$

is integrable but it does not have any antiderivative.

Proof. Since f is monotone, then f is integrable on $[0,1]$ by virtue of Theorem 2.4.5.

Assume now that there exists an antiderivative F of f on $[0,1]$. Thus, on the one hand, $F'(1) = f(1) = 1$, but on the other hand, L'Hôpital's rule yields

$$F'(1) = \lim_{x \to 1^-} \frac{F(x) - F(1)}{x - 1} = \lim_{x \to 1^-} F'(x) = \lim_{x \to 1^-} f(x) = 0,$$

a contradiction. □

Observe that the integral of the function f given in Example A.5.1 is null. In fact, $L(f, P) = 0$ for any partition P of $[0, 1]$.

Example A.5.2. The function

$$F(x) = \begin{cases} x^2 \sin(1/x^2) & \text{if } 0 < x \leq 1 \\ 0 & \text{if } x = 0 \end{cases}$$

is differentiable on $[0, 1]$ but F' is not integrable on $[0, 1]$.

Proof. The chain rule and the usual results for products and quotients of differentiable functions show f is differentiable at each $x \neq 0$ and

$$F'(x) = 2x \sin\left(\frac{1}{x^2}\right) - \frac{2}{x} \cos\left(\frac{1}{x^2}\right).$$

Moreover, since $\sin(1/x^2)$ is bounded, Proposition 1.2.9 yields

$$F'(0) = \lim_{x \to 0^+} \frac{F(x) - F(0)}{x} = \lim_{x \to 0^+} x \sin(1/x^2) = 0,$$

hence $F(x)$ is differentiable on the entire interval $[0, 1]$. However, F' is not integrable because it is not bounded. □

The derivative F' given in Example A.5.2 is integrable on every compact interval contained in $(0, 1]$, so the integrability of F' on $[0, 1]$ fails because of its bad behavior at the sole point $x = 0$. But there are bounded derivatives which are not integrable on any compact interval. Examples of such functions are difficult and exceed the scope of this book. They can be found in [Kharazishvili (1996)].

For a perfect relation between integration and antidifferentiation it is necessary to use an integration theory more general than that of Riemann such as the integral of Henstock–Kurzweil: given any differentiable function $F : [a, b] \longrightarrow \mathbb{R}$, its derivative is always integrable in the sense of Henstock–Kurweil and its integral on $[a, b]$ is equal to $F(b) - F(a)$. A very accessible article about this subject is [Bartle (1966)].

A.6 Geometrical representation of complex numbers and the n-th complex roots of unity

Every complex number $z = a + bi$ can be identified with the point (a, b) of $\mathbb{R} \times \mathbb{R}$. In this way, the number 1 is identified with the point $(1, 0)$, the number i with $(0, 1)$ and the whole set $\mathbb{C}$ with the plane $\mathbb{R} \times \mathbb{R}$, which is why this representation is called *representation of the complex plane*.

For every $z = a + bi \in \mathbb{C}$, two notions are derived from this geometrical representation: the *modulus of* z is the real number $|z| := \sqrt{a^2 + b^2}$, that is, the length of the fixed vector $\overrightarrow{Oz}$; and if $z \neq 0$, the *argument of* z is the angle $\arg z := \alpha \in [0, 2\pi)$

that carries the fixed vector $\overrightarrow{O1}$ over $\overrightarrow{Oz}$ counterclockwise (see picture below). For $z = 0$, no argument is assigned.

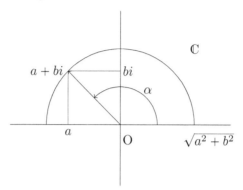

Any complex number z can be also given in *polar form* as

$$z = |z|(\cos \alpha + i \sin \alpha) \qquad (A.12)$$

where $\alpha = \arg z$ and $|\cos \alpha + i \sin \alpha| = 1$ by virtue of (3.36). In a more formal way, the argument α of a complex number $a + bi$ whose modulus is 1 can be identified with any real number α such that $a = \cos \alpha$ and $b = \sin \alpha$. It easily follows from formulas (3.36), (3.30) and (3.35) that for every $k \in \mathbb{Z}$ there exists a unique $\alpha \in [2k\pi, 2(k+1)\pi)$ satisfying (A.12).

The polar representation is very useful to calculate the product of complex numbers. Indeed, given $z_1 = |z|(\cos \alpha_1 + i \sin \alpha_1)$ and $z_2 = |z_2|(\cos \alpha_2 + i \sin \alpha_2)$, statements (i) and (ii) in Proposition 3.5.2 yield

$$z_1 z_2 = |z_1||z_2|[\cos \alpha_1 \cos \alpha_2 - \sin \alpha_1 \sin \alpha_2 + i(\cos \alpha_1 \sin \alpha_2 + \sin \alpha_1 \cos \alpha_2)]$$
$$= |z_1||z_2|[\cos (\alpha_1 + \alpha_2) + i \sin (\alpha_1 + \alpha_2)]$$

$$(A.13)$$

that is, the modulus of $z_1 z_2$ is the product of the moduli of z_1 and z_2, and its argument is the sum of the arguments of z_1 and z_2.

The *n-th roots of unity* are defined as the complex roots of the polynomial $x^n - 1$ which are exactly

$$\cos \left(\frac{2k\pi}{n} \right) + i \sin \left(\frac{2k\pi}{n} \right), \qquad k = 0, 1, 2, \ldots, n - 1$$

such as follows from (A.13).

The n-th roots of unity admit a simple geometrical representation thanks to the notion of argument. For instance, the roots of $x^3 - 1$ are

$$\cos 0 + i \sin 0 = 1,$$
$$\cos(2\pi/3) + i \sin(2\pi/3) = (-1 + \sqrt{3}i)/2,$$
$$\cos(4\pi/3) + i \sin(4\pi/3) = (-1 - \sqrt{3}i)/2,$$

and can be geometrically represented as

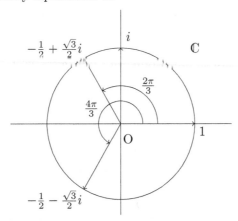

This procedure can be also applied to find the solutions of the equation

$$x^n - z_0 = 0 \qquad (A.14)$$

where z_0 is a fixed, non-null, complex number and n is a positive integer. Indeed, denoting $\alpha_0 := \arg z_0$, the reader may realize that

$$\sqrt[n]{|z_0|}\left[\cos\left(\frac{\alpha_0}{n} + \frac{2\pi k}{n}\right) + i\sin\left(\frac{\alpha_0}{n} + \frac{2\pi k}{n}\right)\right], \qquad k = 0, 1, \ldots, n-1 \qquad (A.15)$$

are the n solutions to (A.14).

A.7   Long division of polynomials

Throughout this section, let $\mathbb{K}$ denote one of the fields $\mathbb{R}$ or $\mathbb{C}$. Given a pair of polynomials $P(x)$ and $D(x)$ over $\mathbb{K}$ such that $\deg P(x) > \deg D(x)$, there exists another pair of polynomials $Q(x)$ and $R(x)$ over $\mathbb{K}$ such that $P(x) = Q(x)D(x) + R(x)$ and $\deg R(x) < \deg P(x)$. The polynomials $P(x)$ and $D(x)$ are respectively called the *dividend* and *divisor*, and the polynomials $Q(x)$ and $R(x)$ are the *quotient* and *remainder*. The algorithm that yields the quotient and remainder for a pair of given polynomials $P(x)$ and $D(x)$ is very like the usual algorithm for the division of real numbers and works as follows: let

$$P(x) = a_n x^n + a_{n-1} x^{n-1} + \ldots + a_1 x + a_0$$
$$D(x) = d_k x^k + \ldots + d_1 x + d_0$$

with $a_n \neq 0 \neq d_k$ and $m := n - k > 0$. The coefficients of the quotient

$$Q(x) = c_m x^m + \ldots + c_1 x + c_0$$

are determined as follows. First we first pick $c_m := a_n/d_k$ so that the degree of $P_1(x) := P(x) - c_m x^m D(x)$ is smaller than $\deg P(x)$. If $\deg P_1(x) < \deg D(x)$, then we take $Q(x) := c_m x^m$ and $R(x) := P_1(x)$; otherwise, we choose c_{m-1} so that the degree of $P_2(x) := P_1(x) - c_{m-1} x^{m-1} \cdot D(x)$ is smaller than $\deg P_1(x)$ (c_{m-1}

may be null). If $\deg P_2(x) < \deg D(x)$, then we stop this procedure by taking $Q(x) := c_m x^m + c_{m-1} x^{m-1}$ and $R(x) := P_2(x)$; otherwise, we choose c_{m-2} so that the degree of $P_3(x) := P_2(x) - c_{m-2} x^{m-2} D(x)$ is smaller than $\deg P_2(x)$ and so on. This procedure ends when we get a polynomial $P_l(x)$ such that $\deg P_l(x) < \deg D(x)$, which obviously happens after at most m steps, so $1 \leqslant l \leqslant m$.

An example will illustrate this simple procedure for the case $\mathbb{K} = \mathbb{R}$. Let

$$P(x) = 5x^4 - 3x^2 + x - 1 \quad \text{and} \quad D(x) = x^2 - 2x + 1.$$

Place $P(x)$ under a squared angle and $D(x)$ on the left as in Diagram (A.16). It is convenient to put $0 \cdot x^n$ in its corresponding place when $a_n = 0$ or $d_n = 0$. The quotient shall be placed right above the squared angle. Obviously, the degree of $Q(x)$ must be two, so $Q(x) = c_2 x^2 + c_1 x + c_0$. First we take $c_2 = 5$ and the polynomial $5x^2 D(x)$ is placed just below $P(x)$, so we may calculate

$$P_1(x) := P(x) - 5x^2 D(x) = 10x^3 - 8x^2 + x - 1$$

whose degree is smaller than that of $P(x)$. Next, we take $c_1 := 10$ so that

$$P_2(x) := P_1(x) - 10x D(x) = 12x^2 - 9x - 1$$

and $\deg P_2(x) < \deg P_1(x)$. Next, we take $c_0 := 12$, so $P_3(x) := P_2(x) - 12 D(x) = 15x - 13$. As $\deg P_3(x) < \deg D(x)$, the algorithm stops by taking $Q(x) = 5x^2 + 10x + 12$ and $R(x) = P_3(x)$.

			$5x^2$	$+10x$	$+12$		
x^2	$-2x$	$+1$	$5x^4$	$+0 \cdot x^3$	$-3x^2$	$+x$	-1
			$5x^4$	$-10x^3$	$+5x^2$	$+0 \cdot x$	$+0$
				$+10x^3$	$-8x^2$	$+x$	-1
				$+10x^3$	$-20x^2$	$+10x$	$+0$
					$+12x^2$	$-9x$	-1
					$+12x^2$	$-24x$	$+12$
						$+15x$	-13

$$(\text{A.16})$$

The procedure just described in the example above admits some shortcuts that the attentive reader will find out in the following Diagram.

$$5x^2 \quad +10x \quad +12$$

$$
\begin{array}{ccc|cccccc}
x^2 & -2x & +1 & 5x^4 & +0\cdot x^3 & -3x^2 & +x & -1 \\
& & & \underline{5x^4} & -10x^3 & +5x^2 & & \\
& & & & +10x^3 & -8x^2 & +x & \\
& & & & \underline{+10x^3} & -20x^2 & +10x & \\
& & & & & +12x^2 & -9x & -1 \\
& & & & & \underline{+12x^2} & -24x & +12 \\
& & & & & & +15x & -13 \\
\end{array}
$$

Let us give an example to illustrate the case $\mathbb{K} = \mathbb{C}$: Let $P(z) = z^3 + iz^2 + z + 1$, $D(z) = z^2 + i$ and by means of the long division algorithm, let us obtain the polynomials $Q(z)$ and $R(z)$ over $\mathbb{C}$ for which $P(z) = Q(z)D(z) + R(z)$ and $\deg R(z) < \deg D(z)$.

$$z \quad +i$$

$$
\begin{array}{cc|cccc}
z^2 & +i & z^3 & +iz^2 & +z & +1 \\
& & \underline{z^3} & +0 & +iz & \\
& & & +iz^2 & +(1-i)z & +1 \\
& & & \underline{+iz^2} & +0\cdot z & -1 \\
& & & & (1-i)z & +2 \\
\end{array}
$$

hence $z^3 + iz^2 + z + 1 = (z^2 + i)(z + i) + (1 - i)z + 2$.

A.8 Synthetic division of polynomials

Let $\mathbb{K}$ denote either $\mathbb{R}$ or $\mathbb{C}$. Given a polynomial $P(x) = a_0 x^n + a_1 x^{n-1} + \ldots + a_{n-1} x + a_n$ over $\mathbb{K}$ and given a number $c \in \mathbb{K}$, consider the finite sequence $(b_k)_{k=0}^n$

given by

$$b_0 := a_0$$

$$b_k := a_{k-1} - cb_{k-1}, \quad 1 \leqslant k \leqslant n.$$

Then $B(x) := b_0 x^{n-1} + b_1 x^{n-2} + \ldots + b_{n-2} x + b_{n-1}$ and b_n are respectively the quotient and the remainder of the division of $P(x)$ by $x-c$ as the reader may prove. Thus, if $b_n = 0$ then c is a root of $P(x)$ and $P(x) = (x-c)B(x)$.

There is a simple procedure called *synthetic division* that yields the coefficients b_k quickly. Arrange the coefficients a_k in line and the number c as indicated in the Diagram below.

The coefficients of the quotient $B(x)$ (in blue) are obtained as follows: first, the coefficient a_0 is moved under the black line and then we let $b_0 := a_0$; next, the product of b_0 by c is put right under the coefficient a_1 and the addition $b_1 := a_1 + cb_0$ is put under the black line. Next, the product cb_1 is put right under a_2 and the addition $b_2 := a_2 + cb_1$ is put under the black line, and so on.

Example A.8.1. Factorize the polynomial $P(x) = 2x^4 + x^3 - 7x^2 - 3x + 3$ in irreducible form over $\mathbb{R}$:

Since the coefficients of $P(x)$ are integers, Proposition 3.2.6 shows the only possible rational roots of $P(x)$ are ± 1, ± 3, $\pm 1/2$ and $\pm 3/2$. By the algorithm of the synthetic division applied for $c = -1$, we obtain

$$
\begin{array}{r|rrrrr}
 & 2 & 1 & -7 & -3 & 3 \\
-1 & & -2 & 1 & 6 & -3 \\
\hline
 & 2 & -1 & -6 & 3 & \boxed{0}
\end{array}
\tag{A.17}
$$

which means that -1 is a root of $P(x)$ and $P(x) = (x+1)(2x^3 - x^2 - 6x + 3)$. Applying again the same algorithm to $2x^3 - x^2 - 6x + 3$ for the choice $c = 1/2$, we get

$$
\begin{array}{r|rrrr}
 & 2 & -1 & -6 & 3 \\
1/2 & & 1 & 0 & -3 \\
\hline
 & 2 & 0 & -6 & \boxed{0}
\end{array}
$$

which means that $P(x) = (x+1)(x-1/2)(2x^2 - 6)$. And as the roots of $2x^2 - 6$ are $-\sqrt{3}$ and $\sqrt{3}$, then $2x^2 - 6 = 2(x+\sqrt{3})(x-\sqrt{3})$, which lets us factorize $P(x)$ in irreducible form as $P(x) = 2(x+1)(x-1/2)(x+\sqrt{3})(x-\sqrt{3})$.

A.9 Completing squares

The technique of completing squares is very useful in handling second degree polynomials. It consists in rewriting a given polynomial $P(x) = ax^2 + bx + c$ with $a > 0$ as $(\alpha x + \beta)^2 + \gamma$ with $\alpha > 0$; in the case $\gamma \neq 0$, a further reduction yields $P(x) = \gamma[(\alpha' x + \beta')^2 + 1]$ if $\gamma > 0$ or $P(x) = |\gamma|[(\alpha' x + \beta')^2 - 1]$ if $\gamma < 0$, where $\alpha' := \alpha/\sqrt{|\gamma|}$ and $\beta' := \beta/\sqrt{|\gamma|}$.

Three examples are sufficient:

Example A.9.1. Complete the square in $x^2 - 4x + 4$.

Solution. The identity

$$x^2 - 4x + 4 = (\alpha x + \beta)^2 + \gamma = \alpha^2 x^2 + 2\alpha\beta x + \beta^2 + \gamma$$

produces the following string of equations:

$$1 = \alpha^2 \qquad\qquad (A.18)$$
$$-4 = 2\alpha\beta \qquad\qquad (A.19)$$
$$4 = \beta^2 + \gamma. \qquad\qquad (A.20)$$

Since α is always chosen positive, (A.18) gives $\alpha = 1$, and then (A.19) gives $\beta = -2$. Finally, (A.20) yields $\gamma = 0$. Thus, $x^2 - 4x + 4 = (x-2)^2$, a perfect square. □

Example A.9.2. Complete the square in $x^2 + x + 1$.

Solution. Let us write down

$$x^2 + x + 1 = (\alpha x + \beta)^2 + \gamma = \alpha^2 x^2 + 2\alpha\beta x + \beta^2 + \gamma$$

for certain coefficients α, β and γ. That produces the following string of equations:

$$1 = \alpha^2 \qquad\qquad (A.21)$$
$$1 = 2\alpha\beta \qquad\qquad (A.22)$$
$$1 = \beta^2 + \gamma. \qquad\qquad (A.23)$$

Thus, (A.21) yields $\alpha = 1$; once α is known, (A.22) gives $\beta = 1/2$, and finally, (A.23) gives $\gamma = 1 - 1/4 = 3/4$. Therefore,

$$x^2 + x + 1 = \left(x + \frac{1}{2}\right)^2 + \frac{3}{4} = \frac{3}{4}\left[\left(\frac{2}{\sqrt{3}}x + \frac{1}{\sqrt{3}}\right)^2 + 1\right] \qquad (A.24)$$

as desired. □

Example A.9.3. Complete the square in $25x^2 - 10x - 1$.

Solution. Writing
$$25x^2 - 10x - 1 = (\alpha x + \beta)^2 + \gamma$$
$$= \alpha^2 x^2 + 2\alpha\beta x + \beta^2 + \gamma$$
we get the following string of equations:
$$25 = \alpha^2 \tag{A.25}$$
$$-10 = 2\alpha\beta \tag{A.26}$$
$$-1 = \beta^2 + \gamma. \tag{A.27}$$
Thus, solving consecutively (A.25), (A.26) and (A.27), we manage $\alpha = 5$, $\beta = -1$, and $\gamma = -2$. Therefore,
$$25x^2 - 10x - 1 = (5x - 1)^2 - 2 = 2\left[\left(\frac{5}{\sqrt{2}}x - \frac{1}{\sqrt{2}}\right)^2 - 1\right].$$

$\square$

A.10   The Fundamental Theorem of Algebra

The celebrated *Fundamental Theorem of Algebra* plays a central role in factorization of polynomials and therefore, in antidifferentiation of fractions of polynomials. It owes its name to historical reasons, when polynomials were the main subjects sought by algebraists. However, all its known proofs use some Real or Complex Analysis. The demonstration given in this section needs only Real Analysis of one variable and the notions of modulus and argument of a complex number, so it should be accessible to any potential reader of this book.[2]

We give the name *two variable polynomial of coefficients* $(a_{kl})_{k=1 l=1}^{n\ \ n}$ to the expression
$$\sum_{1 \leqslant k,l \leqslant n} a_{kl}x^k y^l \tag{A.28}$$
where each a_{kl} is a real number.   Such a polynomial defines a function $f \colon \mathbb{R} \times \mathbb{R} \longrightarrow \mathbb{R}$ which maps each pair (x, y) to the real value $f(x, y)$ given by (A.28).

For each $y \in \mathbb{R}$, let $f_y \colon \mathbb{R} \longrightarrow \mathbb{R}$ denote the function that maps each x to $f_y(x) := f(x, y)$; and for each $x \in \mathbb{R}$, let $f^x \colon \mathbb{R} \longrightarrow \mathbb{R}$ be the function that maps every $y \in \mathbb{R}$ to $f^x(y) := f(x, y)$.

Given a pair of compact intervals I and J, the Heine–Cantor theorem proves that for every $y \in J$, the restriction $f_y|_I$ is uniformly continuous. Moreover, if $f(x, y)$ is a two variable polynomial, more can be said:

Proposition A.10.1. *Let* $f(x, y) := \sum_{1 \leqslant k,l \leqslant n} a_{kl}x^k y^l$ *be a two variable polynomial, let I and J be a pair of compact intervals, and for every $y \in J$ let $f_y \colon I \longrightarrow \mathbb{R}$ be the function given by $f_y(x) := f(x, y)$.*

---

[2]A simple generalization of Theorem 2.2.9 shows that if $\mathbb{R}^2$ is endowed with the usual metric, K is a compact subset of $\mathbb{R}^2$ and f is a continuous function from K into $\mathbb{R}$, then $f(K)$ is also compact. The reader who knows this version of Weierstrass' theorem for real-valued functions of two real variables may safely skip Propositions A.10.1 and A.10.2.

Then, given $\varepsilon > 0$, there exists $\delta > 0$ such that for every $y \in J$ and every pair of numbers x and z in I such that $|x - z| < \delta$, we have $|f_y(x) - f_y(z)| < \varepsilon$.

Proof. Fix $\varepsilon > 0$. Let $M := \max\{|a_{kl}|: 1 \leqslant k, l \leqslant n\}$. Since J is bounded, there exists $H > 0$ such that $|y^l| \leqslant H$ for all $1 \leqslant l \leqslant n$ and all $y \in J$. Moreover, by Theorem 2.2.13, each function $p_k(x) := x^k$ ($1 \leqslant k = 1 \leqslant n$) is uniformly continuous on I so for $\varepsilon' := \varepsilon/(MHn^2)$, there exists $\delta > 0$ satisfying $|x^k - z^k| < \varepsilon'$ for all $1 \leqslant k \leqslant n$ and all x and z in I such that $|x - z| < \delta$.

Thus, given any $y \in J$ and any pair of elements x and z in I such that $|x - z| < \delta$,

$$|f_y(x) - f_y(z)| \leqslant \left| \sum_{1 \leqslant k, l \leqslant n} a_{kl}(x^k - z^k)y^l \right|$$

$$\leqslant MHn \sum_{1 \leqslant k \leqslant n} |x^k - z^k| \leqslant MHn^2\varepsilon' = \varepsilon$$

as we wanted to prove. □

Proposition A.10.2. *Given a pair of compact intervals I and J, every two variable polynomial reaches a minimum on $I \times J$.*

Proof. Let $f(x, y) := \sum_{1 \leqslant k, l \leqslant n} a_{kl}x^k y^l$ be a two variable polynomial and let $M := \max\{|a_{kl}|: 1 \leqslant k, l \leqslant n\}$. As both I and J are compact, we can pick a constant H so that $|x|^k < H$ and $|y|^k < H$ for all $x \in I$, all $y \in J$ and all $1 \leqslant k \leqslant n$. Thus

$$|f(x, y)| \leqslant \sum_{1 \leqslant k, l \leqslant n} |a_{kl}||x|^k|y|^l \leqslant n^2 MH$$

hence $f(I \times J)$ is bounded and therefore, there exists $\iota := \inf f(I \times J) \in \mathbb{R}$.

Take a sequence (x_n, y_n) in $I \times J$ so that $f(x_n, y_n) \xrightarrow[n]{} \iota$. Propositions 1.3.2 and 1.3.3 allow us to assume that $x_n \xrightarrow[n]{} x_0$ and $y_n \xrightarrow[n]{} y_0$ for some $x_0 \in I$ and some $y_0 \in J$. We claim that $f(x_0, y_0) = \iota$. Otherwise, if the result fails, we may take $\varepsilon := (f(x_0, y_0) - \iota)/4 > 0$.

For each $y \in J$, let $f_y: I \longrightarrow \mathbb{R}$ be the function that maps every $x \in I$ to $f_y(x) := f(x, y)$; and for each $x \in I$, let $f^x: J \longrightarrow \mathbb{R}$ denote the function that sends every $y \in J$ to $f^x(y) := f(x, y)$. Two applications of Proposition A.10.1 provide a number $\delta > 0$ such that

$$|f_y(u) - f_y(v)| < \varepsilon \tag{A.29}$$

for all $y \in J$ and all $(u, v) \subset I \times I$ with $|u - v| < \delta$, and

$$|f^x(u') - f^x(v')| < \varepsilon \tag{A.30}$$

for all $x \in I$ and all $(u', v') \subset J \times J$ with $|u' - v'| < \delta$.

Take a natural number m so that $|x_m - x_0| < \delta$, $|y_m - y_0| < \delta$ and $0 \leqslant f(x_m, y_m) - \iota < \varepsilon$. Thus, (A.29) and (A.30) yield

$$|f(x_m, y_m) - f(x_0, y_0)| \leqslant |f(x_m, y_m) - f(x_0, y_m)| + |f(x_0, y_m) - f(x_0, y_0)|$$

$$= |f_{y_m}(x_m) - f_{y_m}(x_0)| + |f^{x_0}(y_m) - f^{x_0}(y_0)| < 2\varepsilon,$$

and therefore,

$$\varepsilon > |\iota - f(x_m, y_m)|$$
$$\geqslant |\iota - f(x_0, y_0)| - |f(x_0, y_0) - f(x_m, y_m)| \geqslant 4\varepsilon - 2\varepsilon = 2\varepsilon,$$

a contradiction. That proves that $f(x_0, y_0) = \iota$, and the proof is done. $\square$

Proposition A.10.3. *Given a polynomial $p(z)$ over $\mathbb{C}$ with real coefficients, there is a pair of two variable polynomials over $\mathbb{R}$, $f_1(x, y)$ and $f_2(x, y)$, such that $p(x+yi) = f_1(x, y) + f_2(x, y)i$ for all pairs of real numbers x and y.*

Proof. Let $p(z) = \sum_{k=0}^n a_k z^k$ with $a_k \in \mathbb{R}$ for all k. Given $r \in \mathbb{R}$, let $\lfloor r \rfloor$ denote the largest integer q such that $q \leqslant r$.

For every pair of real numbers x and y and every natural number m, bearing in mind that $i^{2k} = (-1)^k$ and $i^{2k+1} = (-1)^k i$ for all $k = 0, 1, 2, \ldots$, Newton's binomial formula gives

$$(x + yi)^m = \sum_{k=0}^m \binom{m}{k} x^{m-k} y^k i^k = \left[\sum_{0 \leqslant k \leqslant \lfloor m/2 \rfloor} \binom{m}{2k}(-1)^k x^{m-2k} y^{2k} \right] +$$

$$+ \left[\sum_{1 \leqslant k \leqslant \lfloor m/2 \rfloor + 1} \binom{m}{2k-1}(-1)^{k-1} x^{m-2k+1} y^{2k-1} \right] i$$

$$=: R_m(x, y) + J_m(x, y) \cdot i.$$

Note that $R_m(x, y)$ and $J_m(x, y)$ are two variable polynomials over $\mathbb{R}$. Subsequently,

$$p(x + yi) = \sum_{k=0}^n [a_k R_k(x, y) + a_k J_k(x, y)i] = \sum_{k=0}^n a_k R_k(x, y) + i \sum_{k=0}^n a_k J_k(x, y)$$

and as the coefficients a_k are real, the expressions $f_1(x, y) := \sum_{k=0}^n a_k R_k(x, y)$ and $f_2(x, y) := \sum_{k=0}^n a_k J_k(x, y)$ define two variable polynomials over $\mathbb{R}$ satisfying $p(x + yi) = f_1(x, y) + f_2(x, y) \cdot i$ as desired. $\square$

There are many proofs for the Fundamental Theorem of Algebra, some of them very sophisticated. The proof given here follows this argument: given a polynomial $p(z)$ over $\mathbb{C}$ whose coefficients are all real and $\deg p(z) > 1$, and given a point $(x_0, y_0) \in \mathbb{R}^2$ such that $|p(x_0 + y_0 i)| > 0$, there exists a line r in the plane $\mathbb{R}^2$ passing through (x_0, y_0) and containing points (x_1, y_1) and (x_2, y_2) as close to (x_0, y_0) as we please for which $|p(x_1 + y_1 i)| < |p(x_0 + y_0 i)| < |p(x_2 + y_2 i)|$. This means that if $|p(z)|$ reaches a minimum on a point $z_0 \in \mathbb{C}$ then $|p(z_0)| = 0$ and therefore, z_0 is a root of $p(z)$.[3]

Theorem A.10.4 (Fundamental Theorem of Algebra). *Every polynomial over $\mathbb{R}$ has a root in $\mathbb{C}$.*

---

[3]This argument is an adaptation to polynomials of a classical theorem of Complex Analysis known as the Maximum Modulus Principle.

Proof. Assume the result fails. Then there exists a polynomial over $\mathbb{R}$

$$p(x) = \alpha_n x^n + \alpha_{n-1} x^{n-1} + \ldots + \alpha_1 x + \alpha_0$$

with $\alpha_n \neq 0$ and $|p(z)| > 0$ for all $z \in \mathbb{C}$.

Choose $r > 1$ large enough so that $r^n |\alpha_n|/2 > |p(0)|$ and $n \max_{1 \leqslant i \leqslant n-1} |\alpha_i|/r < |\alpha_n|/2$ and fix a compact interval $I := [-\alpha, \alpha]$ such that $|x + yi| > r$ for all $(x, y) \notin I \times I$. Clearly, if $(x, y) \notin I \times I$ and $z = x + yi$,

$$|p(z)| \geqslant |z|^n \left(|\alpha_n| - \frac{|\alpha_{n-1}|}{|z|} - \ldots - \frac{|\alpha_0|}{|z|^n} \right) \geqslant |z|^n \frac{|\alpha_n|}{2} > |p(0)|.$$

Therefore,

$$\inf_{(x,y) \in I \times I} |p(x + yi)| = \inf_{(x,y) \in \mathbb{R} \times \mathbb{R}} |p(x + yi)|. \tag{A.31}$$

Moreover, by Proposition A.10.3, there is a pair of two variable polynomials $f_1(x, y)$ and $f_2(x, y)$ over $\mathbb{R}$ such that $p(x + yi) = f_1(x, y) + f_2(x, y)i$, so

$$f(x, y) := |p(x + yi)|^2 = f_1(x, y)^2 + f_2(x, y)^2$$

is a two variable polynomial over $\mathbb{R}$. Next, Proposition A.10.2 provides a point $(x_0, y_0) \in I \times I$ such that $f(x_0, y_0) \leqslant f(x, y)$ for all $(x, y) \in I \times I$. In combination with (A.31), and denoting $z_0 := x_0 + y_0 i$, this produces $|p(z_0)| = \min_{z \in \mathbb{C}} |p(z)|$.

Regarding $p(z)$ as a polynomial over $\mathbb{C}$, Proposition 3.2.3 allows us to expand $P(z)$ as $p(z) = \sum_{j=0}^n c_j (z - z_0)^j$. Thus $p(z_0) = c_0 \neq 0$, and if k is the smallest natural number for which c_j is not null, then

$$p(z) = c_0 + c_k (z - z_0)^k + c_{k+1}(z - z_0)^{k+1} + \ldots + c_n (z - z_0)^n.$$

Consider the polynomial over $\mathbb{C}$ given by $q(z) := c_0 + c_k (z - z_0)^k$. Note that if $|z - z_0| < 1$ then

$$\left| \frac{p(z) - q(z)}{(z - z_0)^{k+1}} \right| < |c_{k+1}| + \ldots + |c_n| =: M.$$

Next, let $\varphi := \big(\arg(-c_0) - \arg c_k \big)/k$ and denote

$$z_r := z_0 + r(\cos \varphi + i \sin \varphi), \ 0 < r < 1.$$

Since the argument of the product of two complex numbers is the sum of their arguments (see (A.13)), it readily follows that the argument of $c_k (z_r - z_0)^k$ is $\arg(-c_0)$. Thus, regarding the complex numbers c_0 and $c_k (z_r - z_0)^k$ as vectors of $\mathbb{R}^2$, they keep opposite directions. Moreover, for every

$$0 < r < \min\{1, (|c_0|/|c_k|)^{1/k}, |c_k|/M\}$$

since $r < \min\{1, (|c_0|/|c_k|)^{1/k}\}$, the modulus of c_0 is larger than that of $c_k (z_r - z_0)^k$, and therefore the modulus of $c_0 + c_k (z_r - z_0)^k$ is $|c_0| - |c_k| r^k$. Thus

$$|p(z_r)| \leqslant |q(z_r)| + r^{k+1} \left| \frac{p(z_r) - p(z_0)}{r^{k+1}} \right|$$

$$\leqslant |c_0 + c_k (z_r - z_0)^k| + M r^{k+1}$$

$$= |c_0| - |c_k| r^k + M r^{k+1} = |c_0| - r^k (|c_k| - Mr)$$

and this last inequality and the condition $0 < r < |c_k|/M$ yield $|p(z_r)| < |c_0| = |p(z_0)| = \min_{z \in \mathbb{C}} |p(z)|$, a contradiction which allows us to conclude that $P(z)$ vanishes on some $z \in \mathbb{C}$. $\qquad\square$

Corollary A.10.5. *Each polynomial over* $\mathbb{C}$ *has a root in* $\mathbb{C}$.

Proof. Given a polynomial $P(z)$ over $\mathbb{C}$, the function $Q(z) := P(z) \cdot \overline{P(\overline{z})}$ is a polynomial over $\mathbb{C}$ whose coefficients are all real numbers. Then, by Theorem A.10.4, there exists $z_0 \in \mathbb{C}$ such that $Q(z_0) = 0$. Thus, $P(z_0) = 0$ or $P(\overline{z_0}) = 0$ so at least one of the numbers z_0 or $\overline{z_0}$ is a root of $P(z)$. $\square$

The notions and notation introduced in this section can also be used to manage the following auxiliary results:

Proposition A.10.6. *If $f(x, y)$ and $g(x, y)$ are a pair of two variable polynomials and $f(x, y) = g(x, y)$ for all $(x, y) \in \mathbb{R}^2 \backslash \{(x_0, y_0)\}$ then $f(x_0, y_0) = g(x_0, y_0)$.*

Proof. By hypothesis, given any $y \in \mathbb{R} \backslash \{y_0\}$,

$$f_y(x) = g_y(x), \qquad \text{for all } x \in \mathbb{R} \backslash \{x_0\}.$$

Since f_y and g_y are polynomials over $\mathbb{R}$, they are continuous on $\mathbb{R}$ and subsequently,

$$f_y(x_0) = \lim_{x \to x_0} f_y(x) = \lim_{x \to x_0} g_y(x) = g_y(x_0).$$

We have just proved that $f^{x_0}(y) = g^{x_0}(y)$ for all $y \in \mathbb{R} \backslash \{y_0\}$. Applying again the same continuity argument,

$$f^{x_0}(y_0) = \lim_{y \to y_0} f^{x_0}(y) = \lim_{y \to y_0} g^{x_0}(y) = g^{x_0}(y_0)$$

that is, $f(x_0, y_0) = g(x_0, y_0)$. $\square$

Proposition A.10.7. *If $P(z)$ and $Q(z)$ are a pair of polynomials over $\mathbb{C}$ such that $P(z) = Q(z)$ for all $z \in \mathbb{C} \backslash \{z_0\}$, then $P(z_0) = Q(z_0)$ too.*

Proof. It is a direct consequence of Propositions A.10.3 and A.10.6. $\square$

A.11 Roots of polynomials of degree greater than two

Polynomials are not simple functions: they have produced very tough questions whose solutions have been pursued for centuries. We are speaking of factoring polynomials, which plays an important role in antidifferentiation. The solution of this question for the second degree polynomials over the reals was already known to the classic hellenic world and even before, but the discovery of a general solution by radicals[4] for the polynomials of third and fourth degree was only achieved through the sixteenth to eighteenth centuries by Del Ferro, Tartaglia, Cardano, Ferrari, Descartes, Lagrange, Abel, etc. (see Chapter 4 in [Aleksandrov (1999)]); and in the nineteenth century, Galois confirmed the suspicions of Lagrange and Abel that there is not any formula that yields the roots in radical form of a general polynomial of degree greater than four. The description of the work of the mentioned

---

[4]Solving a polynomial by radicals consists in finding its roots by means of algebraic methods like those used for quadratic polynomials in Proposition 3.2.5.

mathematicians goes far beyond the scope of this book, but a gentle introduction to the subject can be found in [Bewersdorff (2006)].

Everything needed for the factorization of polynomials occurring throughout this book is covered by Proposition 3.2.6, the algorithms of division of polynomials given in Sections A.7 and A.8 and the formula for the roots of a quadratic polynomial given in Proposition 3.2.5.

Although there exist factorization formulas for polynomials of third and fourth degree, their use is rather unpractical, but for the sake of completeness, they are included in the following. The reader interested in their proofs may consult [Bewersdorff (2006)] or [Aleksandrov (1999)]. More information about polynomials may be found in [Barbeau (1989)].

Cubic polynomials:

Given a polynomial $P(x) = x^3 + ax^2 + bx + c$ over $\mathbb{C}$, let

$$L = \frac{a^3}{27} - \frac{ab}{6} + \frac{c}{2},$$
$$M = \frac{a^2}{9} - \frac{b}{3}$$

and calculate the two complex roots of $x^2 - (L^2 - M^3) = 0$, which will be denoted x_1 and x_2.

Next, by means of (A.15), solve the equations

$$z^3 + L - x_i = 0, \ i = 1, 2 \tag{A.32}$$
$$z^3 + L + x_i = 0, \ i = 1, 2 \tag{A.33}$$

and let $\{\mu_j : 1 \leqslant j \leqslant 6\}$ and $\{\nu_j : 1 \leqslant j \leqslant 6\}$ be the respective sets of solutions of (A.32) and (A.33). Among all pairs (μ_i, ν_j), select the only three pairs

$$(\mu_{i_k}, \nu_{j_k}), \quad k = 1, 2, 3$$

which satisfy the identity $\mu_{i_k} \nu_{j_k} = M$. The roots of $P(x)$ are exactly

$$z_k := -\frac{a}{3} + \mu_{i_k} + \nu_{j_k}, \ k = 1, 2, 3.$$

We notice that if $P(x)$ is a polynomial over $\mathbb{R}$, then L and M are real numbers and only nine pairs (μ_i, ν_j) need be checked.

Quartic polynomials:

Given $P(x) = x^4 + ax^3 + bx^2 + cx + d$ with a, b, c and d complex numbers, let

$$p := b - \frac{3}{8}a^2$$
$$q := c - \frac{1}{2}ab + \frac{1}{8}a^3$$
$$r := d - \frac{1}{4}ac + \frac{1}{16}a^2 b - \frac{3}{256}a^4$$

and let z_i, $i = 1, 2, 3$ be the three solutions of the cubic equation

$$z^3 + \frac{p}{2}z^2 + \frac{p^2 - 4r}{16}z - \frac{q^2}{64} = 0.$$

Solve the three quadratic equations $z^2 - z_i = 0$ for $i = 1, 2, 3$ and let $\{\lambda_1, \lambda_2\}$, $\{\mu_1, \mu_2\}$ and $\{\nu_1, \nu_2\}$ be their respective sets of solutions. Select the triples $(\lambda_{i_l} . \mu_{j_l}, \nu_{k_l})$ that satisfy the identity

$$\lambda_{i_l} \mu_{j_l} \nu_{k_l} = q/8.$$

The solutions to $P(x) = 0$ are

$$-\frac{a}{4} + \lambda_{i_l} + \mu_{j_l} + \nu_{k_l},$$

which concludes this laborious procedure.

A.12 Decomposition of fractions of polynomials

As seen in Chapter 6, the methods for finding the primitive of a rational function are based on the decomposition of any fraction of polynomials into a sum of simpler fractions which admit immediate antiderivatives. The long division algorithm decomposes any fraction $P(x)/Q(x)$ of polynomials over $\mathbb{R}$ into $P(x)/Q(x) = C(x) + P_0(x)/Q(x)$ where $C(x)$ and $P_0(x)$ are polynomials over $\mathbb{R}$ such that $\deg P_0(x) < \deg Q(x)$. Since $C(x)$ admits an immediate antiderivative, the problem is reduced to the decomposition of fractions of polynomials of negative degree.

This section requires the use of polynomials over $\mathbb{C}$. If $P(x)$ and $Q(x)$ are a pair of polynomials over $\mathbb{R}$ with $Q(x) \not\equiv 0$, the fraction $P(x)/Q(x)$ can be regarded as a fraction of polynomials over $\mathbb{R}$ but also as a fraction over $\mathbb{C}$. In the first case, the domain of $P(x)/Q(x)$ is $\mathbb{R} \backslash R_Q$ and in the second case, its domain is $\mathbb{C} \backslash C_Q$ where R_Q and C_Q are respectively the set of all real roots of $Q(x)$ and of all complex roots of $Q(x)$. In this second case we will say that $P(x)/Q(x)$ is a fraction of polynomials over $\mathbb{R}$ in spite of the fact that $P(x)/Q(x)$ is a complex-valued function on $\mathbb{C} \backslash C_Q$. Certainly, this is an abuse of language but the context will clear any ambiguity.

The partial fraction decomposition for fractions $P(x)/Q(x)$ is achieved in Theorem A.12.3 for the case where all the roots of $Q(x)$ are simple, and in Theorem A.12.15 if $Q(x)$ possesses some multiple root. Theorem A.12.18 is often a useful shortcut to Theorem A.12.15.

Proposition A.12.1. *Let $\mathcal{P}_{n-1}$ denote the complex vector space of all polynomials over $\mathbb{C}$ whose degree is at most $n - 1$, and let $\{a_i\}_{i=1}^n$ be a finite set of pairwise different complex numbers. Then the system $\{\prod_{i \neq j}(x - a_i)\}_{j=1}^n$ is a basis of $\mathcal{P}_{n-1}$.*

Proof. Let us denote $Q_j(x) := \prod_{i \neq j}(x - a_i)$ for all $1 \leqslant j \leqslant n$. It was already proved in Proposition 3.2.1 that $\mathcal{P}_{n-1}$ is a vector space over $\mathbb{C}$ whose dimension is n, so we only need to see that the system $\{Q_j(x)\}_{j=1}^n$ is linearly independent. Assume then that the linear combination $Q(x) = \sum_{j=1}^n \lambda_j Q_j(x)$ is identically null. As $Q_i(a_j) = 0$ if and only if $i \neq j$, we get $0 = Q(a_j) = \lambda_j Q_j(a_j)$ hence $\lambda_j = 0$. That proves that $\{Q_j(x)\}_{j=1}^n$ is linearly independent, and therefore, a basis of $\mathcal{P}_{n-1}$. $\square$

Throughout this section, given a complex polynomial $Q(x)$ with $\deg Q(x) \geqslant 1$, the set of all fractions $P(x)/Q(x)$ of polynomials over $\mathbb{C}$ with $\deg P(x) < \deg Q(x)$ will be denoted by $\mathcal{W}_Q$. As said before, $\mathcal{P}_{n-1}$ denotes the vector space of all polynomials over $\mathbb{C}$ of degree at most $n-1$. Clearly, $\mathcal{W}_Q$ is a complex vector space, and since $\{z^k\}_{k=0}^{n-1}$ is a basis of the vector space $\mathcal{P}_{n-1}$, it readily follows that $\{z^k/Q(z)\}_{k=0}^{n-1}$ is a basis of $\mathcal{W}_Q$. Thus, the dimension of $\mathcal{W}_Q$ is n.

Proposition A.12.2. *Let $\{a_i\}_{i=1}^n$ be a collection of pairwise different complex numbers and let $Q(x) := \prod_{i=1}^n (x - a_i)$. Then the set $\{1/(x - a_i)\}_{i=1}^n$ is a basis of $\mathcal{W}_Q$.*

Proof. By Proposition A.12.1, the polynomials $Q_j(x) := \prod_{i \neq j}(x - a_i)$ for $1 \leqslant j \leqslant n$ form a basis of $\mathcal{P}_{n-1}$, hence, given any complex polynomial $P(x)$ with $\deg P(x) < \deg Q(x)$, there exists a unique collection $\{\lambda_i\}_{i=1}^n \subset \mathbb{C}$ such that $P(x) = \sum_{i=1}^n \lambda_i Q_i(x)$. Now it is immediate that

$$\frac{P(x)}{Q(x)} = \sum_{i=1}^n \lambda_i \frac{Q_i(x)}{Q(x)} = \sum_{i=1}^n \frac{\lambda_i}{x - a_i}$$

for all $x \in \mathbb{C} \backslash \{a_i\}_{i=1}^n$, concluding the proof. $\qquad \square$

The following result is the partial fraction decomposition for fractions $P(x)/Q(x)$ such that all the roots of $Q(x)$ are simple.

Theorem A.12.3. *Let $\{a_i\}_{i=1}^m \subset \mathbb{R}$ and $\{c_i\}_{i=1}^n \subset \mathbb{C} \backslash \mathbb{R}$ be two finite sets of pairwise different numbers and for every $1 \leqslant i \leqslant n$, let μ_i and ν_i be a pair of real numbers so that $x^2 + \mu_i x + \nu_i = (x - c_i)(x - \overline{c_i})$. Consider the polynomial $Q(x)$ whose irreducible factorizations over $\mathbb{R}$ is*

$$Q(x) = \prod_{i=1}^m (x - a_i) \cdot \prod_{i=1}^n (x^2 + \mu_i x + \nu_i).$$

Then, given a polynomial $P(x)$ over $\mathbb{R}$ such that $\deg P(x) < \deg Q(x)$, there exist four unique sets $\{A_i\}_{i=1}^m \subset \mathbb{R}$, $\{C_i\}_{i=1}^n \subset \mathbb{C}$, $\{M_i\}_{i=1}^n \subset \mathbb{R}$ and $\{N_i\}_{i=1}^n \subset \mathbb{R}$ such that

$$(i) \qquad \frac{P(x)}{Q(x)} = \sum_{i=1}^m \frac{A_i}{x - a_i} + \sum_{i=1}^n \frac{\overline{C_i}}{x - \overline{c_i}} + \sum_{i=1}^n \frac{C_i}{x - c_i}$$

$$(ii) \qquad = \sum_{i=1}^m \frac{A_i}{x - a_i} + \sum_{i=1}^n \frac{M_i x + N_i}{x^2 + \mu_i x + \nu_i}$$

and both (i) and (ii) are valid for all $x \in \mathbb{C} \backslash C_Q$.

Proof. (i) By Proposition A.12.1 there exist three unique collections of complex numbers $\{A_i\}_{i=1}^m$, $\{B_i\}_{i=1}^n$ and $\{C_i\}_{i=1}^n$ such that

$$P(x) = \sum_{i=1}^m A_i \cdot \prod_{j \neq i}(x - a_j) \cdot \prod_{j=1}^n (x - \overline{c_j})(x - c_j)$$

$$+ \sum_{i=1}^n B_i \cdot \prod_{j=1}^m (x - a_j) \cdot \prod_{j \neq i}(x - \overline{c_j}) \cdot \prod_{j=1}^n (x - c_j)$$

$$+ \sum_{i=1}^n C_i \cdot \prod_{j=1}^m (x - a_j) \cdot \prod_{j=1}^n (x - \overline{c_j}) \cdot \prod_{j \neq i}(x - c_j). \tag{A.34}$$

Using this last expression to evaluate $P(x)$ at each of its roots, we get

$$P(a_k) = A_k \cdot \prod_{j \neq k}(a_k - a_j) \cdot \prod_{j=1}^n (a_k - \overline{c_j})(a_k - c_j) \tag{A.35}$$

$$P(\overline{c_k}) = B_k \cdot \prod_{j=1}^m (\overline{c_k} - a_j) \cdot \prod_{j \neq k}(\overline{c_k} - \overline{c_j}) \cdot \prod_{j=1}^n (\overline{c_k} - c_j) \tag{A.36}$$

$$P(c_k) = C_k \cdot \prod_{j=1}^m (c_k - a_j) \cdot \prod_{j=1}^n (c_k - \overline{c_j}) \cdot \prod_{j \neq k}(c_k - c_j). \tag{A.37}$$

As each a_k is real, then $P(a_k) \in \mathbb{R}$ and $(a_k - \overline{c_j})(a_k - c_j) \in \mathbb{R}\backslash\{0\}$ for all $1 \leqslant j \leqslant n$, so equation (A.35) proves that $A_k \in \mathbb{R}$ for all $1 \leqslant k \leqslant m$.

Moreover, as the coefficients of $P(x)$ are real, then $P(\overline{c_k}) = \overline{P(c_k)}$, and comparing (A.36) with the conjugate of (A.37), it is immediate that $B_k = \overline{C_k}$ for all $1 \leqslant k \leqslant n$.

(ii) Let us reorder the addends of the decomposition given in (i) as

$$\frac{P(x)}{Q(x)} = \sum_{i=1}^m \frac{A_i}{x - a_i} + \sum_{i=1}^n \left[\frac{\overline{C_i}}{x - \overline{c_i}} + \frac{C_i}{x - c_i} \right]. \tag{A.38}$$

For $M_i := 2\Re C_i$ and $N_i := -2(\Re C_i \Re c_i + \Im C_i \Im c_i)$, a few calculations yield

$$\frac{\overline{C_i}}{x - \overline{c_i}} + \frac{C_i}{x - c_i} = \frac{M_i x + N_i}{x^2 + \mu_i x + \nu_i} \tag{A.39}$$

and plugging (A.39) into (A.38), statement (ii) is obtained.

The uniqueness of the coefficients A_i, M_i and N_i is a consequence of the uniqueness of the coefficients A_i and C_i given in statement (i). $\qquad\square$

The following remark readily follows from (A.35), (A.36) and (A.37).

Remark A.12.4. All the coefficients A_k and C_k in Theorem A.12.3 are non-null if and only if $P(x)/Q(x)$ is irreducible.

Dealing with polynomials with multiple roots requires the use of the derivative of fractions of polynomials over $\mathbb{C}$.

Definition A.12.5. Given a polynomial $P(z) = \sum_{k=0}^n a_k z^k$ over $\mathbb{C}$ with $n \geqslant 1$, its *formal derivative* is defined as the polynomial $P'(z) := \sum_{k=1}^n k a_k z^{k-1}$ over $\mathbb{C}$. If $n = 0$, the formal derivative of $P(z)$ is defined as $P'(z) :\equiv 0$.

Proposition A.12.6. *Given a pair of polynomials $P(z)$ and $Q(z)$ over $\mathbb{C}$, the following properties hold:*

(i) $P'(z) \equiv 0$ if and only if $P(z)$ is constant;
(ii) $(P + Q)'(z) = P'(z) + Q'(z)$;
(iii) $(P \cdot Q)'(z) = P'(z)Q(z) + P(z)Q'(z)$;
(iv) if $P(z) = \alpha \prod_{i=1}^{n}(z - a_i)^{n_i}$ is the irreducible factorization of $P(z)$ over $\mathbb{C}$ then
$$P'(z) = \alpha \sum_{i=1}^{n}\left[n_i(z - a_i)^{n_i - 1} \cdot \prod_{j \neq i}(z - a_j)^{n_j}\right].$$

Proof. (i) It is a direct consequence of the fact that $\{z^k\}_{k=0}^n$ is a basis for the linear space $\mathcal{P}_n$ established in Proposition 3.2.1.

(ii) Trivial.

(iii) If $\deg P(z) = 0$ or $\deg Q(z) = 0$, the proof is trivial. In case $P(x) = az^n$ and $Q(z) = bz^m$ with $n \geqslant 1$ and $m \geqslant 1$, then the proof is also straightforward. The general case now follows easily.

(iv) It is an immediate consequence of (iii). □

By analogy with its real counterpart, the formal derivative of a fraction of complex polynomials is defined as follows:[5]

Definition A.12.7. *Given a pair $P(z)$ and $Q(z)$ of polynomials over $\mathbb{C}$, the formal derivative of $P(z)/Q(z)$ is defined as the fraction of polynomials over $\mathbb{C}$ given by*
$$\left(\frac{P(z)}{Q(z)}\right)' := \frac{P'(z)Q(z) - P(z)Q'(z)}{Q(z)^2}.$$

Proposition A.12.8. *Given polynomials $P_1(z)$, $P_2(z)$ and $Q(z)$ over $\mathbb{C}$ with $\deg P_i(z) < \deg Q(z)$ for $i = 1$ and $i = 2$, we have*
$$\left(\frac{P_1(z)}{Q(z)} + \frac{P_2(z)}{Q(z)}\right)' = \left(\frac{P_1(z)}{Q(z)}\right)' + \left(\frac{P_2(z)}{Q(z)}\right)'.$$

Proof. It is immediate from Proposition A.12.6. □

The roots of a polynomial over $\mathbb{C}$ and of its derivative are linked as shown in the following:

Proposition A.12.9. *Let $P(z)$ be a polynomial over $\mathbb{C}$. If $a \in \mathbb{C}$ is a root of $P(x)$ of multiplicity n then a is a root of $P'(x)$ of multiplicity $n - 1$.*

Proof. Let $P(z) = \alpha \prod_{i=1}^{m}(z - a_i)^{k_i}$ be the irreducible factorization over $\mathbb{C}$ of $P(z)$. Proving the result for the root a_1 will be sufficient.

---

[5]There exists a meaningful notion of differentiability for complex-valued functions whose corresponding derivative for fractions of polynomials over $\mathbb{C}$ coincides exactly with the formal derivative introduced in Definitions A.12.5 and A.12.7 (see [Bak (1993)]). The reader acquainted with that genuine derivative for complex-valued functions may safely skip everything between Definition A.12.5 and Proposition A.12.8.

Following Proposition A.12.6 (iv), the formal derivative of $P(z)$ is

$$P'(z) = \alpha \sum_{i=1}^{m} k_i (z - a_i)^{k_i - 1} \prod_{j \neq i} (z - a_j)^{k_j},$$

and pulling out the factor $(z - a_1)^{k_1 - 1}$, we manage

$$P'(z) = \alpha (z - a_1)^{k_1 - 1} \left[k_1 \prod_{j \neq 1} (z - a_j)^{k_j} + Q(z) \right] \qquad \text{(A.40)}$$

where $Q(z) := (z - a_1) \sum_{i=2}^{m} k_i (z - a_i)^{k_i - 1} \prod_{j \neq 1, i} (z - a_j)^{k_j}$. As $Q(z)$ vanishes at $z = a_1$ but $\prod_{j \neq 1} (z - a_j)^{k_j}$ does not, it follows from (A.40) that a_1 is a root of $P'(z)$ of multiplicity $k_1 - 1$. □

Corollary A.12.10. *Let $P(z)$ be a polynomial over $\mathbb{C}$. If $0 = P(a) = P'(a) = \ldots = P^{(n)}(a)$ and $0 \neq P^{(n+1)}(a)$ then a is a root of $P(z)$ of multiplicity n.*

Proof. Immediate from Proposition A.12.9. □

The fractions $1/(z - a_i)$ of Statement (iii) in the following proposition are to be understood as complex-valued functions defined on $\mathbb{C} \backslash C_Q$.

Given a subset A of a vector space V over $\mathbb{K}$ (where $\mathbb{K}$ is either $\mathbb{R}$ or $\mathbb{C}$), the vector subspace spanned by A in V is denoted $\text{span}\{A\}$.

Proposition A.12.11. *Let $Q(z)$ be a polynomial over $\mathbb{C}$ whose irreducible factorization is $Q(z) = \prod_{i=1}^{n} (z - a_i)^{k_i}$ and assume $k_i > 1$ for some index i. Let $S(z) := \prod_{i=1}^{n} (z - a_i)^{k_i + 1}$ and let $\mathcal{U} := \text{span}\{1/(z - a_i)\}_{i=1}^{n}$. Let ∂ be the map defined on the complex vector space of all fractions of polynomials over $\mathbb{C}$ which sends each $P(x)/Q(x)$ to $\big(P(x)/Q(x)\big)'$. Then:*

(i) the map ∂ is linear and injective on $\mathcal{W}_R$ for every polynomial $R(x)$ over $\mathbb{C}$;
(ii) $\partial(\mathcal{W}_Q) \subset \mathcal{W}_S$;
(iii) $\mathcal{W}_S = \partial(\mathcal{W}_Q) \oplus \mathcal{U}$.

Proof. (i) and (ii) The linearity of ∂ follows from Propositions A.12.6 and A.12.8. In order to prove that ∂ is injective on $\mathcal{W}_Q$, take any non-null fraction $P(z)/Q(z) \in \mathcal{W}_Q$, let α be the leading coefficient of $P(z)$, let $m := \deg P(z)$ and let $p := \deg Q(z)$. As $m < p$, the degree of $P'(z)Q(z) - P(z)Q'(z)$ is $m + p - 1$ and its leading coefficient is $\alpha(m - p)$. Hence $P'(z)Q(z) - P(z)Q'(z)$ is not identically null and neither is $\partial\big(P(z)/Q(z)\big)$. That proves the injectivity of ∂ on $\mathcal{W}_Q$.
To see that $\partial\big(P(z)/Q(z)\big) \in \mathcal{W}_S$, denote

$$Q_i(z) := \prod_{j \neq i} (z - a_j) \quad \text{for all } 1 \leqslant i \leqslant n.$$

Since $\partial\big(P(z)/Q(z)\big)$ is

$$\frac{P'(z)Q(z) - P(z)\big[\sum_{i=1}^{n} k_i (z-a_1)^{k_1} \cdot \ldots \cdot (z-a_i)^{k_i-1} \cdot \ldots \cdot (z-a_n)^{k_n}\big]}{Q(z)^2}$$

$$= \frac{P'(z)\cdot(z-a_1)\cdot\ldots\cdot(z-a_m)\quad P(z)\big[\sum_{i=1}^{n} k_i Q_i(z)\big]}{(z-a_1)^{k_1+1}\cdot\ldots\cdot(z-a_n)^{k_n+1}} \tag{A.41}$$

it is straightforward that $\partial\big(P(z)/Q(z)\big) \in \mathcal{W}_S$.

(iii) First, we shall prove that $\partial(\mathcal{W}_Q) \cap \mathcal{U} = \{0\}$. Indeed, if $\partial(\mathcal{W}_Q) \cap \mathcal{U} \neq \{0\}$ then there exists a fraction $P(z)/Q(z) \in \mathcal{W}_Q$ whose image by ∂ is a non-null linear combination $\sum_{i=1}^{n} \lambda_i/(z-a_i) \in \mathcal{U}$. Thus,

$$\frac{\lambda_1}{z-a_1} + \ldots + \frac{\lambda_n}{z-a_n} = \frac{R(z)}{\prod_{i=1}^{n}(z-a_i)}$$

where $R(z)$ is a non-null polynomial over $\mathbb{C}$ such that $\deg R(z) < n$.

Since $\deg P(z) < \deg Q(z)$, at least one of the roots a_i of $Q(z)$ is either a root of $P(z)$ of multiplicity smaller than k_i or it is not a root of $P(z)$. Without loss of generality, assume that is the case for a_1. Thus $Q(z) = (z-a_1)^{k_1}Q_0(z)$ and $P(z) = (z-a_1)^{m_1}P_0(z)$ where $0 \leqslant m_1 < k_1$ (the value $m_1 = 0$ is reserved for the case when a_1 is not a root of $P(z)$) and $P_0(x)$ and $Q_0(x)$ are a pair of polynomials such that $Q_0(a_1) \neq 0$ and $P_0(a_1) \neq 0$. As

$$\partial\left(\frac{P(z)}{Q(z)}\right) = \frac{P'(z)Q(z) - P(z)Q'(z)}{Q(z)^2} = \frac{R(z)}{\prod_{i=1}^{n}(z-a_i)}, \tag{A.42}$$

defining

$$T(z) := \big[m_1 P_0(z) + (z-a_1)P_0'(z)\big]Q_0(z)$$
$$- \big[k_1 Q_0(z) + (z-a_1)Q_0'(z)\big]P_0(z)$$

a few simple calculations yield

$$P'(z)Q(z) - P(z)Q'(z) = (z-a_1)^{m_1+k_1-1}T(z) \tag{A.43}$$

so plugging (A.43) into (A.42) and removing denominators in the resulting expression we obtain

$$(z-a_1)^{m_1+k_1-1}T(z) = R(z) \cdot \prod_{i=1}^{n}(z-a_i)^{2k_i-1}, \tag{A.44}$$

which holds for all $z \in \mathbb{C}$ by virtue of Proposition A.10.7. Clearly, the multiplicity of the root a_1 of the polynomial on the right side of (A.44) is at least $2k_1 - 1$. But $T(a_1) = (m_1 - k_1)P_0(a_1)Q_0(a_1) \neq 0$, hence the multiplicity of the root a_1 on the left side of (A.44) is $m_1 + k_1 - 1$ exactly, so it is strictly smaller than $2k_1 - 1$, a contradiction. This proves $\partial(\mathcal{W}_Q) \cap \mathcal{U} = \{0\}$.

Now, the identity $\partial(\mathcal{W}_Q) \oplus \mathcal{U} = \mathcal{W}_S$ can be easily proved by computing $\dim \partial(\mathcal{W}_Q) + \dim \mathcal{U} = \dim \mathcal{W}_S = n + \sum_{i=1}^{n} k_i$. $\qquad\square$

The two following propositions are preparatory for the partial fraction decomposition of Theorem A.12.15.

Proposition A.12.12. *Let $Q(z)$ be a polynomial over $\mathbb{C}$ whose irreducible factorization is $Q(z) = \prod_{i=1}^{n}(z - a_i)^m$. Then the system*

$$\frac{1}{(z - a_i)^j}, \quad 1 \leqslant j \leqslant m, \quad 1 \leqslant i \leqslant n$$

is a basis of the vector space $\mathcal{W}_Q$.

Proof. For every $1 \leqslant j \leqslant m$, let $Q_j(z) := \prod_{i=1}^{n}(z - a_i)^j$ and $\mathcal{W}_j := \mathcal{W}_{Q_j}$.
By virtue of Proposition A.12.11,

$$\mathcal{W}_{m+1-j} = \partial(\mathcal{W}_{m-j}) \oplus \mathcal{W}_1, \quad \text{for all } 1 \leqslant j \leqslant m - 1. \tag{A.45}$$

A consecutive application of identity (A.45) for $j = 1, 2, \ldots, m - 1$ yields

$$\mathcal{W}_m = \partial(\mathcal{W}_{m-1}) \oplus \mathcal{W}_1$$
$$= \partial(\partial(\mathcal{W}_{m-2}) \oplus \mathcal{W}_1) \oplus \mathcal{W}_1 = \partial^2(\mathcal{W}_{m-2}) \oplus \partial(\mathcal{W}_1) \oplus \mathcal{W}_1$$
$$= \ldots\ldots\ldots\ldots\ldots\ldots\ldots\ldots\ldots\ldots\ldots\ldots\ldots\ldots$$
$$= \partial^{m-1}(\mathcal{W}_1) \oplus \partial^{m-2}(\mathcal{W}_1) \oplus \ldots \oplus \partial^2(\mathcal{W}_1) \oplus \partial(\mathcal{W}_1) \oplus \mathcal{W}_1. \tag{A.46}$$

Since $\partial^j(1/(z-a_i)) = (-1)^j j!/(z-a_i)^{j+1}$ and ∂ is injective, Proposition A.12.2 shows that $\{1/(z - a_i)^{j+1}\}_{i=1}^{n}$ is a basis of the vector space $\partial^j(\mathcal{W}_1)$ for $j = 0, 1, \ldots, m-1$, and in turn, (A.46) yields $\{1/(z - a_i)^{j+1}\}_{i=1\,j=0}^{n\,\,m-1}$ is a basis of $\mathcal{W}_m$. $\square$

Proposition A.12.13. *Let $Q(z)$ be a polynomial over $\mathbb{C}$ whose irreducible factorization is $Q(z) = \prod_{i=1}^{n}(z-a_i)^{k_i}$, where $(k_i)_{i=1}^{n}$ is non-decreasing and $k_n \geqslant 2$. Then the set*

$$\mathcal{A} := \left\{ \frac{1}{(z - a_i)^j} : 1 \leqslant j \leqslant k_i, \ 1 \leqslant i \leqslant n \right\}$$

is a basis of the vector space $\mathcal{W}_Q$.

Proof. Let $m := \max\{k_i : 1 \leqslant i \leqslant n\}$ and let $Q_0(z) := \prod_{i=1}^{n}(z - a_i)^m$. Clearly, $\mathcal{W}_Q$ is a linear subspace of $\mathcal{W}_{Q_0}$, and $\mathcal{A} \subset \mathcal{W}_Q$, Proposition A.12.12 proves that $\mathcal{A}$ is a set of linearly independent vectors. Moreover, $\dim \mathcal{W}_Q = \sum_{i=1}^{n} n_i$, so $\mathcal{A}$ is a basis of $\mathcal{W}_Q$. $\square$

Proposition A.12.14. *Given a natural number l and a pair of numbers $C \in \mathbb{C}$ and $z \in \mathbb{C}\backslash\mathbb{R}$, there exists a unique finite collection $\{F_k(x)\}_{k=1}^{l}$ of polynomials over $\mathbb{R}$ of degree 1 such that*

$$\frac{\overline{C}}{(x - \overline{z})^l} + \frac{C}{(x - z)^l} = \sum_{k=1}^{l} \frac{F_k(x)}{(x^2 + \mu x + \nu)^l}, \quad x \in \mathbb{C}\backslash\{z, \overline{z}\}$$

where μ and ν are real numbers such that $(x - \overline{z})(x - z) = x^2 + \mu x + \nu$.

Proof. Given the real numbers $S := 2\Re C$ and $T := -2(\Re C \Re z + \Im C \Im z)$, we have

$$\frac{\overline{C}}{x - \overline{z}} + \frac{C}{x - z} = \frac{Sx + T}{x^2 + \mu x + \nu} \tag{A.47}$$

and for $l > 1$,

$$\frac{\overline{C}}{(x - \overline{z})^l} + \frac{C}{(x - z)^l} = \frac{(-1)^l}{(l - 1)!} \partial^{l-1} \left(\frac{\overline{C}}{x - \overline{z}} + \frac{C}{x - z} \right)$$

$$= \frac{(-1)^{l-1}}{(l - 1)!} \partial^{l-1} \left(\frac{Sx + T}{x^2 + \mu x + \nu} \right). \tag{A.48}$$

Next, note that for any natural number $k \geqslant 2$,

$$\partial \left(\frac{Sx + T}{(x^2 + \mu x + \nu)^{k-1}} \right)$$

$$= \frac{S(x^2 + \mu x + \nu)^{k-1} - (k - 1)(Sx + T)(x^2 + \mu x + \nu)^{k-2}(2x + \mu)}{(x^2 + \mu x + \nu)^{2k-2}}$$

$$= \frac{S}{(x^2 + \mu x + \nu)^{k-1}} - (k - 1)\frac{x^2 + \mu x + \nu}{(x^2 + \mu x + \nu)^k}2S$$

$$+ (k - 1)\frac{(S\mu - 2T)x + 2S\nu - T\mu}{(x^2 + \mu x + \nu)^k}$$

$$= \frac{(3 - 2k)S}{(x^2 + \mu x + \nu)^{k-1}} + (k - 1)\frac{(S\mu - 2T)x + 2S\nu - T\mu}{(x^2 + \mu x + \nu)^k}$$

$$= \frac{F_1(x)}{(x^2 + \mu x + \nu)^{k-1}} + \frac{F_2(x)}{(x^2 + \mu x + \nu)^k} \tag{A.49}$$

where $F_1(x)$ and $F_2(x)$ are polynomials over $\mathbb{R}$ with degree at most one.

Successive applications of (A.49) on (A.48) yield polynomials $F_{ij}(x)$ over $\mathbb{R}$ of degree at most one satisfying

$$\frac{\overline{C}}{(x - \overline{z})^l} + \frac{C}{(x - z)^l} = (-1)^{l-1}\partial^{l-1} \left(\frac{Sx + T}{x^2 + \mu x + \nu} \right)$$

$$= \partial^{l-2} \left(\frac{F_{11}(x)}{x^2 + \mu x + \nu} + \frac{F_{12}(x)}{(x^2 + \mu x + \nu)^2} \right)$$

$$= \partial^{l-3} \left(\frac{F_{21}(x)}{x^2 + \mu x + \nu} + \frac{F_{22}(x)}{(x^2 + \mu x + \nu)^2} + \frac{F_{23}(x)}{(x^2 + \mu x + \nu)^3} \right)$$

$$= \cdots\cdots\cdots\cdots\cdots\cdots$$

$$= \partial \left(\sum_{k=1}^{l-1} \frac{F_{l-2,k}(x)}{(x^2 + \mu x + \nu)^k} \right) = \sum_{k=1}^{l} \frac{F_{l-1,k}(x)}{(x^2 + \mu x + \nu)^k}.$$

The uniqueness of the polynomials $F_{ij}(x)$ is a consequence of the uniqueness of the polynomials $F_i(x)$ occurring at (A.49). $\square$

Theorem A.12.15. *Let $\{a_i\}_{i=1}^m \subset \mathbb{R}$ and $\{c_i\}_{i=1}^n \subset \mathbb{C} \backslash \mathbb{R}$ be a pair of collections of pairwise different numbers and let $Q(x)$ be a polynomial over $\mathbb{R}$ whose irreducible factorization over $\mathbb{C}$ is*

$$Q(x) = \prod_{i=1}^m (x - a_i)^{m_i} \cdot \prod_{i=1}^n (x - \overline{c_i})^{n_i} (x - c_i)^{n_i}.$$

For every $1 \leqslant i \leqslant n$, let μ_i and ν_i be a pair of real numbers such that $x^2 + \mu_i x + \nu_i = (x - \overline{c_i})(x - c_i)$. Assume that at least one of the powers m_i or n_i is greater than one.

Then for every polynomial $P(x)$ over $\mathbb{R}$ such that $\deg P(x) < \deg Q(x)$ there exist five unique subsets $\{A_{il}\}_{l=1}^{m_i}{}_{i=1}^{m} \subset \mathbb{R}$, $\{A'_{il}\}_{l=1}^{m_i}{}_{i=1}^{m} \subset \mathbb{R}$, $\{C_{il}\}_{l=1}^{n_i}{}_{i=1}^{n} \subset \mathbb{C}$, $\{M_{il}\}_{l=1}^{n_i}{}_{i=1}^{n} \subset \mathbb{R}$ and $\{N_{il}\}_{l=1}^{n_i}{}_{i=1}^{n} \subset \mathbb{R}$ such that

$$(i) \quad \frac{P(x)}{Q(x)} = \sum_{i=1}^{m} \sum_{l=1}^{m_i} \frac{A_{il}}{(x - a_i)^l} + \sum_{i=1}^{n} \sum_{l=1}^{n_i} \left[\frac{\overline{C_{il}}}{(x - \overline{c_i})^l} + \frac{C_{il}}{(x - c_i)^l} \right]$$

$$(ii) \qquad = \sum_{i=1}^{m} \sum_{l=1}^{m_i} \frac{A'_{il}}{(x - a_i)^l} + \sum_{i=1}^{n} \sum_{l=1}^{n_i} \frac{M_{il} x + N_{il}}{(x^2 + \mu_i x + \nu_i)^l}.$$

Proof. (i) By Proposition A.12.13, there exist unique real coefficients A_{il} and unique complex coefficients B_{il} and C_{il} such that

$$\frac{P(x)}{Q(x)} = \sum_{i=1}^{m} \sum_{l=1}^{m_i} \frac{A_{il}}{(x - a_i)^l} + \sum_{i=1}^{n} \sum_{l=1}^{n_i} \left[\frac{B_{il}}{(x - \overline{c_i})^l} + \frac{C_{il}}{(x - c_i)^l} \right], \quad x \in \mathbb{C} \backslash C_Q \quad (A.50)$$

where C_Q denotes the set of all complex roots of $Q(x)$.

Let $\mathbf{m} := \max\{m_i, n_j\}_{i=1}^{m}{}_{j=1}^{n}$. The proof will be done as soon as we demonstrate for each $0 \leqslant k \leqslant \mathbf{m}$ that

$$A_{i, m_i - k} \in \mathbb{R} \quad \text{for all } 1 \leqslant i \leqslant m \text{ and all } 0 \leqslant k \leqslant m_i - 1, \quad (A.51)$$

$$B_{i, n_i - k} = \overline{C_{i, n_i - k}} \quad \text{for all } 1 \leqslant i \leqslant n \text{ and all } 0 \leqslant k \leqslant n_i - 1. \quad (A.52)$$

Let us prove (A.51) and (A.52) for $k = 0$:

Multiply both sides of (A.50) by $Q(x)$, so that

$$P(x) = \sum_{i=1}^{m} \sum_{l=1}^{m_i} A_{il}(x - a_i)^{m_i - l} \prod_{j \neq i}(x - a_j)^{m_j} \prod_{j=1}^{n}(x - \overline{c_j})^{n_j}(x - c_j)^{n_j} \quad (A.53)$$

$$+ \sum_{i=1}^{n} \sum_{l=1}^{n_i} B_{il}(x - \overline{c_i})^{n_i - l} \prod_{j=1}^{m}(x - a_j)^{m_j} \prod_{j \neq i}(x - \overline{c_j})^{n_j} \prod_{j=1}^{n}(x - c_j)^{n_j}$$

$$+ \sum_{i=1}^{n} \sum_{l=1}^{n_i} C_{il}(x - c_i)^{n_i - l} \prod_{j=1}^{m}(x - a_j)^{m_j} \prod_{j=1}^{n}(x - \overline{c_j})^{n_j} \prod_{j \neq i}(x - c_j)^{n_j}.$$

By Proposition A.10.7, the identity (A.53) holds for all $x \in \mathbb{C}$, so evaluating the polynomials of (A.53) at each root of $Q(x)$ we obtain

$$P(a_i) = A_{i m_i} \cdot \prod_{j \neq i}(a_i - a_j)^{m_j} \cdot \prod_{j=1}^{n}(a_i - \overline{c_j})^{n_j}(a_i - c_j)^{n_j}, \quad (A.54)$$

$$P(\overline{c_i}) = B_{i n_i} \cdot \prod_{j=1}^{m}(\overline{c_i} - a_j)^{m_j} \cdot \prod_{j \neq i}(\overline{c_i} - \overline{c_j})^{n_j} \prod_{j=1}^{n}(\overline{c_i} - c_j)^{n_j}, \quad (A.55)$$

$$P(c_i) = C_{i n_i} \cdot \prod_{j=1}^{m}(c_i - a_j)^{m_j} \cdot \prod_{j=1}^{n}(c_i - \overline{c_j})^{n_j} \prod_{j \neq i}(c_i - c_j)^{n_j}. \quad (A.56)$$

Clearly, as a_i and the coefficients of $P(x)$ are real numbers, working out A_{im_i} in (A.54), we get $A_{im_i} \in \mathbb{R}$ for all $1 \leqslant i \leqslant m$; and as $P(\overline{c_i}) = \overline{P(c_i)}$, identifying (A.55) with the conjugate of (A.56) we get $B_{in_i} = \overline{C_{in_i}}$ for all $1 \leqslant i \leqslant n$. We have just proved (A.51) and (A.52) for $k = 0$. For $k = 1$, consider the polynomial over $\mathbb{R}$ given by

$$Q_1(x) := \prod_{i=1}^{m}(x - a_i)^{m_i - 1} \cdot \prod_{i=1}^{n}[(x - \overline{c_i})(x - c_i)]^{n_i - 1}.$$

Since each A_{im_i} is real, and with the help of Proposition A.12.14, we know there exists a polynomial $P_1(x)$ with real coefficients such that

$$\frac{P_1(x)}{Q_1(x)} = \frac{P(x)}{Q(x)} - \sum_{i=1}^{m}\frac{A_{im_i}}{(x - a_i)^{m_i}} - \sum_{i=1}^{n}\left[\frac{\overline{C_{in_i}}}{(x - \overline{c_i})^{n_i}} + \frac{C_{in_i}}{(x - c_i)^{n_i}}\right].$$

Thus, by (A.50),

$$\frac{P_1(x)}{Q_1(x)} = \sum_{i=1}^{m}\sum_{l=1}^{m_i-1}\frac{A_{il}}{(x - a_i)^l} + \sum_{i=1}^{n}\sum_{l=1}^{n_i-1}\left[\frac{B_{il}}{(x - \overline{c_i})^l} + \frac{C_{il}}{(x - c_i)^l}\right], \qquad x \in \mathbb{C}\backslash C_Q \quad \text{(A.57)}$$

so $\deg P_1(x) < \deg Q_1(x)$, and if the argument applied on $P(x)/Q(x)$ is applied on the fraction $P_1(x)/Q_1(x)$ of (A.57) we get

$$A_{i,m_i-1} \in \mathbb{R}, \qquad \text{for all } 1 \leqslant i \leqslant m \text{ with } m_i \geqslant 2$$
$$B_{i,n_i-1} = \overline{C_{i,n_i-1}} \qquad \text{for all } 1 \leqslant i \leqslant n \text{ with } n_i \geqslant 2.$$

Now, (A.51) and (A.52) are already proved for $k = 0$ and $k = 1$. Repeating this procedure $\mathbf{m} - 2$ more times, (A.51) and (A.52) are proved for all $0 \leqslant k \leqslant \mathbf{m}$.

(ii) It is immediate from statement (i) and from Proposition A.12.14. $\qquad \square$

Remark A.12.16. In general, $A_{il} \neq A'_{il}$ in Theorem A.12.15.

For the last theorem, we need an auxiliary result:

Lemma A.12.17. *Let $P(z)$ and $Q(z)$ be a pair of polynomials over $\mathbb{C}$ such that $\deg P(z) < \deg Q(z)$ and all the coefficients of $Q(z)$ are real. Then all the coefficients of $P(z)$ are real if and only if the coefficients of $S(z) := P'(z)Q(z) - P(z)Q'(z)$ are also real.*

Proof. If the coefficients of $P(z)$ are real, it is straightforward that the coefficients of $S(z)$ are real too.

For the converse, assume some coefficient of $P(z)$ is not real. Let

$$P(z) = a_m z^m + \ldots + a_1 z + a_0$$
$$Q(z) = b_n z^n + \ldots + b_1 z + b_0$$

so that $a_m \neq 0$, $b_n \neq 0$ and $m < n$.

If $a_n \in \mathbb{C} \backslash \mathbb{R}$, then it is straightforward that the leading coefficient of $S(z)$ is $(m-n)a_m b_n \in \mathbb{C} \backslash \mathbb{R}$.

If $a_n \in \mathbb{R}$, let l be the largest index for which $a_l \in \mathbb{C} \backslash \mathbb{R}$. Let $P_2(z) := \sum_{k=l+1}^{m} a_k z^k$ and let $P_1(z) := P(z) - P_2(z)$. Thus, by linearity,

$$S(z) = \partial(PQ)(z) = \partial(P_2 Q)(z) + \partial(P_1 Q)(z). \tag{A.58}$$

As proved before, since all the coefficients of $P_2(z)$ are real, the coefficients of $\partial(P_2 Q)(z)$ are real too, and as the leading coefficient of P_1 is not real, neither is the leading coefficient of $\partial(P_1 Q)(z)$. Therefore, it follows from (A.58) that the $(l+n-1)$-th coefficient of $S(z)$ is not real, and the proof is done. $\qquad \square$

Theorem A.12.15 admits the following shortcut also known as Hermite's method. It is very useful for solving primitives of rational functions.

Theorem A.12.18. *Consider a pair of real polynomials whose irreducible factorizations over $\mathbb{R}$ are*

$$Q(x) = \prod_{i=1}^{m}(x-a_i)^{m_i} \cdot \prod_{i=1}^{n}(x^2 + \mu_i x + \nu_i)^{n_i}$$

$$Q_0(x) = \prod_{i=1}^{m}(x-a_i)^{m_i-1} \cdot \prod_{i=1}^{n}(x^2 + \mu_i x + \nu_i)^{n_i-1}$$

where at least one of the natural numbers m_i or n_i is greater than 1.

Then, given a polynomial $P(x)$ over $\mathbb{R}$ such that $\deg P(x) < \deg Q(x)$, there exists a unique polynomial $P_0(x)$ over $\mathbb{R}$ such that $\deg P_0(x) < \deg Q_0(x)$, and there exist unique subsets of real numbers $\{A_i\}_{i=1}^{m}$, $\{M_i\}_{i=1}^{n}$ and $\{N_i\}_{i=1}^{n}$ such that $\deg P_0(x) < \deg Q_0(x)$ and

$$\frac{P(x)}{Q(x)} = \partial\left(\frac{P_0(x)}{Q_0(x)}\right) + \sum_{i=1}^{m}\frac{A_i}{x-a_i} + \sum_{i=1}^{n}\frac{M_i x + N_i}{x^2 + \mu_i x + \nu_i}.$$

Proof. By Theorem A.12.15, there exist unique real coefficients A_{ik}, M_{ik} and N_{ik} such that

$$\begin{aligned}
\frac{P(x)}{Q(x)} &= \sum_{i=1}^{m}\frac{A_{i1}}{x-a_i} + \sum_{i=1}^{n}\frac{M_{i1}x + N_{i1}}{x^2 + \mu_i x + \nu_i} \\
&\quad + \sum_{i=1}^{m}\sum_{l=2}^{m_i}\frac{A_{il}}{(x-a_i)^l} + \sum_{i=1}^{n}\sum_{l=2}^{n_i}\frac{M_{il}x + N_{il}}{(x^2 + \mu_i x + \nu_i)^l}.
\end{aligned} \tag{A.59}$$

By virtue of Proposition A.12.11(iii), the expression in the second line of (A.59) is the formal derivative of a fraction $P_0(x)/Q_0(x)$ where $P_0(x)$ is a polynomial with $\deg P_0(x) < \deg Q_0(x)$. Moreover, as the coefficients M_{ik} and N_{ik} are real, it follows from Lemma A.12.17 that the coefficients of $P_0(x)$ are real too. The proof is fulfilled by letting $A_i := A_{i1}$, $M_i := M_{i1}$ and $N_i := N_{i1}$. $\qquad \square$

A.13 Cardinality of sets

Roughly speaking, the *cardinal* of a set is the number of elements it possesses. Although the notion of cardinality is intuitive to a certain extent, its formalization is not simple at all. A gentle introduction to the subject is [Halmos (1960)]. In the following, we intend to explain the notions of finite cardinal and infinite cardinal with special emphasis on distinguishing the sets with countable infinite cardinal from the rest of infinite sets, but avoiding all sort of mathematical technicalities.

A set A is said to be *finite* if either it is empty or there exists $n \in \mathbb{N}$ and a bijection $\varphi \colon \{1, 2, \dots, n\} \longrightarrow A$; in the first case, the cardinal of A is zero, and in the second case, it is n. If A is not finite then A is said to be *infinite* (more formally, it is said that the cardinal of A is infinite), in which case A possesses at least as many elements as $\mathbb{N}$ does, hence there exists an injective function $f \colon \mathbb{N} \longrightarrow A$.

At this point, it is natural to ask if all infinite sets have exactly the same quantity of elements as $\mathbb{N}$. If an infinite set A has the same number of elements as $\mathbb{N}$, then there exists a bijection $f \colon \mathbb{N} \longrightarrow A$, A is said to be *countably infinite* (or *countable* for short) and the cardinal of A is denoted $\aleph_0$.

The simplest example of a countable set is $\mathbb{N}$ itself. The set $\mathbb{Z}$ is also countable: it is sufficient to realize that the function $\phi \colon \mathbb{N} \longrightarrow \mathbb{Z}$ given by $\phi(2n-1) := n-1$ and $\phi(2n) := -n$ for $n \in \mathbb{N}$ is a bijection that maps the odd natural numbers onto the non-negative integers and the even natural numbers onto the negative integers.

Roughly speaking, an infinite set is countable if its number of elements is small enough that its elements can be enumerated in a list. In the example provided by the bijection $\phi \colon \mathbb{N} \longrightarrow \mathbb{Z}$ given before, that list is

$$
\begin{aligned}
1^{\text{st}} &: \quad 0, \\
2^{\text{nd}} &: -1, \\
3^{\text{d}} &: \quad 1, \\
4^{\text{th}} &: -2, \\
5^{\text{th}} &: \quad 2, \\
&\cdots \quad \cdots
\end{aligned}
$$

Obviously, a subset of a countable set is either finite or countable. Given a finite set A, its only subset with the same cardinal as A is A itself. However, a countable set contains countable proper subsets. This is the case with $\mathbb{N}$ and $\mathbb{Z}$.

Proposition A.13.1. *If A and B are two countable sets, then $A \times B$ is also countable.*

Proof. Since A and B are countable, their elements can be listed as

$$
\begin{aligned}
A &= \{a_1, a_2, a_3, \dots\} \\
B &= \{b_1, b_2, b_3, \dots\}.
\end{aligned}
$$

Thus, the set $A \times B$ can be arranged in an infinite table as

$A \times B$	a_1	a_2	a_3	a_4	· · ·
b_1	(a_1, b_1)	(a_2, b_1)	(a_3, b_1)	(a_4, b_1)	· · ·
b_2	(a_1, b_2)	(a_2, b_2)	(a_3, b_2)	(a_4, b_2)	· · ·
b_3	(a_1, b_3)	(a_2, b_3)	(a_3, b_3)	(a_4, b_3)	· · ·

so, if we follow the finite diagonals of the table above in descending order starting at the shortest one, we can countably list all elements of $A \times B$ as

$$(a_1, b_1), \ (a_2, b_1), \ (a_1, b_2), \ (a_3, b_1), (a_2, b_2), \ (a_1, b_3), \ldots$$

which gives a bijection between $\mathbb{N}$ and $A \times B$, proving that $A \times B$ is countable. $\square$

A very important consequence of the last proposition is the following:

Corollary A.13.2. *The set of all rational numbers is countable.*

Proof. Since both $\mathbb{Z}$ and $\mathbb{N}$ are countable, then Proposition A.13.1 gives that $\mathbb{Z} \times \mathbb{N}$ is countable, and as the set $\mathbb{Q}$ can be identified with an infinite subset of $\mathbb{Z} \times \mathbb{N}$, then it follows that $\mathbb{Q}$ is also countable. $\square$

This last result is surprising because $\mathbb{N}$ is sparsely distributed in the real line while there are rational numbers everywhere in $\mathbb{R}$; indeed, $\mathbb{N}' = \varnothing$ but $\mathbb{Q}' = \mathbb{R}$.

However, $\mathbb{R}$ is uncountable. Otherwise, the elements of $\mathbb{R}$ could be countably listed as

$$r_1, r_2, r_3, \ldots; \tag{A.60}$$

thus, given each real number r_i, and letting

$$0.r_i(1) \, r_i(2) \, r_i(3) \ldots$$

be the decimal expansion of its fractional part, we can construct the real number α whose decimal expansion is

$$\alpha := 0.\alpha_1 \, \alpha_2 \, \alpha_3 \ldots$$

where $\alpha_i \neq r_i(i)$ for all i. Clearly, α is a real number which has not been listed in (A.60), a contradiction that shows that $\mathbb{R}$ is uncountable.

Bibliography

Aleksandrov, A.D. (1999). Aleksandrov, A.D., Kolmogorov, A.N., and Laurentiev, M.A. (eds.) (1999). Mathematics: its Content, Methods and Meaning. Dover (New York).

Barbeau, E.J. (1989). Polynomials. Problem books in math. Springer.

Bartle, R.G. (1966). Return to the Riemann Integral. *American Math. Monthly* **103**, pp. 625–632.

Bak, J. (1993). Complex Analysis. Undergraduate texts in Mathematics, Springer (New York).

Bewersdorff, J. (2006). Galois Theory for Beginners. Amer. Math. Soc. Student Math. Library.

Bronstein, M. (1997). Symbolic integration. I. Transcendental functions. Springer.

Brychkov, Yu.A., Marichev, O.I., and Prudnikov, A.P. (1998). Handbook of Integral and Series. CRC.

Bruckner, A.M. (1978). Differentiation of Real Functions. Springer (Berlin Heidelberg).

Ciesielski, K.C., Seoane-Sepulveda, J.B. (2018). Differentiability versus continuity: restrictio and extension theorems and monstrous examples. *Bull. Amer. Math. Soc.* **56**, 2 pp. 211–260.

Chakraborty, K., Kanemitsu, S., Kuzumaki, T. (2016). Quick introduction to Complex Analysis. World Scientific (Singapore).

Fulks, W. (1961). Advanced Calculus. An introduction to Analysis. John Wiley and sons Inc. (New York, London).

Halmos, P.R. (1960). A naïve set theory. Univ. series in undergraduate texts, D. Van Nostrand Company Inc. (Princeton).

Kharazishvili, A.B. (2000). Strange functions in real analysis. Marcel Dekker (Basel).

Petruska, G., Laczkovich, M. (1974). Baire 1 functions, approximately continuous functions and derivatives. *Acta Math. Acad. Sci. Hungaricae* **25**, 1–2 pp. 189–212.

Risch, R.H. (1969). The problem of integration in finite terms. *Transacations of the American Mathematical Society* **76** pp. 167–189.

Risch, R.H. (1970). The solution of the problem of integration in finite terms. *Bullettin of the American Mathematical Society* **139** pp. 605–608.

Ross, K.A. (1986). Elementary Analysis: the theory of Calculus. Springer (New York).

Index

CPSIA information can be obtained
at www.ICGtesting.com
Printed in the USA
LVHW102137160920
666264LV00004B/6